REGULATING MEDIA

THE GUILFORD COMMUNICATION SERIES

Editors

Recent Volumes

REGULATING MEDIA

The Licensing and Supervision
of Broadcasting in Six Countries

WOLFGANG HOFFMANN-RIEM

THE GUILFORD PRESS
New York   London

© 1996 The Guilford Press
A Division of Guilford Publications, Inc.
72 Spring Street, New York, NY 10012

Printed in the United States of America

This book is printed on acid-free paper.

Last digit is print number: 9 8 7 6 5 4 3 2 1

Library of Congress Cataloging-in-Publication Data

Hoffmann-Riem, Wolfgang.
    Regulating media : the licensing and supervision of broadcasting in
six countries / Wolfgang Hoffmann-Riem.
       p.   cm. — (The Guilford communication series)
    Includes bibliographical references and index.
    ISBN 1-57230-029-9
    1. Broadcasting—Law and legislation.   2. Television
stations—Licenses.   3. Radio stations—Licenses.   I. Title.
II. Series.
K4325.4.H64   1996
343.09'94—dc20
[342.3994]   96-14095
                                          CIP

Acknowledgments

This study could not have been realized without the support of a wide range of people. Although it is impossible to thank individually all those who were involved in discussions or who provided information or constructive criticism, a special thanks is due to Mark Armstrong, Melbourne; Jay Blumler, Leeds; Peter Humphreys, Manchester; Andre Lange, Strassbourg; Eli Noam, New York; Tony Prosser, Glasgow; Don LeDuc, Milwaukee; Bill Melody, Melbourne; and Richard Yanda, Montreal. Among my colleagues, I am particularly grateful to Martin Eifert, Bernd Holznagel (who is entirely responsible for the chapter on Canada), Margarete Schuler-Harms, Stefan Engels, and Wolfgang Schulz, as well as Paul Aliferis and Carol Semple, who translated most of the study and proofread those parts I wrote in English. It goes without saying that responsibility for any shortcomings of the finished work is entirely mine.

Contents

PART II

A Comparative Study
of Broadcasting Regulation and Supervision

CHAPTER 7

Justifications for the Regulation
and Supervision of Broadcasting

CHAPTER 11

Patterns of Practical Supervisory Action   **324**

CHAPTER 12

Assessment   **335**

Abbreviations

AAC Advertising Advisory Committee (UK)

ABA Australian Broadcasting Authority

ABC American Broadcasting Companies, Inc. (U.S.)

ABC Australian Broadcasting Corporation (public broadcaster/Australia)

ABCB Australian Broadcasting Control Board

ABT Australian Broadcasting Tribunal

AfP *Archiv für Presserecht* (journal/Germany)

ALC Advertising Liaison Commitee (UK)

AM amplitude modulation

ARD Arbeitsgemeinschaft der öffentlichrechtlichen Rundfunkanstalten der Bundesrepublik Deutschland (Association of Public Broadcasters/Germany)

ARTE public broadcaster (Germany–France)

ATV 10 Associated Television 10 (private broadcaster/Australia)

A 2 La Société Nationale de Télévision en Couleur "Antenne 2" (public broadcaster/France); now France 2

AUSSAT Australian domestic satellite system

Austl.L.R. Australian Law Review

BayVerfGH Bayerischer Verfassungsgerichtshof (court/Germany)

BayVerfGHE Endsheidungssammlung des Bayerischen Verssaungsgerichtshof (selection of court decisions/Germany)

BayVerwGH Bayerischer Verwaltungsgerichtshof (court/Germany)

BBC British Broadcasting Corporation (public broadcaster)

BBG Board of Broadcasting Governors (Canada)

BCC Broadcasting Complaints Commission (UK)

BCN Broad Band Communications Network (U.S.)

BLM Bayerische Landeszentrale für Neue Medien (Bavarian State Authority for New Media)

BSB British Satellite Broadcasting Consortium

BSC Broadcasting Standards Council (UK)

BSEG Broadband Services Expert Group (Australia)

CAB   Canadian Association of Broadcasters

CANCOM   Canadian Satellite Communications, Inc.

C.F.R.   Code of Federal Regulations (statute law/U.S.)

CBC   Canadian Broadcasting Corporation (public broadcaster)

CBDF   Canadian Broadcast Development Fund

CBS   Columbia Broadcasting System, Inc. (U.S.)

CDU   Christlich Demokratische Union (political party/Germany)

cert.   certiorary

Cir.   Circuit

CIS   Commonwealth of Independent States (former Soviet Union)

CLT   Compagnie Luxembourgoise de Télédiffusion

CNCL   Commission Nationale de la Communication et des Libertés (France)

CNN   Cable News Network (U.S.)

CRTC   Canadian Radio–Television and Telecommunications Commission

CSA   Conseil Supérieur de l'Audiovisuel (France)

CSU   Christlich Soziale Union (political party/Germany)

CTR   Comités Techniques Radiophonique (France)

CTV   Canadian Television Ltd. (private broadcaster)

D2 MAC   D2 = technical term; MAC = multiplexed analogue components (standard of transmission for satellite broadcasting in Europe)

DBS   direct broadcast satellite

DC   Décision du Conseil Constitutionel (France)

DLM   Direktorenkonferenz der Landesmedienanstalten (Germany)

DOC   Department of Communications (Canada)

DSF   Deutsches Sport Fernsehen (private broadcaster/Germany)

DVBl.   *Deutsches Verwaltungsblatt* (journal/Germany)

EC   European Community

ECHR   European Convention on Human Rights

EEC   European Economic Community

EEO   equal employment opportunity

epd/KiFu   *Evangelischer Pressedienst/Kirche und Rundfunk* (journal/Germany)

F.2d   *Federal Reporter,* Second Series (journal/U.S.)

F.3d   *Federal Reporter,* Third Series (journal/U.S.)

F.C.C.   *FCC Records, First Series*

F.C.C.2d   *FCC Records, Second Series*

F.C.C.Rcd.   *FCC Records, Third Series*

F.Supp.   *Federal Supplement* (journal/U.S.)

FACTS   Federation of Australian Commercial Television Stations

FAG   Fernmeldeanlagengesetz (statute law/Germany)

FARB   Federation of Australian (Commercial) Radio Broadcasters

FCC   Federal Communications Commission (U.S.)

FM   frequency modulation

FP   La Société Française de Production et de Création Audiovisuelles (France)

FR 3   La Société Nationale de Télévision "France-Régions 3" (public broadcaster/France); now France 3

France 2   se A 2

France 3   se FR 3

France 5   private broadcaster (France)

FRG   Federal Republic of Germany

FTC   Federal Trade Commission (U.S.)

GATT   General Agreement on Tariffs and Trade

GDR   German Democratic Republic

GLOBAL private broadcaster (Canada)

HAM Hamburgische Anstalt für Neue Medien (State Media Authority of Hamburg)

HMSO Her Majesty's Stationery Office

IBA Independent Broadcasting Authority (UK)

ILR Independent Local Radio (UK)

IRA Irish Republican Army

ISDN Integrated Services Digital Network

ITA Independent Television Authority (UK)

ITB Interstate Treaty on Broadcasting in Unified Germany

ITC Independent Television Commission (UK)

ITCA Independent Television Companies Association (UK)

ITN Independent Television News (UK)

ITV Independent Television (UK)

ITVA Independent Television Association (UK)

J.O. *Journal Officiel* (France)

LAR Landesanstalt für das Rundfunkwesen Saarland (State Media Authority of Saarland)

LfK Landesanstalt für Kommunikation Baden–Württemberg (State Media Authority of Baden–Württemberg)

LfR Landesanstalt für Rundfunk Nordrhein–Westfalen (State Media Authority of North Rhine–Westphalia)

LMA local marketing agreement (U.S.)

LMG BW Landesmediengesetz Baden-Württemberg (statute law/Germany)

LPR Hessische Landesanstalt für Privaten Rundfunk (State Media Authority of Hessen)

LPR Landeszentrale für Private Rundfunkveranstalter/Rheinland–Pfalz (State Media Authority of Rhineland–Pfalz)

LPTV low-power TV

LRA Landesrundfunkausschuss für Sachsen–Anhalt (State Media Authority of Saxony–Anhalt)

LRZ Landesrundfunkzentrale Mecklenburg-Vorpommern (State Media Authority of Mecklenburg–Vorpommern)

M 6 Métropole Télévision (private broadcaster/France)

MABB Medienanstalt Berlin–Brandenburg (State Media Authority of Berlin–Brandenburg)

MCA Media Council of Australia

MCS multichannel service

MDS multipoint distribution service/system

MMDS multichannel, multipoint distribution service

MSO multiple-system operator

MTV Music Television (private broadcaster/U.S.)

n-tv News Television (private broadcaster/Germany)

NAB National Association of Broadcasters (U.S.)

NARB National Advertising Review Board (U.S.)

NBC National Broadcasting Company (U.S.)

NII National Information Infrastructure (U.S.)

NJW *Neue Juristische Wochenzeitschrift* (journal/Germany)

NTCA National Television Companies Association

NVwZ *Neue Zeitschrift für Verwaltungsrecht* (journal/Germany)

OLG Oberlandesgericht (court/Germany)

ORTF Office de Radiodiffusion Télévision Française (France)

OVG Oberverwaltungsgericht (court/Germany)

PBS Public Broadcasting Service (U.S.)

PCS Personal Communications Services (U.S.)

PEG Channels channels for public, educational, and government use (U.S.)

Pro 7 (Sieben) private broadcaster (Germany)

PTT Post Telephone and Telegraph

R. Rex/Regina

R.R.2d *Pike & Fisher Radio Regulations,* Second Series (journal/U.S.)

RAI Radiotelevisione Italiana (public broadcaster/Italy)

RCA Radio Corporation of America

Rcd. record

Reg. register; registration; registration cases; registry; regulation

RPI retail price index

RPS Radio Program Standards (Australia)

RTL Radio Télé-Luxembourg (private broadcaster/Germany)

RTL2 Radio Télé-Luxembourg 2 (private broadcaster/Germany)

RuF *Rundfunk und Fernsehen* (journal/Germany)

S.Ct. *Supreme Court Reporter* (journal/U.S.)

SAT 1 private broadcaster (Germany)

SBS Special Broadcasting Service (Australia)

2d second series

SED Sozialistische Einheitspartei Deutschlands (East German Communist Party)

SFP La Société Française de Production et de Création Audiovisuelles (France)

SLM Sächsische Landesanstalt für Privaten Rundfunk und neue Medien (State Media Authority of Saxony)

SMATV satellite master antenna television

SOP Service d'Observation des Programmes (France)

SRF La Société Nationale de Radiodiffusion "Radio France" (public radio broadcaster/France)

STV subscription TV (Australia)

TBS Turner Broadcasting System

TDF Etablissement Public de Télédiffusion Française (France)

Tele 5 private broadcaster/Germany

TF 1 Télévision Française 1 (private broadcaster/France)

3 Sat public broadcaster (Germany)

TLR Thüringer Landesanstalt für Privaten Rundfunk (State Media Authority of Thuringia)

TNT Turner Network Television

TPC Trade Practices Commission (Australia)

TPS Television Program Standards (Australia)

TSN The Sports Network (private broadcaster/Canada)

TVA Télé Diffuseurs Associées (broadcaster/Canada)

TV-a.m. private broadcaster/UK

TWW The West of England and Wales (UK)

U.S. *United States Reports* (journal/U.S.)

U.S. App *United States Reports, Appendix* (journal/U.S.)

U.S.C. United States Code

U.S.C.A. United States Code Annotated

ULR Unabhängige Landesanstalt für das Rundfunkwesen/Schleswig–Holstein (State Media Authority of Schleswig–Holstein)

USSB United States Satellite Broadcasting

VerfGH Verfassungsgerichtshof (court/Germany)

VerwGH Verwaltungsgerichtshof (court/Germany)

VG Verwaltungsgericht (court/Germany)

VHF very high frequency

Vox private broadcaster (Germany)

WARC World Administrative Radio Conference

ZDF Zweites Deutsches Fernsehen (public broadcaster/Germany)

ZUM *Zeitschrift für Urheberrecht* (journal/Germany)

REGULATING MEDIA

Introduction

1. BROADCASTING AND ITS REGULATION

The history of broadcasting is about three-quarters of a century old. In this relatively short period of time, broadcasting—both radio and television—has undergone a huge technological transformation. As a system of communication, broadcasting has always been a reflection of changing political and economic circumstances. Thus the medium of broadcasting offers a particularly good platform for observing the interaction of the various factors of technological, economic, and political development. Through its programming, broadcasting reflects those issues that society generates for mass communication or that are of greatest interest. It also reflects which issues are "socially acceptable" as a result of filtering either by state censorship or by the broadcasting company itself. As a conveyer of the communication requirements of others, broadcasting functions as a medium of communication; however it also actively influences the information supply of the respective society in which it operates and thus plays a proactive role—is in effect a key factor—in the communication supply itself. Consequently, in modern information societies, broadcasting plays a significant role in communication infrastructure.

As a result, broadcasting, since its infancy, has enjoyed the intense interest and solicitous attention of governmental authorities and legislatures. This has often resulted in close state guardianship or at least in subtle attempts by governmental authorities to influence programming, the selection of personnel, sources of financing, and so forth. Many countries, especially in Eastern Europe, Asia, Africa, and South America, have as yet been unable to decide whether to free broadcasting from state control. Others, especially in North

America, Western Europe, and Australia, have a longer tradition of independent broadcasting or have recently managed to reduce or completely remove the broadcasting industry's ties to government and the powers that support them, particularly the political parties.

Reducing state dependence most certainly does not mean that broadcasting should be released into a power vacuum. Other institutions—for example, companies, as well as other organizations—are anxious to use broadcasting to promote their own interests. Their aim is to exert influence upon the information storehouse of citizens. Above all, however, they aim to influence the values and attitudes of the public. Their motives vary; they may be seeking political influence, or they may wish to promote the sales of goods and services that have nothing to do with broadcasting.

Spurred by technological innovations—especially those in transmission technology as well in production—the broadcasting industry has proved to be a rapidly expanding growth area with a high profit potential. Economic analyses of production costs and their relation to earnings opportunities, especially analyses of disproportionate growth as listener numbers rise and the markets extend accordingly, explain why companies that are internationally oriented and multimedially interlocked have particularly good prospects of economic success. A high level of international and multimedial concentration is virtually a structural necessity and goes hand in hand with the media market. Nonetheless, there are still opportunities for smaller companies to find and utilize market niches in the field of communications.

Technological, economic, and political developments in broadcasting have already been outlined a number of times. Without going into great detail here, it can be pointed out that the technological transformation in production and distribution, the struggle to use the (assumed) communicative power of broadcasting, and, above all, the prospect of economic success have, from the very beginning, been the decisive factors in the development of broadcasting. It is the interaction of all these factors that has shaped broadcasting's rapid development over the last few decades. It is not the transformation itself, however, that is the subject of the following discourse but, rather, the influence of the state upon its development. Of particular interest is the reaction of governmental authorities to changed circumstances in broadcasting and telecommunications.

In every country in the world, government has attempted to regulate the development of broadcasting, especially after it has experienced an initial phase of rapid development. This movement to regulation has also occurred in countries where other parts of the media, notably the press, have not been closely regulated. It has been particularly prevalent in countries where the media sector was once closely monitored by the state, for example in totalitarian states.

Regulation usually took place in the form of legislation. Even in former totalitarian states, the law was used as a legitimizing vehicle. Recently, however, regulation by law has been supplemented and replaced with "regulation" by market forces. This trend applies not only to Western countries but also to Eastern countries as well. In Western countries—for example, nations of Western Europe, North America, and Australia—a trend toward deregulation, or better toward reregulation, has been evident, especially in the last few years, as a reaction to changed circumstances.

In the Western European countries, opening up the market to competition represents a significant change, as there had previously been a monopoly of so-called public broadcasting in most of these countries. However in the United States, Canada, and Australia, broadcasting was the responsibility of the private commercial sector from its very inception. In these countries, public broadcasting emerged later, originally in the form of educational broadcasting. Such a duopoly of public and private broadcasting first began to develop at the beginning of the 1930s in Australia and shortly thereafter in North America. It did not reach Europe until 1954 when the British broadcasting system was set up.

The legal framework of broadcasting and the supervision of its rules are the subject of the following analyses. The questions are posed as to how and with what prospect of success the development of the broadcasting system and the conduct of those media companies within that system can be influenced by legal regulation. The work goes on to focus more closely on the licensing and supervision of broadcasting. The subjects of the examination are particular institutions (media authorities) that have largely been separated from general state administration. On one hand, these institutions award licenses for distribution of broadcasting. On the other hand, they ensure that broadcasting companies adhere to the existing legal obligations.

2. THE POLITICAL FUNCTION OF INTRODUCING BROADCASTING SUPERVISION

This study examines the effectiveness of law and the supervision of how rules are applied; as such, it is directly connected to the development of broadcasting. The 1980s in particular were characterized in many countries by a trend toward deregulation and reregulation of broadcasting. In Western Europe, the function of broadcasting was becoming increasingly detached from its traditional public service obligations. Although traditional public service broadcasting remained basically in place (e.g., the BBC in Britain, the RAI in Italy, and the ARD and ZDF in Germany), it was joined by commercial broadcasting companies. Funded by the private sector, these companies traded in the free market. To some extent, as in Great Britain, commercial broadcasters had already been in existence, in the form of Independent Radio and Television, which was organized as a monopoly. However the new regulation in 1990 gave this sector of the broadcasting system a new quality, a direct result of the introduction of competition.

In all Western European countries, these periods of radical change resulted at times in intense political controversies about the future of broadcasting. In these debates critics raised such issues as fears about the loss of program quality, the public service concept, and the threat to vulnerable values. Others stressed the advantages of increased choice and the prospect of satisfying consumer demand to a greater extent. Both sides pointed—from their respective viewpoints—to experiences in countries outside of Europe, particularly in the United States. Those who were skeptical about the new system in Europe were mollified by the idea that any potential risks could be kept to a minimum through statutory regulation and supervision. However it seemed less important how those instruments created in the United States were actually applied and what difficulties and successes the supervisory body, the Federal Communications Commission (FCC), had experienced. It was not the practical experience of broadcasting supervision in the United States but, rather, the legal range of these instruments that aroused the most interest in Western Europe. Using this material as a starting point, all countries in Western Europe that were going through radical change in broadcasting created special legal control mechanisms and supervisory bodies for private broadcasters.

The resulting regulation of a private sector broadcasting system encompassed instruments and procedures that are also applied in other areas, in which goods and services can be bought or sold on the open market. However, certain rules were created specifically for the broadcasting industry as well. Broadcasting companies were liable to much stricter regulation than most other commercial companies.

The creation of broadcasting regulation in this way was an important precondition for the introduction of private sector broadcasting. The supervisory body set up by the state served as a guarantee that broadcasting would develop in a responsible manner—or at least not proliferate uncontrollably. The reference to a legal framework and the possibility of supervision of broadcasting companies were assets when it came to politically authorizing the new development; in some countries, such as Germany, the political forces against the introduction of private broadcasting could not otherwise have been overcome. Ever since the introduction of private broadcasting, broadcasting supervision has aided political authorization. This is especially evident if difficulties arise in protecting what are traditionally considered "vulnerable" values, such as diversity or personal dignity. Situations such as this usually provoke discussion about possible changes in the legal framework and about reform of broadcasting supervision.

3. STUDY METHODS

The following study attempts to trace the effect licensing and supervision have upon private broadcasting. It is not our aim to produce a comprehensive compilation of individual rules in the various countries. This study, rather, aims to look for typical spheres of supervision and to examine how the supervisory bodies operate within them. In particular, it will consider whether the supervisory bodies succeed in achieving the objectives set down by law. There will be no attempt made to examine the various broadcasting orders in terms of their efficiency, as this would presuppose a common normative level or standard that does not exist. Irrespective of the recognition of certain common principles—for example, freedom of communication—there exist in the different broadcasting systems widely differing appraisals on the subject of realizing objectives. Since a mandatory common standard does not exist, the following chapters will attempt merely to measure the regulation of broadcasting and those related activities by the objectives that are recognized in each of the broadcasting systems.

In order to gain some insight into the possibilities and difficulties presented, a comparison of broadcast regulation in selected countries will be made. On one hand industrialized nations with a long tradition of private sector broadcasting and broadcasting supervision (the United States, Australia, and Canada) were chosen; selected on the other hand were leading Western European countries that equate broadcasting with the public service concept (Britain, France, and Germany). Thus, the line of proceeding is inductive: An examination of the various broadcasting systems will highlight both common elements and differences. The study will select individual regulations and developments, but a comprehensive or detailed description is not intended. Nor is the study intended to be a comparison of the actual state of regulation in these countries. It is rather a close examination of the various phases of broadcasting regulation to establish what kind of regulatory action is taken where and which tools are typically used. As a result the studies of the various countries are not standardised according to a set form. Rather, we have attempted to push the characteristics of the various broadcasting systems into the foreground. In this way, the main points of distinction are set accordingly, and the diversity of possible regulation options becomes evident.

The analysis is based upon materials that are accessible to the public, for example, annual reports of regulatory bodies and academic publications. Further sources of information are discussions in which facts and opinions were sought. Unfortunately, no systematic, empirical research on broadcasting supervision has been undertaken in any of the countries examined, although studies have been made of individual aspects of the subject. On the basis of available material, then, no definitive statements can be made, only appraisals that may be plausible to a greater or lesser degree. The comparison of the various broadcasting systems, which extends over a period of time and covers different countries, serves to increase the level of plausibility. Indeed, such comparison uncovers structures that indicate there are tools, procedures, and effects that are typical of broadcasting supervision. The aim of this work is to expose these structures; the selected details as such are merely for illustrative purposes.

Part I consists of chapters about the various broadcasting systems. Each chapter begins with a short look into the development and structure of the broadcasting system in question. There follows an outline of how broadcasting supervision is organized, which areas it covers, and how it fits into the network of the various interested political institutions. The focus is mainly on regulations that are typical of the respective country or where the analyses promise significant insights. Part II takes a transnational view and consists of a systematic depiction of broadcasting supervision procedures and the experiences that have been gained.

A comparison of countries in this way is both difficult and risky. For an author—or at least for the author of this study—it is easier to understand his own national and social system, and consequently there is a great temptation to consider other countries and societies from this perspective. Thus, there exists the danger that those readers who have grown up knowing other systems may not find their perspective adequately represented in the study.

It is also difficult to gain access to data as well as to achieve any real understanding of how various functions are linked, and this is not only because of the language barrier. Particular problems present themselves when the systems being analyzed are in a state of rapid transformation, as is the case at present in the broadcasting industry. This constituted a distinct obstacle in studies comparing countries. Just when one country study was completed and a new one begun, the situation would suddenly change in the previous one. This was the case in all six of the broadcasting systems compared here. The chapters, particularly those in Part I, were updated several times before they could be sent to the publisher. By the time this work is in print, there will no doubt have been further changes. As it is not the aim of this study to describe the up-to-the-minute situation but, rather, to outline developments by example and to seek out basic structures, it is of little consequence if certain details are no longer totally up-to-date.

This study has been done at a time in which the broadcasting systems in those countries examined are faced with a new challenge. As a result of technological developments, in particular the possibility of digital compression, the problem of scarcity of transmission possiblities will soon be a thing of the past. Thus, a large part of the justification for the regulation of broadcasting may soon no longer apply. This study demonstrates that those objectives that regulation sought to promote can retain their relevance even when the problem of scarcity is overcome. As a result, it is highly probable that broadcasting regulation will remain in existence, although its justification will have to be changed somewhat and the objectives and tools adjusted accordingly. In any case, it can be assumed, for the foreseeable future, that broadcasting as a mass medium will continue to exist and that the opportunities to individualize the media supplies and to use the media will in fact be limited. Even the possibilities of interactivity, which are expanding more and more as a result of technological advancements, will presumably be used only marginally in the foreseeable future. It will most certainly not result in the majority of consumers having to do without primarily passive reception.

Thus this study is based upon the supposition that although future broadcasting regulation will be faced with new tasks and will undergo many changes, those typical structural elements described here and, above all, the problem of effective regulation, will remain.

PART I

Broadcasting
and Its Supervision in
Selected Broadcasting
Systems

CHAPTER 1

The United States

The pioneer of the modern broadcasting system worldwide was the United States. However, it was not just that the United States initiated a number of technological innovations or that in the United States these could be implemented for the first time in a large setting. Rather, the commercial and organizational forms adopted for U.S. broadcasting also served as a model for many other countries. Moreover, America's broadcast programming is now dominant throughout the world.

It should be noted, however, that this de facto pioneering role applies only to market-financed broadcasting. The United States has been unsuccessful at developing a noncommercial broadcasting system comparable to European public broadcasting. The U.S. media order rests atop the powerful pillar of commercial broadcasting; the thin stake beside this—public broadcasting—makes no pretense to serve as an analogous foundation of the media system, though it has an important supplementary function.

As a result, the United States cannot claim to be a model for a dual broadcasting system with two equal pillars but rather a model for how private commercial broadcasting can work. To this extent, the American broadcasting system has served as a pacesetter for many industrialized countries as well as for some developing nations. But in central Europe, there continue to be substantial reservations about the functioning of the marketplace and thus about commercial broadcasting. The concept of broadcasting as a trustee of the public is still strongly embedded in European cultural tradition and political rhetoric. In day-to-day media politics, however, much of the groundwork is laid in a manner similar to that in the United States. Apparently, the U.S. media system considers the principle of free speech as a transnational "commercial logic" that runs its course irrespective of legal, cultural, and political traditions. For this reason, the American example is particularly instructive for comparative media studies.

The media order in the United States has also been a pioneer with respect to the relationship between media markets and governmental supervision. The reasons for, and tools of, U.S. supervision have a number of parallels in the more recent British, French, Canadian, Australian, and even German media systems. In addition, the de- and reregulatory policies practiced in the United States since the early 1980s are finding an increasing number of adherents in Europe. The concept of liberalization is a virtually unavoidable consequence of a commercial, internationally networked media system (see Hoffmann-Riem, 1985: 528ff.). It not only benefits established private broadcasters but can also help prevent supervisory authorities from having to admit to being only moderately successful with forms of traditional broadcasting regulation.

In many respects, the experiences of the FCC reveal a bottom line of limited success. Although the FCC has played a major role in shaping the American broadcasting system, it has consistently had to yield in the face of limitations that a commercial media system (necessarily) places on governmental supervision in trying to protect some vulnerable societal values. As a result, it comes as no surprise that the FCC has become one of the main driving (though also driven) forces behind deregulation.

In order to be able to understand the supervisory dilemma facing the FCC, it is first necessary to describe the U.S. broadcasting system (Sections 1–3). After the position and duties of the FCC (Section 4) are identified, individual problem areas are used to show the direction indicated by the experiences of the FCC and the consequences it has drawn from these (Section 5). Sections on new electronic media follow (Sections 6 and 7). Finally, an effort is made to sketch the political arena in which the FCC acts (Section 8).

1. DEVELOPMENT AND STRUCTURE OF THE U.S. BROADCASTING SYSTEM

1.1. Commercial Broadcasting

In contrast to broadcasting in Europe, broadcasting in the United States from its inception exhibited a private form of organization. It was only after the "chaos in the ether" had led to mutual blockades, that the federal government assumed a regulatory responsibility—first (in 1927) with the Federal Radio Commission and subsequently (in 1934) with the FCC.

The use of cable technology in the United States also began quite early (since 1948). Yet as late as 1979, less than 10% of all U.S. homes with television had cable. In the 1980s an explosive cable TV expansion took place. At the beginning of the 1990s, roughly 60% of all households with television subscribed to cable services. There are now almost 3 million satellite dishes that allow households to receive programming distributed by telecommunication satellites. They receive signals that are intended for transmission by cable networks.[1]

Television by direct broadcast satellite (DBS) has remained relatively insignificant thus far. Although the FCC decided to license direct broadcast satellites[2] as early as 1982 and thereafter received a number of applications, DBS has failed to gain a foothold, despite being largely exempt from legal requirements (for the legal broadcasting arrangement, cf. Barnett, 1988: 140f.) or from the possibility of these.[3] The main reason DBS did not meet with success was the excessively high financial burden associated with its introduction, along with great uncertainty as to chances in the market (Teeter & Le Duc, 1992: 415f.). It was ten years before serious plans were again made to introduce DBS. However, late 1990 saw the withdrawal of two major members of the consortium that was formed to develop DBS—Rupert Murdoch and the National Broadcasting Company (NBC)[4]—and the resulting loss of the capital that these two financial giants were to contribute could initially not be recouped completely. Since then, new technology has improved the outlook for DBS. Small, relatively inexpensive receiving equipment has reduced the costs for consumers, and digital compression techniques will enable DBS operators to offer hundreds of channels. As a result, direct satellite broadcasting promises to be more profitable in the future. Two operators (Hughes and Hubbard's United States Satellite Broadcasting—USSB) have sucessfully started operation in 1994. Other companies have also announced plans to deliver DBS service.

In the area of terrestrial television, attempts to expand the market to new forms of distribution have met with only limited success. For example, the basic legal requirements were changed to create transmission opportunities that had previously been either insignificant or used in other ways, such as low-power TV (LPTV) and the multichannel, multipoint distribution service (MMDS), also known as the multipoint distribution service (MDS). The former represents the use of low-power transmitters, traditionally used as translators, as independent stations; the latter refers to television transmitted in the microwave area of the frequency spectrum. Neither completely fulfills the expectations placed in them, and despite large initial interest by broadcasters, they have yet to achieve any great significance (Teeter & Le Duc, 1992: 419ff.). However, it is anticipated that in the future MMDS may enter some markets, though it seems that in the United States cable television plays such a dominant role that new dissemination technologies have difficulty entering the market and their chances in the market are clearly limited. It is debatable whether this will change when new technologies (e.g., digital compression) allow a far bigger number of signals to broadcast via terrestrial frequencies.

Broadcasters in the United States are essentially at liberty as to the financing of their programming. Since the very beginning of the broadcasting system, financing has mainly come from advertising. The term "commercial broadcasting" was coined for those operators financing their programming by advertising, and the public largely accepts this term as a neutral designation for this type of broadcasting. The term can also be used in a broader sense for broadcasting that employs other quasi-advertising forms of financing, such as forms of product placement as well as sponsoring.

Certain broadcasters, especially those who provide cable services,[5] rely on subscriber payments, either exclusively or in addition to advertising. Such payments may be made for receipt of a large package of channels (sometimes referred to as the "basic" service) of individual channels (sometimes termed "premium" services, e.g., for movies). In addition, a separate charge may be made for each program viewed, so-called pay-per-view TV, which has so far been confined mainly to first-run films and spectator sports (e.g., a boxing championship bout).

There are a large number of both radio stations in the United States (roughly 11,000, including more than 1,400 noncommercial broadcasters) and television stations (almost 1,500, including approximately 350 noncommercial and nearly 600 low-power stations). This does not mean, however, that there is a similar diversity in terms of programming.[6] Most commercial television and radio stations belong to so-called networks, that is, companies whose programs are broadcast nationally—these programs are supplemented by locally or regionally oriented programming of network affiliates. The remaining stations are called "independents." In the television sector, one also finds so-called superstations, whose programming likewise is distributed nationally via cable. More specialized offerings are available in cable television, particularly with regard to entertainment (especially film), news and sports channels; such programming is generally transmitted to cable systems by way of telecommunications satellites.

1.2. Public Broadcasting

In the early years of broadcasting, commercial broadcasters were accompanied by others supported and financed by educational institutions (originally taking the form of educational radio). They became the predecessors of public broadcasting, that is, radio and television stations organized privately but not commercially and supported by associations, foundations, universities, communities, and so forth, that have been allocated special frequencies (Lively,

1992: 146). They also received government subsidies. However, "fee financing" was never introduced. Instead, broadcasters now seek voluntary donations from viewers and listeners and from industry. The latter often plays the part of sponsor by financing entire programs in exchange for being named as financial backer. As a result, public broadcasting can serve to boost the image of its sponsors. The Public Broadcasting Service (PBS), a program supplier for public broadcasters, emphasizes high-quality cultural and informational programs as well as minority programming. Public broadcasting has consistently been able to garner only limited attention: Roughly 3% of the entire broadcasting consumption by Americans is attributable to public broadcasting. The public broadcasting system has always been under extreme financial pressure.

PBS will not be considered in further detail in this chapter.

2. BROADCASTING: FIDUCIARY FUNCTION VERSUS MARKET DEVELOPMENT

From the outset, broadcasting in the United States developed under different political and legal conditions than in central Europe. At no time did the federal government have a broadcasting monopoly. This would have conflicted with the specifically American understanding of the basic First Amendment right of freedom of speech and the press,[7] which is conceived as a legitimate defensive right on the part of private citizens against the government.[8] Accordingly, governmental regulation is not there to serve comprehensive, "positive" assurance of a communications system but, rather, to create certain minimum conditions for a functional broadcasting order.

Indeed, broadcasting legislation was initiated in order to put an end to the "chaos in the ether," that is, to enact a kind of set of "traffic regulations." But since frequencies (at least those of terrestrial broadcasting) were scarce, the regulatory duty grew into an area in which considerations of the common good were used to find or justify criteria for allocation. In making allocation decisions as well as in its subsequent supervisory activities, the FCC was bound by the Communications Act of 1934 to observe the common good (i.e., "public interest, convenience and necessity"). Since regulatory power was primarily justified by scarcity of frequencies,[9] the legitimacy of governmental regulations waned with the advent of new means and technologies of transmission. This differs considerably from, for example, the German situation (cf. BVerfGE 57, 295, 320ff.; Hoffmann-Riem, 1985: 534f.).

Since the legislature was motivated by a perceived concern of "public interest, convenience and necessity" (i.e., for the common good), the Communications Act of 1934 largely lacks rules for positively structuring the broadcasting order and the (programming) conduct of broadcasters. Accordingly, there are no specific statutory norms regarding broadcasters' conduct, programming requirements, and so forth. At the same time, however, the Communications Act does empower the FCC to enact general rules for the common good and, based on these, to make the appropriate licensing decisions (see in particular 47 U.S.C. 303, 307ff.). This authorization has been broadly interpreted in case law to mean that rules relating to programming are also permissible, insofar as these are necessary to ensure that despite frequency scarcity the manifold interests of viewers and listeners are satisfied. It is against this background that the Supreme Court stressed that broadcasting is a proxy or fiduciary of viewers and listeners.[10] This ensures the right of the public to adequate access to social, political, aesthetic, and moral ideas and experiences. Such requirements necessitate specification by the FCC, a so-called independent agency. However, the Congress monitors which rules are applied and, if need be, it threatens to use its powers under the appropriate statutes as well as informally.

Thanks to political support by the President and the Congress, the FCC was able to deregulate radio and television during the 1980s. The majority of the rules in force were diluted by the FCC in 1981 and 1984, with some being rescinded entirely.[11] This deregulation was made possible by changes in the direction of regulatory philosophy (Horwitz, 1989). A departure was made from the concept of broadcasting as fiduciary, which had been the basis of older case law.[12] In contrast to the situation prevailing in Europe, in the United States the common good as the standard for broadcasting supervision is not defined from the standpoint of cultural policy or democratic freedom; rather, it is understood to be the result of market forces whose effects sufficiently aid in achieving the objectives of common good. Regulation of conduct, therefore, is largely considered to be dysfunctional, since this threatens to curb or impair the effectiveness of key economic factors. The goal of diversity, which continues to be considered important, is considered to have been achieved when there are a sufficient number of providers and offerings. The concept of internally pluralistic diversity of each program (as in Germany) is foreign to the FCC and to the American public in general.

The FCC supports the (worldwide) expansion of the U.S. media industry for reasons of industrial and cultural policy. The formation of powerful media companies and the creation of interlocked, multimedia systems are being gradually promoted. Broadcasting's trustee role thus begins to take more of a back seat so that the greatest possible development of the market can be achieved.

There is an independent set of rules in the United States for cable broadcasting. These federally uniform rules for cable are concerned mainly with preventing communities, which are responsible for licensing cable systems, from enacting inconsistent (i.e., normally stricter) rules (i.e., they can be preempted by federal law). The Cable Communication Policy Act of 1984, marked by the deregulatory philosophy, also underscored the concern for expanding the media industry on a broad scale. However, in October 1992 a comprehensive reregulation statute, the Cable Television Consumer Protection and Competition Act, was enacted, entrusting the FCC with a wealth of new tasks vis-à-vis cable operators. These rules mainly concern economically oriented measures, such as setting the maximum rates for "public interest, convenience and necessity." In addition, cable operators were deprived of economic privileges. They may no longer be shielded from local competition and, for example, now must under certain conditions[13] also pay for the retransmission of programming broadcast by local TV stations. This change was a result of a shift in the balance of power. The cable industry has become firmly established and begun working very profitably, while the networks, and with them terrestrial broadcasting, are now faced with increasing financial difficulties.

The following remarks relate primarily to terrestrial television; some aspects of cable regulation will be dealt with later (see Section 6).

3. COMPETITION AS A PRINCIPLE OF MEDIA POLICY

The conviction that self-regulation by the market is superior to governmental regulation is much more broadly held in the United States than in Europe. Open entry and competition are necessary prerequisites to this concept. The possibility of market failure and externalities is also seen in the broadcasting sector, but many consider this to be less problematic than a failure of governmental regulation. Some critics even stress that the crucial issue is not only market failure but market reach: the

> market constrains the presentation of matters of public interest and public importance, [since] the market brings to bear on editorial and programming decisions factors that might have a great deal to do with profitability or allocation efficiency . . . but little to do with the democratic needs of the electorate. (Fiss, 1987: 787f.; see also Ingber, 1984: 48f.)

Because of the government's licensing monopoly and the remaining broadcasting regulations, as well as complementary antitrust rules, the controlling principle is one of limited regulation of competition. Within the statutory framework, there is much room for economic competition among the various broadcasters. Insofar as these are financed by advertising, the battle for viewers is in effect a struggle to divide up their recipients in order to be able to "sell" as many of them as possible to the advertising industry. Since advertising prices generally are calculated according to audience (viewer/listener) numbers, programming is plainly oriented toward attracting the highest possible ratings. In addition, advertisers prefer special groups of viewers/listeners (e.g., young adults and the well-to-do) to others (e.g., the elderly and minorities with below-average income). For these reasons, advertising financing has a particularly dramatic, indirect effect on the programming conduct of broadcasters.

With regard to programming not financed by advertising—for instance, pay-TV or PBS productions—commercial broadcasters may not compete with these broadcasters for advertising revenues, but they do compete for viewers. Although the operators of per-channel pay-TV are not economically dependent on high ratings for their programs, they do rely on ratings being generally attractive enough to hold on to current subscribers and to recruit new ones. Thus, even though the goals in each case may be somewhat different, economically motivated competition also characterizes the competition over dissemination of programs between pay-TV broadcasters and those financed by advertising. Pay-per-view TV is somewhat different (cf. Baldwin, Wirth, & Zenaty, 1978). But it apparently seems suitable for only a few, particularly attractive programs targeted at a selected market and is not available on a broad scale.

4. THE FCC AS SUPERVISORY AUTHORITY

Supervision of broadcasting in the United States is essentially regulated at the federal level, but state and local governments are additionally involved in franchising and regulatory processes for cable services, creating a bifurcated structure of cable regulation.

The FCC, whose scope of authority covers the entire telecommunications sector (it is not restricted to mass communications) is composed of five commissioners,[14] who are nominated by the President for five-year terms and confirmed by the Senate (see Bittner, 1994: 37f.). Partisan politics are avoided by requiring that no more than three members belong to the same political party. However, the President also determines which commissioner will serve as chairperson. Routine internal administrative duties are performed by a managing director. Substantive groundwork for the agency is undertaken by various bureaus.[15] The Mass Media Bureau is responsible for the FCC duties in all fields of mass communications. It, in turn, is composed of the following departments: audio services, video services, policy, and rules and enforcement. With respect to broadcasting, the FCC is responsible both for decisions on the use of technical facilities and the standards relating to these as well as for the operations of broadcasters. The FCC is required to report to the Congress annually.

As part of its mandate, the FCC is empowered to determine the conditions of licensing and license renewal in the broadcasting sector, including the transmission power of the licensed broadcaster. It can revoke or suspend licenses or issue only a short-term renewal (see Bittner, 1994: 49f.). It may also impose sanctions, such as fines, although it makes an effort to avoid formal sanctions as far as possible. In place of these, it has developed a number of measures less drastic than formal sanctions. These include inquiries or informal objections, as well as warning letters notifying a licensee of violations. The use of nonformalized sanctions is often termed "regulation by raised eyebrows."

The FCC tends to specify its policies by way of supplementary rules or via internal administrative guidelines or authorizations. Further, it uses reports, notices, and primers to keep the public informed of its guiding considerations. In addition, its individual decisions often reveal general rules. But despite such specifications, clearly enforceable standards are often lacking, at least in part part because it is nearly impossible to work in the area of broadcasting regulation without broadly defined, generalized formulas. Moreover, the FCC tends to ensure flexibility through the possibility of exceptions.

The extent of the FCC's supervisory authority differs, depending on which transmission technology is used for broadcasting. This makes clear that the main goal in the United States is not to regulate broadcasting as a mass medium because of its significance for the formation of public opinion. Rather, the underlying principle behind these regulations often has to do with the special disseminational and marketing conditions associated with the type of transmission technology employed; the programming aspects themselves may have an impact, but often only indirectly. For example, traditional terrestrial television was regulated as broadcasting, whereas microwave transmissions by MDS were assigned to the "common carrier" classification. However, because of court decisions limiting its discretion to so categorize, the FCC has also increasingly been moving to orient its regulation—that is, classification in regulatory categories—according to the service, that is, the type of offering.[16] This has enabled some offerings to be reclassified. For instance, in classifying subscription TV (STV), the FCC switched from looking at the content of programming to the question of whether it was directed toward the general public. By referring to the limited target audience, the necessity of special reception equipment, and contractual relationships, the FCC made it possible to redefine STV as nonbroadcasting and thus to exempt it from many broadcasting requirements.[17]

The Communications Act of 1934 is not limited to one particular technology, (e.g., terrestrial broadcasting), though the power to regulate new video media is less clear. The FCC tends to enact different regulations depending on the type of medium, especially the transmission technology. This is partly justified with the goal of giving financially sound, experienced broadcasters an incentive to enter the market in areas with underdeveloped transmission technologies.[18]

The FCC has made use of its regulatory power with varying intensity.

During the 1960s and 1970s, it fought to delay the emergence of cable broadcasting in the interest of terrestrial broadcasters.[19] But at other times—particularly during the more recent phases of de- and reregulation—it has tried to eliminate some of the discrepancies between regulatory theory and reality by adjusting its normative claim.

Over its nearly sixty-year history, the FCC has been able to gain more experience than any other broadcasting regulatory authority. Although we cannot examine its activities in detail here, there is a wealth of literature on this subject (see Cole & Oettinger, 1978; Krasnow, Longley, & Terry, 1982; Levin, 1980; Le Duc, 1973, 1987; Mosco, 1979). Instead, we will look at individual fields of operation, primarily those for which supervisory institutions in other countries are also responsible. The regulatory policy of the FCC is especially illuminating with regard to broadcasting by terrestrial operators. This is the focus of the section that follows.

5. FIELDS OF REGULATION AND SUPERVISION

5.1. The Evaluation and Decision-Making Dilemma in Licensing

Licensing has always been a requirement for broadcasters operating with terrestrial frequencies. With more than 10,000 radio and almost 1,500 television stations in existence, licensing is quantitatively an enormous task in the United States. The licensing procedure is centralized under the auspices of the FCC, which is set up as a federal agency. Indeed, the FCC lacks even regional offices. It is fair to assume that the examination process is simplified for the simple reason of expediting work. This results in an advantage for those who have already surmounted the licensing hurdle: Although they receive only a periodic license (five years for television; seven years for radio), it brings with it a rather firmly protected status. Thus, of the 11,000 applications for license renewal between 1982 and 1989, only 40 were opened to bids from competitors. From the standpoint of media policy, therefore, it is the initial license that is most important.

As is essentially the case in the other countries studied here, the U.S. license is awarded after what are called comparative hearings, in which the various applicants are compared with one another. The FCC has repeatedly noted how problematic these procedures can be.[20] Even among broadcasters (at least licensees), they meet with criticism. Consideration has been given to replacing them with other procedures, such as a lottery or an auction. Indeed, auctioning was orginally used in some marginal services, particularly LPTV und MMDS.[21] In 1993 the commission was authorized for five years to use competitive bidding or auctions to award initial licenses and construction permits when mutually exclusive applications are filed for subscription-based services.[22] The authorization by Congress does not cover auctioning of broadcasting licenses.[23]

In public discussions serious attention has often been given to the concept that a broadcasting license—a right to use a scarce public good—represents an economic commodity, making free allocation difficult to justify economically. Up to now, the FCC has stayed with the procedure of awarding licenses at no charge (for the reasons behind this, see Barnett, 1988: 108; with reference to the various proposals, see Franklin & Anderson, 1990: 774f.). But the less broadcasters are expected to fulfill special public service obligations, the fairer it is to require financial compensation in exchange for the awarding of the license.

Applicants for broadcasting licenses must, first, satisfy certain basic qualifications (i.e., prerequisites and restrictions) and, second, meet criteria for determining rank or priority. The basic qualifications include requirements dealing with citizenship (only U.S. citizens may acquire a license), character, financial qualifications (the ability to sustain a station for ninety days without any additional revenue whatsoever), and familiarity with the community to be served (community ascertainment studies). (For the basic and priority criteria, see Carter, Dee, Gaynes, & Zuckman, 1994: 353ff., 380ff.) The scope of the licensing restrictions also addresses the problem of diversity of media ownership (see Section 5.3).

In addition, FCC selection criteria seek to effect a structural guarantee to ensure that the substance of programming is oriented according to the public interest (for the factors relating to the public interest, see Holsinger, 1991: 431f.). The FCC has never established official minimum norms or requirements for any programming category; instead, prior to deregulation, it used unofficial guidelines as application processing criteria. Some of these were abandoned in 1981 and the remainder in 1984. Therefore, no programming information is required from broadcasting applicants for new stations or from the transferees of existing stations. The stations must simply make available to the public quarterly issues/program lists, reflecting a licensee's most significant programming treatment of various topics, including community issues (see also Section 5.4).

There are, however, a number of rules based on the assumption that the organizational and personnel structures of a broadcaster are reflected in its programming. For this reason, the FCC gives preference to applications from stations whose owners are also involved in the broadcaster's day-to-day activities, since it is believed that a locally anchored owner is more likely to take up local concerns in station programming (Holsinger, 1991: 431). This rule, moreover, helps to curb industry concentration. Furthermore, special preference has been given to applications from minorities and women in order to increase diversity in program offerings. But this policy of promoting women and minorities has of late been strongly questioned by the FCC itself as well as, on several occasions, in court decisions.[24] However, Congress pressured the FCC, ultimately forcing it, via budget legislation, to readopt and continue the policy. The Supreme Court has since held such support for ethnic minorities to be substantially related to the achievement of the important governmental objective of broadcast diversity and has thus ruled it constitutional.[25] Even in areas in which licenses are awarded in lotteries, minority and gender ownership is preferred by according them "weighted" chances[26] (see also Section 5.7).

The decision-making criteria employed are generally vague and in need of interpretation, and for this reason, the underlying premises are of only limited use. Moreover, the decision relies upon prognoses that are often without accurate bases (e.g., with respect to the applicant's financial capabilities). This was particularly true as far as programming commitments were concerned in the period before deregulation. For instance, it has often been observed that broadcasters made fantastic promises—such as giving consideration to local needs in programming, special children's broadcasts, and so forth—that they were later unable to meet (see Cole & Oettinger, 1978: 146ff.; Le Duc, 1987: 47). Apparently, some broadcasters acted with the realistic expectations that pledges made when a decision is to be reached are hard to refute and that even when pledges are not kept, the license would not later be revoked. In fact, revocation of a license or refusal to renew a license has very rarely occurred (for data, see Abel, Clift, & Weiss, 1970: 411ff.; Weiss, Ostroff, & Clift, 1980: 69ff.)—and this usually in the event of violations in nonprogramming matters or a combination of violations, but almost never as a sanction solely for violating programming duties (see also Cole & Oettinger, 1978: 134, 164ff., 253f., 264f.; Chamberlin, 1982: 1069ff.).

Once licensed, a broadcaster generates facts through its activities. It becomes an important journalistic, cultural, political, and economic factor at its location; it makes investments, creates jobs, and normally generates expectations among its audience that it will continue its activities. Even when one broadcaster is criticized, it is uncertain that some other applicant would broadcast "better" programming. With regard to this finding, the FCC has tended to apply stronger criteria for initial licensing than for decisions on license renewal. It has expressly accepted the interest of licensed broadcasters in protection of their status.[27] It has explicitly stated that the reasons for this are the uncertainty regarding the actual fulfillment of competitors' promises, encouragement for broadcasters to improve their service, and structural protection for the broadcasting industry (for a critical appraisal, see McGregor, 1989: 295–301). The courts also accept the protected-status argument, although they sometimes attempt to mitigate the prominence it is accorded.[28]

Even the transfer of a license[29] is subject to only weak review: A broadcaster need only request the transfer, submit the application for examination, and agree to observe FCC regulations. No consideration is given to whether some other applicant might be available that is better qualified than the transferee (47 U.S.C. 310[d]). However, it does not seem that freely negotiable stations have engaged in tactics of short-term profits with quick resale, as some critics have feared. (This, at least, is the result reached by an analysis of sales between 1973 and 1986 by Bates, 1987: 317–23.) The FCC uses the concept of sale as a regulatory issue only with regard to promoting minority ownership in broadcasting stations. In principle, when the FCC has reservations about a licensee's fulfillment of license conditions and for this reason wishes to hold a revocation or nonrenewal hearing, the licensee is prohibited from selling its station as of that time. The only exception to this rule is the so-called distress sale policy, whereby under certain conditions approval will be given to a sale to a group with significant minority ownership.[30] Yet this possibility has had only a marginal effect.[31]

In deciding on renewal applications, the FCC relies, on one hand, on information provided by the broadcaster and, on the other, on programming complaints by third parties and the submissions of competitors for the license. It operates under the legal presumption that the licensed broadcaster is satisfying the needs of the audience in its reception area.[32] Submission of contrary evidence showing violations has recently been made more difficult. Moreover, for license renewals, a simplified procedure has been introduced—the so-called postcard renewal—for which a small amount of information on foregoing broadcasting operations suffices.[33] Only 5% of licensees are randomly selected for the purpose of closer examination (Holsinger, 1991: 434). The FCC does not conduct systematic program monitoring, which would be almost impossible in light of the large number of broadcasters and the FCC's relatively small staff. The earlier duty of broadcasters to maintain program logs has been replaced by the requirement to submit quarterly reports indicating the programs the broadcaster believes to represent "significant treatment [of] community issues."[34] The FCC decision to relax the requirements still further was overturned by the courts, because, among other reasons, this did not guarantee a sufficient information basis for a possible challenge to the license renewal by third parties.[35]

The process of FCC deregulation and subsequent control and correction by the courts first took place in the radio sector and then later in television. This reveals a recurrent pattern of the FCC, first deregulating radio and thereafter applying the deregulation to television. In this way, the radio sector became a sort of experimental area.

It is evident that at the moment the FCC is satisfied with having only a small amount of data on which to make its decisions, whereas previously, the FCC had worked with more extensive documentation. However, it is still able to conduct complex hearings in order to gain a more thorough resolution of the facts. (For the hearings procedure, see Carter et al., 1994: 375ff. For hearings procedures in the past, see Cole & Oettinger, 1978: 213ff.) Affected community groups may now also participate in the licensing procedure, for example, by filing a "petition to deny" and thus calling for hearings, which in turn enables them to examine the submitted material closely and, if need be, question it and make it public.[36] Citizens' groups have very often participated in the application process. Public-interest law firms have stepped in, representing minority groups or women's groups who criticize applicants as being unresponsive to their needs. Since hearings can result in both financial and image costs for a broadcaster, such groups can pose a considerable threat potential. The FCC and the broadcasting industry view these complaints with reservation, since political interest groups can use them to exert pressure on programming changes (Teeter & Le Duc, 1992: 377f.).

Licensed broadcasters tend to object to the fact that competitors for a license can be given preference over them solely on the basis of promises; broadcasters claim that consideration must also be given to all that they have done to develop viable broadcasting. The FCC has agreed that the situation of a broadcaster who has been in the public eye for a long time is different from that of a competitor who makes pledges for the future that cannot be examined for their reliability. At the same time, the FCC has found that a licensee often becomes a political power factor, one that cannot easily be pushed around once the license expires. As a result, it has only been able to decide on nonrenewal of licenses in the event of grievous violations but not solely on the assumption that a competitor might, for example, broadcast even better programming (see, e.g., Lively, 1992: 175).

The recognition of a relatively protected status for licensed broadcasters is related to the reduction of substantive programming requirements placed on them: The lower the standards of conduct, the less chance there is that the FCC will orient its decisions according to such standards. Nevertheless, the possibility that the license may be awarded to a competitor is a foreboding idea for licensees.

The suspicion is occasionally raised that some companies intentionally take part in the licensing process without any likelihood of being awarded one. Their motivation is to scare existing licensees into paying cash settlements in exchange for withdrawal of these applications.[37] A number of years ago, Congress felt compelled to limit such payments to the actual expenses incurred in filing the spurious application. However, this policy also caused serious competitors to avoid reaching agreements, a situation contrary to the law's intended policy of encouraging conflict resolution through settlement (Lively, 1992: 172). Ultimately, the cap on costs was relaxed, and the FCC enacted new rules to address the problem more effectively.[38] Now, prior FCC approval is required before any negotiations between a license holder and a challenger or competing applicant can be resolved. Such approval will be measured against the public interest and in essence only will be granted when possible payments do not exceed opponents' legitimate, prudent expenses and no other concessions are made. This is intended to offer adequate protection against abusive proceedings by both civic groups as well as potential competitors. Initial studies indicate that this is in fact happening (Teeter & Le Duc, 1992: 378f.).

Uncertainties during the licensing phase have to do with the virtual lack of clear, recognized standards for issuing and renewing licenses. It therefore comes as no surprise that the FCC has been looking for ways to alleviate the decision-making dilemma and also to reduce the risk of abuse. As a result, a practically unstoppable trend can be observed in the United States toward more and more protected status for the licensee. In other countries, there are similar developments. But protected status leads to situations where the possibilities decrease for compelling broadcasters to adhere to legal requirements and to their pledges: The sanction risk is reduced. This in turn amounts to de facto deregulation in the area of administrative requirements. As long as the nonfulfillment of expectations or even the violation of rules of conduct cannot be severely sanctioned because of de facto protected status, rules lose their practical validity.[39]

5.2. Disparity between the Object of Supervision and De Facto Programming Responsibility

The "frequency policing" aspect of U.S. broadcasting regulation has meant that a license is awarded for each locally available frequency. As a result, licenses are granted only to local broadcasters. But it quickly became apparent in the United States that the local broadcasting structure collides with economic necessities or incentives that aim at interlocked systems functioning above the local level. Thus, very early—in 1926—the first multistation programming company, a network, arose in the guise of the National Broadcasting Company, and others soon followed. The large network companies dominated first in the radio sector and then in television (for an overview, see Levin, 1980: 276ff.).[40] Then came other, smaller networks, such as those for public broadcasting, as well as so-called community radio (cf. Kleinsteuber, 1988: 132f.; Hardt, 1981: 373ff.). From an editorial as well as economic standpoint, the most important radio and television stations are affiliated with networks. Network companies in the United States do not limit themselves to acquiring programming but also sell commercials and advertise their programming. Therefore, they have a large impact on the editorial and economic development of broadcasters and at the same time enable the latter to work with a small number of personnel. Basically, U.S. networks do not themselves produce their programming; rather, they contract with producers for it or purchase it on the open market. Independent stations rely on an independent supply of programming, though they often buy and air reruns of network programs.

Due to the dominant role of the network companies, a media structure has developed for which the legislature's regulatory concept is not prepared: The local broadcaster, licensed and supervized by the FCC, is dependent on another company whose program offerings it can hardly influence and for which it can accept only formal programming responsibility. However, this other company—the network company—is not subject to any licensing or supervision. The gap between legal regulatory structure and the practice of de facto programming provision leads to a weakening of control, since the controlling body can at best exert indirect pressure on the programming company, namely, only through the rules of conduct and supervisory measures directed at local broadcasters. Since the network companies are also themselves owners of local broadcasting stations, they are directly affected by FCC rules in this role.

The FCC has thus far not found a way to submit the networks directly to its supervision. Many observers believe that an important reason for the basic failure of the FCC's former efforts to enforce programming requirements for broadcasters lies in the network structure. However, the FCC has developed instruments for, above all, limiting the market power of the networks and thereby countering the earlier threat of complete domination of the broadcasting market by the networks. In so doing, it works on the relationship of the networks with both their affiliates and their program producers, that is, the emergence of a software market, and also on the role of the network companies as broadcast licensees. The FCC forbids certain clauses in affiliates' contracts with the networks, particularly those that place the affiliates at their network's mercy. These are the so-called network rules. For example, contracts may not forbid the affiliates from rejecting network programming they consider to be unsatisfactory or prevent them from broadcasting programs from other networks.[41] Particularly important is the "prime-time access rule," by which limitations regarding the share of network programming during prime time are imposed on network-owned or network-affiliated stations in the fifty largest television markets.[42] In practical effects, the prime-time access rule means that three hours of the four-hour prime-time period are to be devoted to network-produced or network-distributed entertainment programs, half an hour to network news programming, and the remaining half-hour to either nonnetwork-produced entertainment programs or special documentary or public-affairs features (Carter, Dee, et al., 1994: 437).

Regarding the relationship between the networks and program producers, the FCC tried to prevent the networks from exploiting their market-dominant position by banning ownership interests in broadcast programming produced by others. Accompanying this stricture were restrictions on in-house production of programming and on operating as a syndicator of nonnetwork programming, which the FCC hoped would ensure a more diverse program production and distribution market. In this respect, the object was to combat vertical concentration. These rules were generally categorized as "fin–syn" (financial–syndication) rules. Because of the increasing economic difficulties faced by the three traditional networks,[43] the FCC bowed to considerable pressure from the broadcasting industry to abolish the fin–syn rules (see, e.g., *Broadcasting*, 15 April 1991: 40; 15 July 1991: 29) and relaxed them.[44] But as has often been the case, courts have intervened here as well,[45] and the FCC amended the rules in accordance with the court order.[46]

The networks are no longer prevented from entering into prime-time television entertainment programming production and syndication business. The only remaining restriction is that even if the networks own syndication rights to prime-time entertainment programming, they are prohibited from actively being the syndication agent. There are no longer any restrictions on networks' holding a financial interest in any type of network television programming or owning syndicated rights in such network programming. The only remaining regulation to separate production from distribution deals with nonnetwork first-run programs: The networks are prohibited from owning financial interests or syndicated rights for such programs unless the network itself produced them solely in-house, and here as well, the network's syndicational role must be passive (Carter, Dee, et al., 1994: 439f.). These restrictions will sunset in 1995 unless the FCC finds that they should be continued.

The approaches taken by the FCC have not been able to alter the fact that a considerable amount of journalistic and economic power in the broadcasting sector lies in the hands of the networks (cf. Levin, 1980: 297) and that the FCC has only limited means of controlling this power. Therefore, a supervisory deficit is perforce built into the system structurally. Supervision is in danger of being satisfied with fictions of licensed broadcasters' true programming responsibility.

5.3. Combating Concentration

As media economists have proved, broadcasting is a branch of the economy that is subject to strong pressure of concentration. At the same time, however, concentration is considered to be a threat to diversity. In order to surmount this problem, the United States has long recognized, in addition to the general antitrust rules, special limitations on concentration in the broadcasting sector. The main thrust of these regulations targets the awarding of the license. In the foreground are two regulatory types of restrictions: on multiple ownership and on cross-ownership (for details, see Carter, Dee, et al., 1994: 362ff.; Ginsburg, Botein, & Director, 1991: 180ff.; Head & Sterling, 1990: 452ff.). Without doubt, these rules have contributed to a limitation on the level of concentration. They are also enforceable, although at times with considerable difficulty.

That these rules have had an effect is clear, since there has always been a jump in concentration whenever they were relaxed or rescinded.[47] This effectiveness is also indicated by the fact that large media companies either fight them[48] or make an effort to sidestep them via circumvention or by applying for exceptions, and so forth. Nevertheless, the way the FCC battles concentration shows how difficult the task is. This can be seen in the development of these rules, which can be described here only in brief.[49]

One striking feature of this development is that, in view of the pressure to concentrate, the FCC constantly saw itself forced to leave loopholes or to relax the rules. From the mid-1980s and until the early 1990s, permissible multiple ownership was governed by the following rules: In a given market, which was essentially defined as the territory of transmission, television ownership was restricted to one station (the duopoly rule), and radio, to one AM and one FM station. In addition to these local rules, multiple ownership at the national level was limited to twelve AM, twelve FM, and twelve TV stations, whereby such TV stations could not reach more than 25% of all television households. The cap could go to 30% if at least 5% was contributed by minority-managed broadcasters. These restrictions were themselves the product of increasing relaxation of regulation. In other words, these limits went from the earlier total of six radio stations to seven and from three TV stations to five and then later to seven, before expanding to the aforementioned restrictions in 1984. Likewise in 1984, additional rules to combat regional concentration, which were introduced in 1975, were eliminated completely. It is also evident however, that in reality, concentration was stronger than that permitted by the prescribed number of stations. But this usually did not result from rule violations but rather from various of the FCC's strategies to ensure regulatory flexibility—primarily with the objective of deflating opposition to the anticoncentration measures. Thus, instead of undertaking preventive measures to combat concentration, the FCC just tended to observe mounting concentration developments. If growing concentration meant that action was called for, the FCC inclined to look at the status quo, that is, not to attack existing concentration even when it conflicted with newly enacted guidelines (thus allowing widespread grandfathering). (For examples in the area of cross-ownership, see Zuckman, Gaynes, Carter, & Dee, 1988: 401, 403.) Moreover, the FCC often made full use of its capacity to allow waivers. In 1992, the FCC undertook a comprehensive realignment of its concentration regulation, concluding that this should be largely abolished or at least relaxed.

Against the background of economic difficulties in the broadcasting field, the FCC first sought to increase permissible multiple ownership in the radio sector to thirty AM and thirty FM stations nationally and, locally, depending on market size, to up to three AM and three FM stations with an audience of not more than 25%.[50] However, Congress reacted with strong opposition to such a broad easing of the regulations, threatening to fix by statute the old 12-12-12 rule (*Television Digest,* 16 March 1992, 2; *Broadcasting,* 16 March 1992: 14), and after congressional pressure on the FCC via budget negotiations (*Communications Daily,* 20 March 1992) and an investigation of the matter by a subcommittee (*Television Digest,* 23 March 1992, 5; Bureau of National Affairs, Daily Report for Executives, 24 March 1992, A12), the FCC opted for more moderate relaxation of the rules.

Permissible multiple ownership at the national level was increased to eighteen AM and eighteen FM radio stations, and since 1994 up to twenty of each. Furthermore, under special circumstances, holdings in an additional three stations are allowed, especially if they are managed by minorities or represent small broadcasters. Multiple ownership at the local level is limited in markets with fifteen or more stations to two AM and two FM stations, whereby any combination may not lead to excessive concentration. This is presumed to be the case with a share in excess of 25% of the listening audience. In markets with less than fifteen stations, ownership of up to three stations is permitted, of which not more than two may be in the same area (AM/FM), as long as concentration amounts to less than 50% of the stations represented in the market.[51] Thus, for the first time in the history of U.S. broadcast regulation, the FCC now permits one licensee to own and program dual AM and FM facilities serving a single audience. Following this reregulation of the radio sector, the FCC is seeking to undertake a corresponding relaxation of the multiple ownership restrictions for television as well.[52] The most important motive behind this relaxation policy is the increase in competing media that has siphoned audiences away from traditional over-the-air broadcasting and increased the total available pool of media voices.

Despite relaxation of the rules, they continue to meet with opposition in wide areas of the media industry. Media history shows that media companies have always found ways to sidestep governmental regulations without committing formal violations. The decisive factor is the economic interest in not being hindered by governmental regulation when pursuing company policies. This can also be observed with respect to efforts to combat concentration. Recent years have seen the emergence of a type of cooperation in broadcasting designed to avoid the multiple-ownership restrictions and to grant stations some of the economic advantages of a larger alliance. By way of so-called local marketing agreements (LMAs) or time brokerage agreements, a radio station "rented" broadcasting time with another station or entered into advertising groups.[53] Under the new concentration rules, the FCC now has assumed jurisdiction over, and regulates, such cooperation to the extent that it, roughly speaking, only allows broadcast programs to be brokered when the brokered station could permissibly be acquired under the concentration rules by the renting station. In particular, stations have also been forbidden from simultaneously broadcasting more than 25% of their programming over another, nonowned station when the latter offers the same type of service (AM/FM) and serves basically the same market.[54]

Also important in combatting concentration—and very complex in their details—are the cross-ownership restrictions, which generally prohibit the operation of VHF television and radio broadcasting in the same area of transmission.[55] The FCC has undertaken substantial curtailments in the rules here as well (see Carter, Dee, et al., 1994: 365f.).

A third regulatory type has to do with diagonal cross-ownership between different media, especially between the press and broadcasting and between cable operators and broadcasters. From the outset (1975), such rules applied to ties between the press and television, but only at the local level. Although it had found no evidence of monopoly abuse, the FCC concluded that the twin goals of economic competition and competition in the marketplace of ideas would be furthered if (future) newspaper–broadcasting combinations were prohibited. Conglomerate concentration above the local level was left untouched (see Kübler, 1982: 38ff.). There are certain restrictions for cross-ownership between cable systems and television stations. For instance, local television broadcasters may not own holdings in cable systems in their area of transmission. But from 1970 onward, networks were also precluded from entering the cable television market by the network–cable cross-ownership rule; however, this was rescinded in the course of the reworking of the concentration rules in 1992.[56] In this way, the FCC fulfilled—at least partially—the networks' wishes. But at the same time it raised fears with network affiliate stations, which became concerned that the networks would bypass them by using cable television to deliver network programming directly to viewers. This shows that regulations run up against the most divergent of interests.

Networks may now operate cable television systems as long as they do not reach nationally more than 10% and locally more than 50% of households with cable television connections. The local limitation does not apply to markets in which the cable system competes with other systems (e.g., MMDS), thereby limiting the threat of acquisition of disseminational power.[57] No such restrictions whatsoever apply to coownership between cable systems and newspaper publishers.[58]

Thus, the history of the FCC's battle against concentration consists of step-by-step retraction of former restrictions. In a media system driven by market philosophy, there are a number of obvious reasons for this pattern. One is the reasoning behind the new FCC rules on multiple ownership in the radio sector.[59] The decisive factors are, on the one hand, mechanisms of the economic market itself and, on the other, the assumptions made with respect to the tying of economic competition to extraeconomic objectives, that is, the general public interest.

The media market, and most particularly the broadcasting market, is characterized by considerable economies of scale. Cooperation in administration, acquisition of advertising, production, programming, and other functions offers substantial cost advantages. This makes a relaxation of the concentration rules appear to be a reasonable means to aid media sectors encountering financial difficulties. Their economic viability is in turn considered necessary and desirable for the simple reason that without them, the assurance of (financially costly) public interest offerings, such as news or minority programs, would be impossible.

However, the policy of increased concentration is normally countered with the objection that consolidation of economic power threatens diversity of content. But as long as several, similarly strong competitors are in the market and the functioning of competition is not threatened by mergers or—when this results in a larger number of robust market participants—is even fostered, this objection thus takes on less significance. With respect to the radio rules, the FCC also points to intermedia competition, which also facilitates such argumentation. Under such circumstances, it can generally be proved and rarely refuted that the viability brought about by concentration is what enables maximum performance in the competition of the fittest. The ever more concentrated market[60] to a certain extent itself legitimizes the argument that market growth is necessary. The more broadcasters or networks compete for expensive series, movies, large news staffs, and so forth, the more important economies of scale become. As a result, market opportunities and market access are accorded primarily to those who are economically viable. Earlier objectives, such as locally anchoring programming, participation by average citizens in broadcasting, or assumptions regarding a technologically conditioned reduction in the barriers to access, are threatening to become lost from sight.

When the concentration process has the effect that, structurally, groups with fewer chances in the economic market (small broadcasters, minorities) are placed at a disadvantage, the FCC tries (at least partially) to compensate by tailoring the concentration rules. Mentioned above was the higher number for permissible multiple ownership, when such membership comprises the participation of minority broadcasters or smaller broadcasters. Furthermore, the FCC introduced hearings to gather proposals for allowing multiple owners to acquire and hold a certain number of shares above and beyond the concentration limits if they conduct an "incubator" program to reduce the barriers to market access for the above-mentioned groups. The requirements placed on such a program are to be defined through flexible FCC policy. The focus has been on components such as technical and organizational assistance, loans and other financial aid, as well as training programs.[61] The basic idea behind these approaches is to place conditions on the economic incentives to promote concentration that alleviates their detrimental consequences. Whether such measures would be effective is certainly doubtful.

Concentration tendencies can be found in other facets as well. Particularly evident are strong diversification by media companies into other economic sectors and a drive by other companies to enter the media field. The public is tensely watching these transactions and new formations. This particularly is the case for the often spectacular changes in the ownership of the large networks, as well as for many other mergers. The media industry is not only in and of itself becoming more concentrated; it has also become increasingly interwoven with the rest of the economy.

Prior to a change in ownership of a license, the FCC must grant its consent. To handle tender offers and proxy contests, the FCC set general guidelines in a policy statement dealing with attempts to change corporate ownership.[62]

Becoming increasingly rare are the special aspects of the industry that once characterized it, such as a media owner's specific, journalistic commitment. Just as broadcasting has become an everyday commodity, the broadcasting company has increasingly become an everyday business. The FCC is rather powerless when it comes to this. Its powers, deriving from the "frequency police" argument, make it more difficult for it to control processes of diversifying concentration. As a result, the FCC keeps its distance. The most likely preventative measure, though weak, that could be employed would be antitrust law.[63] In this way, it has also become apparent in the United States that, upon the retreat of broadcasting law, commercial law is a remaining lever for controlling the market and thus broadcasting as a whole.

5.4. Supervision of Programming Conduct

Since U.S. constitutional law does not recognize the concept of objective basic rights,[64] the legal system has been particularly reluctant to impose direct programming regulations. Preference is given to structural provisions, that is, precautions of a structural nature, which serve to aid indirectly in achieving programming repercussions. The market principle is one such element of structural control: Precautions to ensure this principle thus likewise belong in this context. In the course of deregulation, market control has more and more become the principal element of control in the broadcasting sector.

However, recent developments show that the market result is perceived to be inadequate in protecting vulnerable values. This is recognized to be especially true for programs considered indecent, particularly those shows depicting sex or violence. As a result, more stringent measures are being resorted to. Such attempts, which can also be observed in other countries, can be classified as a kind of "consumer protection" legislation, primarily serving the protection of juveniles. However recently, more stringent measures, such as the encouragement of children's programming, represent a certain break with the market concept, even though it may not be justified to conclude from this that there has been a partial retreat from the market paradigm.[65] Regulation of programming and other policies targeting conduct still meets with strong opposition.

5.4.1. Unofficial Guidelines and the Encouragement of Self-Regulation

Up until the wave of deregulation, there were a variety of guidelines and rules, including those relating to programming, but these were embedded in a "soft" regulatory technique. They were composed in part of resort to so-called self-regulation by the broadcasting industry, that is, to a code of conduct set up by the National Association of Broadcasters (NAB). But this code had to be abandoned, because a court objected to the way it was enacted[66] and since the antitrust arm of the Department of Justice considered it to be the equivalent of an agreement in restraint of trade. (For details on the reversal, see *Broadcasting*, 15 March 1982: 45, 7 January 1985: 204.) In addition, the FCC had "rules of thumb" in the form of administrative practices as well as guidelines that showed under which circumstances the FCC would consider programming conduct positive or negative.

Such programming guidelines were largely abolished in the course of deregulation.[67] This applied to minimum time requirements for news and informational programs and for local broadcasts. The application of such rules of thumb caused a number of definitional problems: For example, what is an informational program? what defines locally connected? and so forth. In addition, it was unclear whether the objectives of program content could be attained in a sensible manner using quantitative values. The FCC's decision to abandon these guidelines was accompanied by its express expectation that this would not change programming conduct appreciably. Apparently, the FCC believed that the quantitative values set forth in its rules did in fact correspond to the economic and editorial interests of broadcasters or that they were so broad and flexible that their application did not persistently collide with these interests.

The technique of resorting to self-regulatory mechanisms was taken up anew with the Television Program Improvement Act of 1990 (47 U.S.C. 303[c]) with respect to restricting violent programs on television.[68] Reflecting heated social discussion and the decrying of violence in the media, the act encourages TV networks, local stations, producers, and the cable industry to enact voluntary guidelines to reduce the negative impact of depictions of violence in television broadcasts, in that it exempts them for 3 years from the antitrust provisions that would normally prevent such action. Initially, the reaction of the TV industry to this law ranged from cool to resistant, since it viewed the market mechanism as basically offering sufficient protection. It also doubted whether, in view of the diverse interests in this highly differentiated and competitive market, a uniform code could even be worked out. In the interim, there have been meetings by the networks and other terrestrial broadcasters,[69] although these have resulted in nothing more than statements of principles and declarations of intent. The cable industry, which is unwilling to cooperate with terrestrial broadcasters for the time being, seems even farther from such a code. This has led to signs of impatience in the political arena, coupled with comments about alternative means of regulation.[70]

It should be mentioned that the FCC has consistently been reluctant to demand regulation of broadcasting formats. It has never established any list of preferred program categories, nor has it opposed changes in a station's format.[71]

5.4.2. The Fairness Doctrine

A particularly illustrative example for the difficulties associated with the supervision of broadcast programming is the so-called fairness doctrine. Since this was the model for the duties of balance and fairness set forth in some Western European countries such as Germany, it is especially worthwhile to study the fate of this doctrine. In substance, the general fairness obligations comprised two main duties. First, it was expected that a broadcaster spend adequate time in its programming on the treatment of controversial matters of general importance. Second, the broadcaster was obligated to undertake such treatment in a fair manner, that is, to give suitable opportunity for the depiction of other views. (For the substance of this doctrine, see Simmons, 1978; Rowan, 1984.) Related to fairness were also requirements to provide equal opportunity for political candidates (47 U.S.C. 315) and some obligations dealing with personal attacks and political editorials. The following remarks focus on the general fairness obligations.

From a legal standpoint, the fairness doctrine was controversial on a number of levels. For instance, there was disagreement about whether it was consistent with the First Amendment. It was also long unclear whether the doctrine was rooted in the Communications Act or only in the FCC's programming requirements dealing with the common good, so that in the latter case the FCC could then abolish it. After a federal Court of Appeals decided in 1987 that the doctrine was essentially not a statutory duty,[72] the FCC largely abrogated it.[73] However, the constitutionality of the doctrine continues to be an open question, as the courts have explicitly refused to rule on this. Accordingly, Congress is not prevented from reintroducing it any time by statute.[74] But despite its future prospects, it is, from a historical standpoint, an important example of the possibilities for supervision of broadcasting.

The enforcement of a rule like the fairness doctrine entails difficulties in a commercial television system. Experiences with such broadcasting systems teach us that, left to their own devices, they tend to deal with controversial topics and to depict controversial opinions only to a limited extent (see, e.g., Barrow, 1975: 673f.; Mallamud, 1973: 115ff.; Malone, 1972: 216f.). Every effort at governmental control conflicts with the editorial and economic interests of broadcasters and thus represents a virtually irresolvable task if the government expects broadcasters to present conflicting views on issues of public importance and to devote a reasonable amount of time to the discussion of controversial issues (part one of the fairness doctrine). Similar difficulties are encountered with the enforcement of the second part of the fairness doctrine, the obligation to provide coverage in a fair and balanced manner, that is, to afford reasonable opportunity for the presentation of opposing viewpoints. The depiction of controversial opinions requires television time and, according to the predominant view, is often less attractive for the majority of viewers than the simplified emphasis of only one viewpoint. In addition, the advertising industry is rarely interested in a program setting that is not characterized by harmony. It therefore comes as no surprise that the large U.S. networks are generally uninterested in widening the spectrum of controversy. At one point, they had even decided not to broadcast any commercials dealing with controversial topics because they would then be obligated to broadcast commercials with contrasting substance.

When the FCC tried to implement the fairness doctrine, it had to work in an area that was not only constitutionally sensitive but also full of conflicts. It did more than just practice restraint in terms of the substance of the doctrine; it also erected large procedural hurdles for fairness complaints by viewers (details can be found in Hoffmann-Riem, 1981b: 178ff., 186ff.; Zuckman et al., 1988: 462ff.). As a result, the number of successful fairness complaints was relatively small. Nearly 10,000 complaints were submitted by citizens annually, of which 99% were not processed because the procedural requirements had not been fulfilled (for details, see Federal Communications Commission, 1984: 74). Only approximately 0.15% of the complaints led to corrective action by the FCC with broadcasters (see also Le Duc, 1988: 59; earlier figures are in Hoffmann-Riem, 1981b: 185f.). However, broadcasters were able to avoid sanctions by declaring themselves willing to redress the situation. Only those unwilling to do so had to expect sanctions—specifically, rather low fines. Violations of the fairness doctrine have almost never been accepted as cause for denying a license renewal. In the 60-year history of broadcasting, there have been only two cases where a license was denied due to systematic fairness violations (see Le Duc, 1988: 60 n. 18).

Yet it would be improper to measure the effectiveness of the fairness doctrine according to such statistics. On the contrary, the doctrine had an impact in a number of ways, even if the FCC did not institute formal proceedings. For reasons of image and publicity alone, U.S. broadcasters tend to be very careful about avoiding any public conflict concerning their programming activities; above all, they go to great lengths to avert the FCC's time-consuming, costly administrative proceedings or even subsequent court actions (see Le Duc, 1988: 61f.). This means that they have monitored themselves for possible fairness violations. In addition, citizens' groups in the United States take active part in monitoring programming, and, as mentioned above (see Section 5.1), they have certain rights of participation in FCC proceedings. As a result, such groups have come up with many ways to activate the public as an informal control authority (see Krasnow et al., 1982: 54ff.; Cole & Oettinger, 1978: 63ff.; Hoffmann-Riem, 1981b: 108f., for further references). Often, local groups are taken by broadcasters "more" seriously than the distant FCC. Thus, there have developed tight networks of communication between local broadcasters and civic groups, which are then used as an avenue of direct yet informal contact for, on the one hand, bringing forward programming requests and complaints and, on the other, working together to come to terms (see Cole & Oettinger, 1978: 228ff.). Among other interests, the networks of informal broadcasting control have addressed the maintenance of fairness and balance (see Rowan, 1984: 72, 81ff.; Le Duc, 1988: 60f.). Comity has often extended so far that broadcasters have sometimes assumed the costs incurred by citizens' groups (see Grundfest, 1977: 85ff; Robinson, 1978a: 231ff.).

It is difficult to estimate the effects of this "extra-FCC" supervision of broadcasting.[75] It is plausible to assume that the success of this informal instrument correlates with the ability of interested parties to organize. Even if a certain degree of one-sidedness is to be expected in group participation, the diversity of the civic groups active in the United States makes it fair to assume that such one-sidedness acts as more of a counterweight than a fortification of the one-sidedness to which a commercial broadcasting system necessarily tends.

This impression is strengthened by the fact that the majority of broadcasters are interested in diminishing the power citizens have to intervene in FCC proceedings, including formal rights of participation, thereby eliminating the potential threat that civic groups have acquired thanks to the development of informal systems of control. In particular, the abolition of the fairness doctrine was a major concern of most parts of the broadcasting industry.

Paradoxically, however, it seems the accomplishments of citizens' groups were used as an argument for repealing the fairness doctrine. In fact, operation of the doctrine can generate "dysfunctional" programming effects, which were cited by its opponents and have been enumerated in detail in the literature.[76] For instance, it is apparent that most broadcasters seek to avoid programming containing controversial subjects. Of course, this fits—as mentioned above— with the general logic of a broadcasting system that for economic reasons is dependent on the largest possible audience and therefore tends toward the uncontroversial middle of the road. However, circumspect programming conduct is reinforced by the view that dealing with controversial matters would trigger obligations under the fairness doctrine, making it necessary to offer broadcasting time for positions that might not interest a broadcaster's audience. The second part of the fairness doctrine thus bears the inherent risk that duties under the doctrine's first part will not be fulfilled. In other words, if controversial issues are ignored, then fairness complaints can hardly be expected, thus leading to no subsequent duties under the second part of the doctrine.

The networks in particular felt forced to proceed cautiously. Although fairness complaints could not be brought against networks directly, they were able to be raised against broadcasters locally. For editorial and economic reasons, networks tend to be interested in seeing that their affiliates broadcast their programs without having to fear subsequent difficulties and that they not be exposed to costly fairness complaints. The networks must therefore take care to present programming that gives no cause for complaints in any region or from any audience. Thus, it seems clear that programming is brought to the least common denominator of tolerable conflict, even with respect to the network structure.

In view of this situation, it is not surprising that in its hearings to abolish the fairness doctrine, the FCC heard from nearly all broadcasters that the doctrine had a chilling effect and prevented them from conducting an open and robust public debate in their programming. This was extensively documented in both the 1985 fairness report[77] and the decision to rescind the fairness doctrine. Moreover, the FCC added that, through the fairness doctrine, preference was given de facto to the expression of established and orthodox opinions, whereas minority positions or unconventional expression was placed at a disadvantage. According to this viewpoint, the regulation of fairness had the opposite effect on program diversity and equal communicative opportunity. However, after the FCC did away with the fairness doctrine, there was no noticeable increase in the treatment of controversial topics (Aufderheide, 1990; another study even pointed out a decline in public affairs programs; *Broadcasting,* 10 April 1989: 75f.).

In evaluating the fairness discussion, however, it must be taken into consideration that it belongs in the larger context of the disputes surrounding the U.S. broadcasting system itself. While the FCC now emerges as an opponent of the fairness doctrine, this is a part of its deregulatory strategy. The fairness doctrine was the prototype for a regulatory style that did not conform to the market and for this reason necessarily represented an alien element in a largely deregulated media setting. It should also be noted that the reference to dysfunctional effects comes with particular intensity from, above all, broadcasters, whose other conduct would not seem to justify the assumption that only the fairness doctrine was preventing them from portraying unpopular or unconventional views in their programming. In contrast, proponents of the fairness doctrine are predominantly groups outside the political mainstream, which apparently see in it an instrument with which to compel at least a minimum amount of opportunity at expression.

In the discussion surrounding the abolition of the fairness doctrine and the efforts to reenact it (also Lively, 1992: 241–253), the arguments used often seem more of a pretext. But the fact remains that the fairness doctrine apparently has an ambivalent potential for programming effects. Since the duties covered by it are difficult to define clearly, some prophesize alarming, chilling effects while others raise unrealistic expectations for diversity of expression.

There is no indication that since the rescinding of the fairness doctrine broadcasters have changed or will change their programming conduct decisively, particularly as the FCC had long discontinued its practical enforcement under de facto deregulation. Moreover, as long as it remains unclear whether Congress will reintroduce the fairness doctrine, it is unlikely that much will change in general programming conduct. But even in the event of definitive abolishment of the doctrine, significant modifications cannot be expected from broadcasters who target a mass audience. Radio broadcasters directed at selected groups or minorities are most apt to gain new latitude.

In any case, in the future, the FCC intends to treat broadcasting comparably to the press, where there are newspapers reflecting minority views. But whether broadcasting's (and especially television's) peculiarities—including its technology, programming and media-specific economics—will permit a sufficiently broad spectrum of diversity remains debatable. It would be an overestimation of the fairness doctrine and thus of supervisory tools if the doctrine were to be given as the cause for the high substantive homogeneity of television offerings in the United States.

5.4.3. Protection of Minors

As with those set forth in other broadcasting systems, rules for the protection of minors are among the most hotly contested programming rules in the United States. Nevertheless, while the goal of protecting minors is not challenged in the United States, the way this objective is to be achieved is all the more so. "Negative" protection in this area is accomplished mainly with anticipatory measures to preempt depictions of violence, pornography, and other morally offensive broadcasts; "positive" protection is sought via obligations to show children's programming (for the development of regulations, see Kunkel & Watkins, 1987: 367).

The problem of violence on television was initially approached in the form of a family viewing policy similar to that in Great Britain. The rules provided that programs during prime time were to be appropriate for viewing by the entire family by keeping them free of violence and sexually explicit material. However, because of the intrusion this would represent on programming decisions, the FCC was uncertain whether it should enact such rules itself, and it put informal pressure on the NAB to adopt appropriate policies in its code of conduct. The FCC then sought indirectly to make these subject to its own evaluation. But for legal reasons, this path was later blocked by the courts, and, as mentioned above, the NAB ultimately had to abandon its code because of antitrust problems.

Even after abandoning the code, however, the networks voluntarily declared that they would adhere to the juvenile protection rules. Apparently, it conformed to their broadcasting philosophy to be able to show the public that they maintained a responsible programming policy. But it is unclear whether the rules had much impact, that is, whether they would have been able to have a lasting influence on conduct. Only during the initial period when they were in force—during 1975, when major public attention was focused on the issue—were clear reactions discernible (cf. Wiley, 1977: 188ff; see also Le Duc, 1987: 68f.). Later studies reveal a considerable increase in the number of violent programs, particularly at times when networks were engaged in sharp competition (see Hoffmann-Riem, 1981b: 168f.). Since broadcast violence probably triggers a strong viewing urge among the public, it was clear that observance of the rules, or good intentions, would take a back seat when ratings battles dictated. An increase in competition—the goal of deregulation policy—makes it even more difficult to implement such plans. This is reflected in studies showing that broadcasters' efforts have slackened off since the NAB code was terminated (cf. Barnett, 1988: 117). It thus is no surprise that Congress took another stab at regulating this area with the TV Program Improvement Act of 1990.

In addition, the fight to keep pornographic and other offensive programs off the air is justified with the argument of protecting minors. From the very beginning, there have been statutes in the United States that provide for a fine and loss of license for a station that broadcasts obscene, indecent, or profane language (as set forth since 1948 in 18 U.S.C. 1464). But up until the 1970s, this area was of little relevance for supervision, particularly since the networks demanded of their affiliates that they observe the relatively strict NAB standards, thereby exercising some prior control over them (see Teeter & Le Duc, 1992: 383ff.). The 1970s saw the emergence of radio stations that took up the topic of sexuality, primarily for talk shows ("topless radio"). Although the prohibition of obscene broadcasts is legally unproblematic in the United States, since this is not protected by the First Amendment,[78] the range of permissible restrictions on programs has even today to be definitively resolved.

Supervisory action initiated by the FCC[79] against these new radio stations in response to numerous complaints against them[80] led to a first Supreme Court ruling,[81] which held (despite strictly limiting the decision to the case at issue) that because of its ability to intrude into the privacy of the home and its easy accessibility to children, broadcasting may be subjected to regulation. Subsequently, the FCC considered relevant only those programs broadcast before

10 P.M. Until 1987 it did not undertake any supervisory steps against broadcasters (see Ginsburg et al., 1991: 538f.), which is likely attributable to the basic reticence to regulate content and the exclusive market orientation.

However, from 1987 onward, the regulation of indecent programs became an ongoing issue in media policy and media law, characterized initially by a tightening of the rules. This corresponded to a new tone being set at the FCC: FCC Chairman Alfred Sikes pushed for stronger enforcement of these FCC regulations (*Broadcasting,* 7 January 1991: 100ff.), and the overall political climate made anything that was deemed indecent basically open to restriction.

In early 1987, the FCC did not just begin once again to assail broadcasters for offensive programming; rather, it also made known that it had broadened its definition of indecency and limited the time for broadcasting of programs deemed indecent to the period between midnight and noon.[82] This policy coincided with the start of the oscillation between regulation and court challenges thereto.[83] While upholding the FCC definition and the power to protect juveniles against such programs, a federal Court of Appeals ruled that the FCC's reasoning for the specific broadcast time rule was insufficient.[84] However, shortly thereafter (and certainly as a reaction to the decision), Congress obligated the FCC by statute to enact rules for introducing a twenty-four-hour ban on indecent programming.[85] These rules were thereupon challenged in court as violating the First Amendment. They were voided, and the FCC was again called upon to enact new rules containing comprehensive reasoning.[86] In order to be compatible with the First Amendment, the regulation of indecent speech had to serve a compelling government interest and be narrowly tailored to that deemed absolutely necessary.[87] The FCC argued that children viewed television at all hours of the day, often had their own television sets, and, moreover, were familiar with the operation of videorecorders, which would permit them to record programs broadcast at any time for later viewing. For this reason, the FCC claimed that the only effective and thus necessary protection of children[88] was a twenty-four-hour ban. A renewed challenge led to repeal of the regulation for not being narrowly tailored, and the matter was returned to the FCC for determination, among other matters, of a time period during which indecent material might be broadcast.[89] In reaction to this ruling, Congress attached a midnight–6-A.M. "safe harbor" to the Public Broadcasting Funding Bill of 1992 (cf. *Broadcasting,* 10 August 1992: 28f.; *Electronic Media,* 31 August 1992: 2). Controversy about a "safe harbor for adults" is still ongoing (Carter, Dee, et al., 1994: 432).

The tightening of the regulation of indecent programs, which the public generally considered positive, was accompanied by an increase in the enforcement of regulations by the FCC.[90] The FCC even levied a record $600,000 fine against Infinity Broadcasting for a wide range of allegedly indecent remarks made by a talk show personality, and it also brought indecency charges against more than a dozen other stations.

There have also been attempts with so-called positive protection of juveniles. Much political support was amassed for the idea of using the extensive television consumption by children and adolescents as an opportunity to offer them programming commensurate with their youth—using television to achieve "prosocial" effects (for this concept, cf. National Institute of Mental Health, Television and Behavior, 1982: 48ff.). Although the FCC has undertaken a number of investigations on positive protection of juveniles, it has often backed down in the face of protests by the broadcasting industry. The original, blanket appeal to broadcasters that they fulfill the interest in children's programming was largely ignored[91] and was later tempered even further by the FCC.[92] The only remaining duty the FCC places on broadcasters is to take into account the concerns of the audience in their regions (i.e., including children and adolescents). However, this duty is difficult to operationalize and has no noticeable effects.

Although commercial television originally produced special children's programs, over time these were shifted to low-ratings periods (e.g., early mornings) and eventually largely eliminated (see Hoffmann-Riem, 1981b: 159f.; Ferall, 1989: 29). The FCC's Children's Task Force determined that the market had failed in the area of children's programming, but this did not motivate the FCC to take any action. In any case, the soft supervisory rules set up by the FCC to monitor market failure have never been able to change this. The reasons for this appear to be of an economic nature: Children's programs are generally less profitable than other programming. Children can be reached with commercial messages through other programs where adults can also be counted on in the audience. The FCC's efforts to strengthen its appeals by pointing out that they are important for licensing decisions have only prompted isolated reactions, such as that films originally intended for adults were redefined as children's programming. A certain percentage of children's programs was obtained by scheduling these shows at early morning hours (such as before school) at broadcast times that are difficult to fill anyway. That this created an incentive for children to view even more television was one of the dysfunctional effects of the response to concerns of juvenile protection.

In the area of cable, special children's offerings have been introduced. However, this was not the product of supervisory action but, rather, economic and editorial considerations. Children's programs are used by cable networks as an advertising tool to entice cable subscribers. As a result, cable television fulfills a certain compensatory function vis-à-vis terrestrial television. Nevertheless, there are clear signs that children's programming is also being scaled back in the cable arena when program time is required for other broadcasts (see Le Duc, 1987: 65).

With the enactment of the Children's Television Act in 1990 (47 U.S.C. 303[a]–[b], 394), however, a new effort was launched to reintensify positive protection of juveniles.[93] For instance, the Act prescribes that when a license is renewed, particular attention must be given to the extent to which the broadcaster has served the educational and informational needs of children in its overall programming, including programming specifically designed to serve such needs. This means that a sort of duty to broadcast children's programming has indirectly achieved the force of statute. In a specifying ruling, the FCC opted for a very broad interpretation of the Act and will largely rely on reporting duties.[94] Thus, a specific number of broadcasting hours were not imposed, nor was it required that all age groups be taken into consideration.[95] Further, the FCC determined that any programs that further children's "positive development in any way, including serving their cognitive/intellectual or social/emotional needs" thereby serve their educational and informational needs, thus electing to take a most vague and generous substantive approach.[96] However, broadcasters must provide documentary evidence as to how they determine the needs of children in their broadcast areas[97] and which particularly significant programs they broadcast for children.

Significantly, the reporting duties also permit control of broadcasters by the public. Action for Children's Television, supported by a number of other organizations, has prepared an informational brochure on the new provisions that also contains tips for approaching broadcasters and a list of licenses due for renewal (The Bureau of National Affairs, *Daily Report for Executives,* 15 May 1991: A-17).

In view of the vague definition of the "educational and informational" needs of children, it is no surprise that broadcasters claim that the usual cartoons and comedies fall under this definition and thus that they meet the requirements of the Children's Television Act (cf. *The New York Times,* 30 September 1992: A1; *Broadcasting,* 5 October 1992: 40f.). Insofar as new children's programs have been developed, 60% of these are shown before 7 A.M. (*The Economist,* 10 October 1992: 31; see also *Broadcasting,* 5 October 1992: 40f.). The FCC is reluctant to evaluate the quality of programming and as a result disempowers itself.

In addition to the requirement of children's programming, the National Endowment for Children's Educational Television Act of 1990 set up a foundation that promotes educationally valuable children's programming (Public Law 101-437-Oct. 18, 1990). This is to take place primarily through production contracts and grants to producers.[98]

5.5. Advertising Restrictions

Advertising financing is such an integral part of the broadcasting system in the United States that there are no prohibitions on commercial interruption, requirements of block advertising, and so forth (with the exception of children's programs). Broadcasters are always at liberty as to how they place their commercials. Although the FCC has never enacted a formal limitation on the amount or type of commercial time, a rule of thumb was in place until 1979 and 1984, respectively, which originally derived from self-imposed standards of the NAB. These restrictions on maximum amounts of advertising (usually, sixteen minutes per hour) were measured so generously that the limits were usually not reached. These restrictions were eliminated as part of the deregulation effort, with the expectation that this would have no effect on advertising conduct. However, the United States has seen the development of advertising forms that would have conflicted with the earlier time restrictions: specifically, the so-called infomercials, or program length commercials, and teleshopping broadcasts.

Apart from several broadcasting-specific restrictions, such as lottery commercials (cf. 18 U.S.C. 1304, 1307; for restrictions on tobacco advertising, see Lively, 1992: 278ff.), permissible commercial content is determined by general advertising law. As in other countries under study here, there is distinct self-regulation in this area by the advertising industry (see the short overview in Carter, Franklin, & Wright, 1991: 400). For instance, in order to become a member of the industry's large association, advertising agencies must accept the Standards of Practice. Advertising organizations, in turn, maintain complaint authorities, which in the case of a complaint, attempt to reach voluntary solutions with the advertising industry. If this is not possible, an investigation is conducted by the National Advertising Review Board (NARB), an establishment composed of various advertising industry organizations. If this body should find a commercial unacceptable, it informs the media and the Federal Trade Commission (FTC). Moreover in the broadcasting sector, networks also function as advertising clearance centers for their affiliates.

Because of the frequent resort to advertising law, control of content in broadcast advertising is mainly the province of the FTC, which is responsible for general advertising rules (for the FTC, see, e.g., Teeter & Le Duc, 1992: 434ff.). In accordance with an agreement between the FTC and the FCC, the latter is responsible only for technical broadcasting questions and for time restrictions (only still in place for children's programming) and specific requirements dealing with the separation of advertising from programming. The entire area of misleading and deceptive advertising falls under FTC jurisdiction (Carter, Dee, et al., 1994: 324). The FCC therefore tends to forward such complaints about forms of advertising to the FTC. Only after the latter has determined that advertising restrictions have been repeatedly violated does the FCC consider consequences of broadcasting law, such as revocation of license.

As a result of regulatory freedom, U.S. broadcasters have used considerable inventiveness in satisfying the wishes of the advertising industry for a program setting conducive to advertising and in keeping consumers in front of their TV sets. The interruption of commercial blocks with brief news updates is only one of many ways of preventing "zapping," or channel switching, and other forms of avoiding commercial messages. Moreover, a variety of substitute forms of concealed advertising, product placement, and so forth, ensure a smooth transition between programming and advertising. The only limit to comingling is general advertising law, which forbids such practices as deceptive advertising. In recent years, the FTC has primarily turned its attention to infomercials, which mix advertising with program elements, and condemned a number of such productions as deceptive, although this has not resulted in general regulations. Apparently, the advertising industry is trusted to regulate itself.

In order to protect juveniles, the former NAB code, which was associated with an FCC Policy Statement,[99] had introduced a number of special advertising restrictions; for instance, the abuse of child trust was to be prevented by prohibiting the host of a children's program from appearing personally in commercials. The wording of this appeal was largely followed, but not its intention (FCC, 1979: vol. 2, 56ff.). For instance, although not specifically covered by the FCC prohibition, commercials by television stars or comic figures popular with children were not avoided, despite the fact that exposing children to such commercials was considered undesirable. In addition, it could be seen that the planned restrictions threatened to cause unwanted side effects. In particular, if the use of "hosts" for commercial purposes was proscribed, then local broadcasters lost the incentive to air children's programs and use the popular hosts to advertise for them. In this way, the already apparent trend toward elimination of existing children's programs was only increased (FCC, 1979: vol. 2, 59).

After the standards were relaxed, new problems arose in the 1980s in relation to programming that functioned like commercials, especially those programs created to promote children's toys and games. When network budget cuts created a gap in children's programming, inexpensive shows were slotted in. These intertwined program content with advertising elements (program-length commercials), raising concerns for the immature audience.

The Children's Television Act of 1990[100] and the FCC rules implementing it[101] impose some advertising restrictions on children's programs. For instance, such programs may only contain a maximum of 12 minutes of advertising per hour during the week and 10.5 minutes per hour at weekends, and additional rules regarding the separation of advertising from programming have been laid down. Furthermore, the linking of programming and commercials through the use of comic figures, and so forth has been made considerably more difficult. The FCC was directed by Congress to determine whether certain cartoons based on toy characters merchandised by advertisers constituted program-length commercials and therefore an automatic violation of the maximum advertising standards.

Although the direct economic disadvantage resulting from the regulation vis-à-vis the de facto status quo[102] was limited, the new regulation spawned strong opposition from the various media organizations (*Broadcasting,* 28 January 1991: 48f.). Just as in the case of the rules for the positive protection of juveniles, during the hearings on the enactment of the FCC specifying rules these groups pushed for the weakest possible regulation (*Broadcasting,* 4 February 1991: 22f., 11 February 1991: 98). They were successful in doing so, particularly with the definition of program-length commercials. The FCC's definition—that these refer only to a "program associated with a product in which commercials for that product are aired"—especially avoids restrictions for shows that are based on popular toys, as long as such programs are not interrupted by commercials for the same product.[103] Mainly because of this definition, Action for Children's Television submitted, among other things, a petition with the FCC for reconsideration, but it was successful only with respect to other points, such as a tightening up of the program-advertising separation rule.[104] The broadcasting industry was more successful, in that it prevailed in gaining a postponement until 1 January 1992 of the advertising restrictions coming into force. These had been scheduled to take effect at the worst possible time, namely, on 1 October 1991, shortly before Christmas.[105]

With respect to advertising restrictions for children's programming, the FCC initially undertook relatively strict monitoring (cf., e.g., *Broadcasting,* 20 April 1992: 24, 11 May 1992: 36ff.), but it will certainly rely mainly on complaints in the long run.[106]

5.6. Teleshopping

By doing away with special advertising rules, the FCC also opened the way in American television for teleshopping (Ferrall, 1989: 20f.) and the complete commercialization of leased access channels (for more on leased access channels, see Section 6). So-called home shopping can be seen today on both terrestrial and cable television in a wide variety of forms operated by various companies.

Because the time restriction for TV advertising was eliminated in 1984, it is not a barrier to the development of teleshopping in the United States to classify it as advertising. Large blocks of programming time can be reserved for it, or new teleshopping broadcasters can be licensed. Since there is no prohibition on combining programming with advertising, it is possible to establish teleshopping as a mixed form of advertising and entertainment. The sales pitch can be integrated into a talk show or tied to game shows or other formats that increase audience attention. This does not even constitute a violation of the special sponsorship identification rules (for more on these rules, see Carter, Dee, et al., 1994: 441f.). The FCC has even decided that home-shopping broadcasting stations serve the public interest and thus can be eligible for mandatory free carriage on local cable systems.[107]

In other words, from a programming perspective, teleshopping does not present a special problem for broadcasting supervision in the United States. This reveals the qualitative leap that deregulation can permit for the concept of broadcasting. It is now possible to release broadcasting completely from its trustee role and to relieve it of its obligations to serve as a democratic forum accompanying the formation of public opinion. In a pure market model, broadcasting is legitimated not by its journalistic accomplishments but, rather, by its capacity of being able to be used for diverse entrepreneurial objectives. However, this means that a specific need for programming-related broadcasting supervision is no longer justified.

5.7. Equal Opportunity Employment

It should also be mentioned that the FCC also uses its regulatory authority to examine whether broadcasters offer equal opportunity in employment (EEO). License applicants are required to adopt and file an affirmative action plan to ensure nondiscrimination against such minority groups as African Americans, Hispanics, Native Americans, and women. The FCC scrutinizes a licensee's EEO efforts when a renewal application is filed, using informal guidelines to compare the percentage of minorities employed by the licensee with the percentage available in the labor force. It emphasizes that the guidelines are neither quotas nor a "safe harbor." The licensee's performance is monitored by requiring stations with five or more full-time employees to file annual employment profiles.

The EEO measures are a part of the arsenal of efforts by which U.S. authorities seek to ensure equal opportunities in the workplace. It is disputable whether their main goal is the achievement of diversity in programming (cf. the analysis of Culver, 1993: 219ff.). In any event, the EEO rules are not limited to persons who might have an influence directly on programming. With respect to diversity in broadcasting employment, the FCC continues to rely on market forces. However, it is considered to be a desirable side effect of EEO policies when more consideration is given to minority interests through the mixed composition of personnel.[108] But when this is not ensured through market forces, there is no realistic chance that programming will be substantively influenced by minority interests.

6. REGULATION OF CABLE

The regulation of cable is an entirely different matter (a short overview can be found in Lively, 1992: 285ff.; Carter, Dee, et al., 1994: 449ff.). The history of cable regulation in the United States has gone through too many phases to be able to be described in detail here. Instead, we address characteristic details and important trends.

Now that 60% of American households receive their television programming over cable—radio is normally not broadcast via cable—this transmission medium has become the crux of the U.S. broadcasting order. Cable was originally used only to transmit programming in remote areas with no access to terrestrial broadcasting. However, it was soon recognized that cable was suitable for providing programming to highly populated cities and offered new broadcasting opportunities there, supplying a motivation for originating specialized cable programming. Its significance for media policy thus became immense.

Since the FCC did not have express jurisdiction over cable, it initially practiced restraint, but in any event it was intent on not influencing existing television broadcasters negatively. The FCC grossly underestimated the significance of this new broadcasting technology. In addition, it initially saw no reason for regulating the relationship between cable operators and local television stations, which were suffering considerably from the import of programming from other markets and the transmission of these over cable (see Le Duc, 1973: 82ff.). However, the FCC later made an effort to slow down the expansion of the cable industry as it became increasingly clear that the economic viability of television broadcasters was being threatened by new competitors.

Only in the mid-1960s, under pressure from the cable industry as well as from television broadcasters, did the FCC begin to enact detailed regulations in this domain. The FCC initially presumed that its jurisdiction was limited to that which was reasonably ancillary to the effective performance of its various responsibilities for the regulation of broadcasting.[109] For their part, the courts tended to restrict such jurisdiction. For instance, the Supreme Court struck down FCC rules that required cable systems to offer separate channels for public, educational, and government use (PEG channels) as well as leased access channels.[110]

The FCC's restraint allowed state and local authorities to jump into the regulatory void, as the latter were involved in the laying of wires under public streets and allowing the use of utility poles along public easements. They used their franchising authority to impose various obligations and restrictions on cable operators, including access channel requirements. In this way a bifurcated jurisdictional scheme emerged. Cable operators began to complain of the burden inflicted by the many rules, which differed considerably from region to region. Mainly they were interested in receiving some sort of status protection for the period after their franchises expired. They also wanted to be exempted to the extent possible from obligations dealing with the content of programming (e.g., advertising restrictions, must-carry obligations)—or even to be treated as common carriers. As well as fighting local rate regulations, they sought refuge under federal regulations, pressuring Congress to enact uniform rules and to preempt state and local authorities in several areas. Congress heeded this call for federal regulation and passed the Cable Communication Policy Act of 1984.

Franchising authority still rested with state and local authorities, while exclusive jurisdiction over signal carriage was exercised by the FCC. Franchising authorities could not regulate cable as a common carrier, nor could they require specific programming services. The conditions for franchise renewal were defined more specifically, resulting in the incumbent franchisee receiving extensive protection (47 U.S.C.A. 546). Now cable operators were protected from regulatory pressure in a way that any broadcaster would envy. However, franchising authorities were granted the power to demand from cable operators that they designate a portion of their channel capacity for public, educational, and governmental use. Cable operators also had to set aside a specific number of channels to be leased for commercial use (leased access channels).

The Cable Communication Policy Act of 1984 was a component of the deregulatory policy. Even though it contained an abundance of regulations, these were in the interest of preempting more stringent rules. Deregulatory goals included the broad removal of the restriction on the rates cable television subscribers had to pay (for a discussion of rate regulation, see Ginsburg et al., 1991: 607ff.). Since nearly all cable operators had local monopolies, the release on the rate cap led to substantial rate increases. From 1984 to 1992, the average rate increased almost three times as fast as the consumer price index. During this period, channel offerings certainly had expanded, but not nearly at this pace. These rate increases—and other hotly debated issues—led in 1992 to the enactment of a new law, the Cable Television Consumer Protection and Competition Act of 1992.

In the conference report by Congress, the goals of this new regulation were described in characteristic terms, as follows:

> It is the policy of the Congress in this Act to:
> (1) promote the availability to the public of a diversity of views and information through cable television and other video distribution media;
> (2) rely on the marketplace, to the maximum extent feasible, to achieve that availability;
> (3) ensure that cable operators continue to expand, where economically justified, their capacity and the programs offered over their cable systems;
> (4) where cable television systems are not subject to effective competition, ensure that consumer interests are protected in receipt of cable service; and
> (5) ensure that cable television operators do not have undue market power vis-à-vis video programers and consumers. (House of Representatives Conference Report 102–862, 14 September 1992: 4f.)

Diversity is seen primarily as a quantitative problem: The expansion of capacity and the resulting availability of a large number of channels are considered to be important objectives. It is to be decided in the marketplace which channels are available. The type of financing method employed in so doing is left to the cable operators and broadcasters. Market failure becomes an issue in only two respects: When cable operators have a monopoly, then, on the one hand, consumer protection has to be ensured, particularly with regard to rates,[111] and, on the other, market adjustments have to be made with respect to video programers. To this extent, exclusive contracts between cable operators and their program suppliers are restricted. Moreover, the Act contains a number of must-carry rules, whose constitutionality is, however, uncertain.[112]

The relationship between cable operators and broadcasting stations is regulated in financial respects as well. Transmission over cable presumes retransmission consent (47 U.S.C. 325, as amended), for which the original broadcaster can demand compensation. This means that not only the transmission of specific cable programming but also that of customary, over-the-air broadcast programming became subject to fees. Up until 1992, this service had been free. This was based on the assumption that transmission over cable would provide additional viewers and higher earnings than from commercial advertisement alone, which itself was sufficient for financing programming. Now, due to the triumph of cable, the broadcasting order has entered a new stage. Strong competition and fragmenting of the audience no longer guarantee traditional broadcasting stations sufficient revenues, so that they should also receive a right to require compensation from cable operators. For the latter, it is important that all programming of interest to consumers be able to be broadcast over cable. In other words, or so runs the assumption, cable operators would thereby have an incentive to pay a fee.

Nevertheless, Congress does not rely solely on market forces. Since it can be expected that cable operators will only pay for sufficiently popular programming, must-carry rules ensure that local commercial television signals will also be transmitted even when the cable operator is not interested in them.[113] The same holds true for the programming of certain noncommercial stations, for whom the rules on the retransmission consent do not apply. Thus they are unable to negotiate compensation.

Although cable is even more a medium of entertainment than terrestrial television, the general parameters of the diversity goal apply to "news and information." However, the Act of 1992 does not contain any provisions to ensure this directly. Yet the conference report on the Act sees a connection between the interest in promoting a diversity of views and the distribution of public television services, recognizing "public television's integral role in serving the educational and informational needs of local communities" (House of Representatives, Conference Report 102–863, 14 September 1992: 2f.). On the other hand, the conference report also emphasizes that local commercial stations are an important source of local news and public affairs programming. The Act therefore ensures the transmission of local programming over cable, not, however, that such programming satisfies true needs.

In particular, many of the rules provided in the Act require specification by the FCC, which has the rather unenviable task of trying to find the resolutions indicated by Congress, yet not detailed in the statute, for conflicts of interest. This likely guarantees that the FCC will have to occupy itself increasingly with cable in the future. Included in this will be the task imposed on the FCC of conducting an investigation regarding the trend toward diverting national, regional, and local sports programming from broadcast television to cable, pay-TV, and pay-per-view (47 U.S.C. 322). The FCC has also been charged with adopting rules regulating indecent cable TV program content (47 U.S.C. 532[h]). The FCC has already fulfilled this responsibility, having adopted new rules giving cable operators, among others, the right to restrict all indecent programming submitted for leased access to one designated channel and to encode that channel.

In this manner, the Act of 1992 follows the trend noticeable elsewhere to provide for content regulations only in exceptional cases, for example, to protect against indecency and pornography. Almost everything else is left to regulation by the market, which is not, however, relied upon completely: must-carry rules and very limited precautions against undue market power take into account the interests of weaker broadcasters. Whether and to what extent video programers offer material that serves the public interest are of no relevance. It is assumed that this is sufficiently ensured through the mechanisms of the market and the limited measures to protect against market failure. The object of broadcasting regulations is the transmission of programming, not its content and its production. Even the rules on indecent programs apply not to their content but, instead, to the manner of their transmission.

Until now in actuality cable operators have almost always had monopolies.[114] It therefore would seem fair to treat them as common carriers (the statement is controversial; cf. Johnson, 1994: 64ff.). However, cable operators have protested vigorously against this. They feel that even the narrow must-carry rules go too far. In defending their interests, cable operators benefited and still benefit from the fact that the Supreme Court has permitted them to rely on the First Amendment,[115] citing their original programming and the exercising of editorial control over which stations or programs to include in their repertoire.

As far as the expansion of cable leads to its being the principal medium of transmission, the editorial discretion of cable operators over what to include in their programming is becoming a growing problem in light of the monopoly positions they enjoy. As a result, the cable industry finds itself caught up in a conflict as to whether the First Amendment guarantee is more important to it than protection for its privileged position as holder of an exclusive right of distribution (cf. Teeter & Le Duc, 1992: 428ff.). As one observer remarked: "To media leaders, 'free speech,' 'deregulation,' and 'marketplace' are not sacred ends in themselves, but simply concepts invoked to realize certain industry objectives; and therefore useful only to the extent that they aid in attaining these objectives" (Le Duc, in Teeter & Le Duc, 1992: 431). However, circumstances can change in structural terms, as is shown below.

7. REGULATION IN A TIME OF EXPANDING NEW ELECTRONIC MEDIA

In addition to cable, there are also a number of other transmission technologies. Mentioned above (Section 1.1) were direct broadcast satellites (DBS), low-power TV (LPTV), multichannel, multipoint distribution services (MMDS) and the Multipoint Distribution Service (MDS).[116] Also to be noted is the satellite master antenna television (SMATV): this is a cable system that does not cross a public right of way and is thus not bound by franchising requirements. Technological developments in the fields of coaxial cable, fiber optics, digital transmission, portable uplinks, integrated services digital network (ISDN), high-definition television, and others create a wide range of alternatives. In addition, the boundaries between various transmission technologies are becoming blurred. As a result, governmental regulation is faced with having to devise complicated justifications. Inconsistencies in standards of regulation in various fields place a burden on regulatory activities. Moreover, many media companies are pressuring to enter a variety of sectors in order to be able to use the advantages of an association of various multimedia actors.

Concentration rules proved to be particularly burdensome. Both the 1984 Cable Act and the 1992 Cable Act inspired restrictions on cable ownership. The rules especially relate to network–cable and telephone–cable cross-ownership.[117] They contradict the plans of many cable television companies as well as many telephone companies to link several services, especially to use cable TV systems to tie together both wireless and wired telephone services and thus to form alliances for regional and even national services in telephone and entertainment (cf. *The New York Times,* Oct. 26, 1994: A1, D7). Although the cross-ownership restrictions have been watered down in many respects (see Common Ownership of Cable Television Systems and National Television Networks, 1992; Telephone–Cable Cross-Ownership, 1992), they are still restrictions. The FCC's aim here is to achieve flexibility in a pragmatic way. The release of video dial tone services for regional Bell holding companies is a good illustrative example. The telephone companies are allowed to offer video distribution on a common carrier basis.[118]

In July 1992 the FCC authorized the seven major regional Bell telephone carriers to provide "video dialtone" service, a decision that allows these companies to deliver television programs through their distribution systems on behalf of unaffiliated television programers without either the phone company or the programer requiring a municipal cable franchise.[119] Unable to ignore the 1992 Cable Act's provision for prohibiting telephone companies from providing cable services, the FCC could only free these companies to a limited extent, granting them the right to deliver the video services on a common carrier basis but not to own or control on their own. This was only a first step. Private policy is likely to prevail in the long run. Combined personal, data, marketing, and mass communication networks will emerge. This situation is certain to raise new questions about the meaning of media concentration, state authority over local telecommunication services, and the First Amendment rights of these conglomerates long before government policymakers have formulated any answers to these questions.

The FCC's decision to confine itself mostly to regulating the transmission of programming and to leave the production and thus the content of such programming to the forces of the market makes it increasingly difficult to justify special broadcasting regulations. Moreover, it is less and less convincing to make the category of regulation dependent on the type of transmission technology employed. At the same time, the regulatory concept developed in the United States contains a natural tendency to respond to the expansion of a transmission technology with more deregulation.

However, there is still considerable political pressure to maintain regulation, mainly from Congress. But sections of the media industry also champion regulations. They see that FCC regulation gives established players a protected status as well as some shelter against new competitors. In addition, they assert that at least the same responsibilities for conflicts of interest as they already bear should be imposed upon their competitors. In this regard, it can be expected that even in times of convergence, media regulators will not become unemployed. The interest of regulators in continued existence matches the desire of the media industry to keep some beneficial regulations intact.

However the market is characterized by a dynamic that makes it difficult to foresee which form of regulation will prevail. The first half of the 1990s has been dominated by particular uncertainty. In view of the globalization of markets, the prospect of being able to build new information highways with the help of new technologies has led to a wealth of new ideas. In speeches to the National Press Club (December, 1993) and to the Television Academy in Los Angeles (January, 1994) Vice-President Gore outlined principles for the development of a national information infrastructure (NII), thus giving it an official blessing. Scenarios for the future are being developed at hundreds of conferences and in countless papers and publications.[120] They deal with the convergence of computer and telecommunication technology and the development of new, interactive possibilities of communication. From the wide range of comments, the following quote is just one example:

> Large network providers such as the telephone companies could provide broadband communications networks (BCNs) that are accessible to general public. Single users or groups of private users will likely own BCNs that connect multiple communications stations in single or multiple locations. BCNs will also support the high speed transmission of integrated voice, data, and video information in digital form, which will integrate the transport of voice, data, and video information and promote two-way interaction between users. These two characteristics transform broadband communications technology into more than the functional sum of the computer, fiber optic, and telephone technologies in more than an extension of the existing publishing, broadcasting, cable, and telecommunications networks. When BCNs interconnect users, the combined technology will change not only the manner in which information is used, but also the manner in which information is communicated. The public will be able to receive, seek out, identify, store, manipulate, compose, alter, filter, and transmit digitized data, print, video, and voice information. Moreover, users will be able to select who receives their information. Electronic communication will no longer be a predominantly passive mode of interaction conducted via one-way, single-format information streams controlled by a limited number of senders. Instead, communication will be an interactive process conducted via two-way, multiple-format information streams controlled by users of the media. BCNs thus have the potential to shift the locus of control over communication from the privileged government-sanctioned media to a greater portion of the public. Most important, however, this new control will allow individuals and groups to become electronic speakers and publishers. (Hammond, 1992: 183f.)

Although a multitude of questions have yet to be answered, many companies are already trying to stake their claim and win a possible share in the new market. Mergers,[121] joint ventures, and strategic alliances are aiming at securing know-how (technology, markets, financing, advertising) and sufficient financial muscle. In several pilot projects, such as the particularly costly "Full Service Network" of Time-Warner in Orlando, options, including the reactions of consumers, are being tested.

Even if many of the plans and ideas currently being discussed should never come to fruition, it seems that a combination of the functions of television, radio, computers, and telephone in the new generation of multimedia communications is assured. Consequently, the question of appropriate regulation becomes an issue. Vice-President Gore outlined five principles for the NII: (1) encouragement of private investment (primarily by removing regulatory impediments); (2) preservation of competition; (3) open access to the network (through uniform standards); (4) universal service (to be assured primarily by the private sector); and (5) flexible and responsive government (cf. Branscomb, 1994: 176). These principles show the "direction of private sector dominance in the design and deployment of the information superhighways" though they "purport to preserve public-sector concerns in its growth and development" (Branscomb, 1994: 176). Exactly how the public interest in individual cases should be maintained remains to be seen. Nor is it yet quite clear which forms a possible state-run regulation should take or which issues it should deal with. As always at times of technological and economic upheaval, powerful companies are demanding above all as much faith as possible in market forces. Others, with an eye to the social impact of the new communications possibilities and the risk of market failure are calling for statutory regulation. However one thing is clear to everyone: Traditional broadcasting regulation cannot be the sponsor.

Many people refer to the prototypical power of the Internet (see Office of Technology Assessment, 1993), a "network of networks" that was founded with state help. In the United States, the construction of the Internet was primarily led by the Defense Department (cf. Grossman, 1994: 32). But because of its spontaneity and dynamic network and user structure, it is hardly accessible to effective statutory regulation. Yet others stress that, irrespective of new services and possible uses, problems similar to those of the past will continue to exist, and therefore statutory regulation will be unrenounceable. It is debatable, then, whether the interconnectivity of the various networks and software can adequately be ensured without a statutory framework.

Another topic under discussion is the objective of universal service (cf. Johnson, 1994: 165ff.). How can it be ensured that TV, telephone, and radio will be available to every citizen, rich and poor alike, irrespective of whether they live in a metropolis or in a rural area? Questions of consumer protection (user sovereignty, privacy, right of reply) and, above all, of access are being discussed (see, Branscomb, 1994). In this regard, one question is whether and to what extent the common carrier conception, which until now was used in the area of telephone, can be retained or even further developed for the new methods of transmission, or whether it should be abandoned (for further information see especially Noam, 1994a). Common carriage is aimed at ensuring the neutrality of access, in spite of monopoly structures or other hurdles. Until now this applied to telephone but not to cable TV. Cable operations were, however, burdened with must-carry obligations. Whether in future the principle of competition will be sufficient to enable adequate access to carriers is the subject of great dispute.

But even if there were access opportunities for all users, there still remains the question of whether all programs that are in the public interest can be supplied (financed) through the market. Such a question leads inevitably to the fundamental debate over whether there should be a body that tries in a paternalistic fashion to satisfy needs instead of relying on the market to satisfy consumer wants. These—like all other questions concerning the broadcast media's place in the market—lead into fundamental discussions about the scope of the First Amendment (cf. Hammond, 1992).

The diversity of the American debate on this subject cannot be reproduced here in detail. However these few lines suffice to make clear that many of the same questions concerning communications crop up in old and somewhat new guises that have accompanied the communications system in the past. At the same time, it is clear that future developments cannot be foreseen with certainty. This is a time of both upheaval and revolutionary development in a new phase of the information society. Indeed, many American observers draw a comparison to the rapid developments of the industrial revolution in the nineteenth century.

8. SUPERVISION OF BROADCASTING IN THE POLITICAL ARENA

Regulation and supervision of broadcasting is part of the political and economic order in the United States. Thus, any analysis of such regulation belongs in the larger context of a study of policy fields. There is a wealth of literature on U.S. governmental regulation, the role of independent federal agencies generally, and the conduct of the FCC in particular. Space does not permit a review here. But one finding deserves to be stressed: An agency such as the FCC is part of a political arena composed of a wide variety of actors, whose potential influence determines the latitude for action (see generally Scharpf, 1970: 63 n. 154). The most important actors alongside the FCC are the different sectors of the broadcasting and telecommunications industries, public interest groups, copyright owners, the courts, the President[122] and the Congress. (For more on these actors, see Krasnow et al., 1982: 33ff; cf. also Cole & Oettinger, 1978: 3ff., 23ff.; Ray, 1990: 149ff. For a general discussion see Wilson, 1980.) In all analyses of FCC policies, it always becomes apparent that there are close networks of communication and cooperation among the various parties.

In five case studies, Krasnow, Longley, and Terry have tried to show how the various groups of players exercized influence over decisions on general regulations. Drawing on concepts of the pluralism theory, they have traced the development of a number of diverse coalitions. They observed that a common pattern at the FCC was that it usually formed coalitions with either the broadcasting industry or with Congress (Krasnow et al., 1982: 280). In many cases, the FCC tried to work out amicable settlements between or among the various actors.

To this extent, the FCC pays attention to potential opposition and attempts to negotiate an enforceable result. Accordingly, FCC policy is virtually always marked by gradual, incremental changes that avoid sharp changes of course.[123] In the efforts to combat concentration (see Section 5.3) grandfathering may serve as an example of this conflict-avoidance strategy. With day-to-day supervisory work as well, the FCC seeks consensus, particularly with broadcasting stations, the networks, cable operators, and relevant industry associations. For years it has given precedence to self-regulation by the industry or even forced this situation with gentle pressure.

A study on "Communications Policy Making at the FCC" (Brotman, 1988) particularly determined that the FCC promotes conflict resolution through negotiation between the affected proponents of various interests, both on the level of rule-making and on that of individual decisions. The FCC supports negotiated self-regulation through trade associations, especially in instances where sensitive First Amendment issues may arise. It has adopted industry guidelines in order to respond to public-interest concerns, while avoiding subsequent First Amendment challenges to adoption of these guidelines. The FCC encourages disputants to reach a mutually satisfactory agreement, and it defers to the solution if it complies with the Commission's mandate. The best-known examples relate to the family viewing policy for television and to controversies surrounding the children's television policy. The FCC often supported broadcaster–citizens' agreements—for instance, in cases involving a petition to deny license renewal or of a transfer and in others where judicial appeal was threatened. Consensus was also sought in the area of cable television in order to integrate cable TV into the communications system and to balance the interests of broadcasters, cable operators, and copyright holders. Congress as well as the White House Office of Telecommunications Policy took part in these attempts to reach a consensus agreement. The FCC did not rely only on the power of rational arguments but also used its ability to threaten sanctions and its bargaining power.

The guidelines of the FCC's policy were developed in the political arena. The President exerted decisive influence in this, not least because of his executive right to nominate FCC commissioners. The deregulation policy of the FCC coincided with the general trend in U.S. policies. However, Congress also made use of its power to threaten and bargain—for instance, in seeking to slow down deregulatory steps. The dispute surrounding the fairness doctrine provides a wealth of examples of this interplay. Congress has actively intervened in other areas in recent years, notably in the fields of protection of juveniles and in cable rates. In this way it has tried to fulfill a supervisory role.

In its relationship to those being supervised, particularly the broadcasters, the FCC has seen itself primarily as a promoter of the established broadcasters and the dominant structures. The constant preference given to licensees over new applicants (see Section 5.1) makes plain the trust placed in rooted structures and thus in proven interactive relationships. Efforts are made to do away as far as possible with reasons for supervisory intervention via informal paths that avoid formal instruments: Its policy of raised eyebrows serves as the epitome of an informal sanction technique. If this does not suffice, the FCC makes use of formal instruments only after a renewed use of informal systems of contact.

Such supervisory work marked by cooperation is defined in part by the special nature of the material—particularly the sensitive issue of potential basic rights' violations inherent in broadcasting supervision. In this regard, the American understanding of freedom of speech and of the press plays a predominant role. The scope of speech is defined broadly, and restrictions on speech require a stricter justification than in the other broadcasting orders studied here. Rhetorically, it is continually asserted that the free speech clause is anchored in the promotion of democracy and the process of discovering truth as well as assuring individual self-fulfillment (cf. Emerson, 1970; Anderson, 1983). The fact that the media are primarily characterized by entrepreneurs (i.e., powerful conglomerates), who wish to receive a good return on their investment, has hardly been recognized in legal doctrine: The dramatic change in the political, economic, and cultural environment in the sphere of the First Amendment has not resulted in any decisive change in the extent of the protection it affords.[124] Courts have often intervened in the development of broadcasting regulation. In this regard, a number of nuances can be observed; in every respect, however, the courts have viewed the First Amendment as a basic right of the communicator (i.e., essentially, the broadcasters). The interest of viewers and listeners in receiving diverse programming content was taken into account indirectly via the protection of the communicators.[125]

The more the broadcasting industry becomes integrated into the general economy and sloughs off uniquely journalistic features in favor of entrepreneurship, the more it seems that regulatory activities will approximate those in other economic sectors and that there will be no real chance to take into account special broadcasting-related public service criteria (Ferrall, 1989: 8ff.). It therefore seems obvious—and is often attempted in the literature—to transfer findings from other policy areas to broadcasting supervision, that is, to use the FCC as a model case for analyzing governmental regulation and supervision generally and to learn about broadcasting regulation through studies of regulations in other fields as well.

NOTES

1. Due to amendments to the Federal Communications Act of 1934, which dealt with the direct reception of satellite signals and were intended, among other things, to protect cable operators against loss of revenues, most of these signals are now scrambled and require a decoder. Cf. Barnett, 1988: 138f.

2. Direct Broadcast Satellites, 90 F.C.C.2d 676 (1982).

3. Cf. *NAB v. FCC,* 740 F.2d 1190 (D.C.Cir. 1984).

4. One of the several reasons Murdoch and NBC cited for their decision to pull out of DBS was the exclusive rights cable multiple system operators (MSOs) exercised over their program channels—MTV, CNN, USA, TBS, TNT—that they refused to allow DBS to deliver. The 1992 Cable Television Consumer Protection and Competition Act empowers the FCC to challenge these exclusive program contracts that discouraged DBS competition.

5. In addition, newer transmission technologies, such as MDS, have now come to encompass pay-TV. However, they are of less significance than cable TV.

6. For an overview of programming strategies and practices, see Eastman, 1993.

7. There is a vast range of literature on the subject. See, for example, Emerson, 1970; Blasi, 1977; Meiklejohn, 1948; Anderson, 1983; van Alstyen, 1984; see also Tedford, 1993.

8. See Hoffmann-Riem, 1981b: 37ff. Le Duc, 1987, especially Chapters 1, 3, and 4, gives an analysis of the interaction between public and private policies that have shaped the nature of broadcast regulation in the United States. He also deals with the corresponding understanding of the First Amendment.

9. See *National Broadcasting Co. v. United States,* 319 U.S. 190, 213, 63 S.Ct. 997, 1943; Carter et al., 1994: 326ff.

10. *Red Lion Broadcasting Co. v. FCC,* 395 U.S. 367, 390, 89, S.Ct. 179 (1969).

11. Cf. Report and Order, In the Matter of Deregulation of Radio, 84 F.C.C.2d 968 (1981); Report and Order, In the Matter of the Revision of Programming and Commercialization Policies, Ascertainment Requirements, Program Log Requirements for Commercial Television Stations, 98 F.C.C.2d 1076 (1984).

12. *Red Lion Broadcasting Co. v. FCC,* 395 U.S. 367, 89, S.Ct. 179 (1969). Nevertheless, the Supreme Court has adhered to the "Red Lion" philosophy; see *Metro Broadcasting, Inc. v. FCC,* 110 S.Ct. 2997 (1990).

13. Local broadcasters who demand carriage (mandatory carriage) receive no payment—generally the smaller, independent stations that the systems would be willing to delete. And those who do negotiate with cable for consent, and are willing to risk the possibility of being deleted, may receive other benefits (e.g., a cable channel of their own to operate) rather than payment.

14. Until 1983 there were seven commissioners.

15. The main bureaus are Mass Media, Common Carrier, Private Radio, and Field Operations (Bittner, 1994: 44).

16. At the same time, however, broadcasters were sometimes allowed to subject themselves to the regulatory system they believed most favorable. The courts held this choice to be unlawful for DBS, and they required that it be classified as broadcasting. *NAB v. FCC,* 740 F.2d 1190 (D.C.Cir. 1984).

17. Subscription Video, 2 F.C.C.Rcd. 1001, 62 R.R.2d 389 (1987) upheld in *National Association for Better Broadcasting v. FCC,* 849 F.2d 665, 64 R.R.2d 1570 (D.C.Cir. 1988). But see *Telecommunications Research and Action Center v. FCC,* 836 F.2d 1349 (D.C.Cir. 1988).

18. See Report and Order, In the Matter of Inquiry into the Development of Regulatory Policy in Regard to Direct Broadcast Satellites for the Period Following the

1983 Regional Administrative Radio Conference, 90 F.C.C.2d 676 (1983) regarding satellite television.

19. Cf. Second Report and Order, Community Antenna Television, 2 F.C.C. 725 (1966). See also Le Duc, 1973; Barnett, 1988: 128ff. The FCC subsequently abandoned this policy; Report, In the Matter of Inquiry into Economic Relationships between Television Broadcasting and Cable Television, 71 F.C.C.2d 632, 713ff. (1979).

20. See Public Notice, Policy Statement on Comparative Broadcast Hearings, 1 F.C.C.2d 393 (1965). See also the case study by Krasnow et al., 1982: 206ff.

21. Cf. lottery selection among applicants, 93 F.C.C.2d 952 (1983). See also Teeter & Le Duc, 1992: 420 f.

22. Omnibus Budget Reconciliation Act of 1993.

23. In 1994, competitive bidding was used for licensing personal communications services (PCS).

24. In the Matter of Reexamination of the Commission's Comparative Licensing, Distress Sales and Tax Certificate Politics Premised on Racial, Ethnic or Gender Classifications, 1 F.C.C.Rcd. 1315 (1986). For an example of a court decision supporting this FCC view, see *Steele v. FCC,* 770 F.2d 1192 (D.C.Cir. 1985). For developments, see Gillmor & Barron, 1990: 719ff.

25. *Metro Broadcasting Inc. v. FCC,* 110 S.Ct. 2997 (1990). Cf. *Broadcasting,* 12 Feb. 1990: 25; Lively, 1992: 168f.

26. 47 U.S.C.A. 309(i)(3). Implementing rule: Amendment of the Commission's Rules to Allow the Selection from Among Certain Competing Applications Using Random Selection or Lotteries Instead of Comparative Hearings, 93 F.C.C.2d 952 (1983). However, the FCC decided not to give preference to women: Amendment of the Commission's Rules to Allow the Selection from Among Certain Competing Applications Using Random Selection or Lotteries Instead of Comparative Hearings, 102 F.C.C.2d 1401 (1985); upheld in *Pappas v. FCC,* 807 F.2d 1019 (D.C.Cir. 1986). Cf. Gillmor & Baron, 1990: 720.

27. See Cowles Broadcasting Inc., 86 F.C.C.2d 993 (1981).

28. Cf. *Central Florida Enterprises, Inc. v. FCC,* 683 F.2d 503 (D.C.Cir. 1982); cert. denied 460 U.S. 1084 (1983).

29. For licenses awarded under the comparative procedure, however, transfer is only possible after one year.

30. This was introduced with Commission Policy Regarding the Advancement of Minority Ownership in Broadcasting, 92 F.C.C.2d 849 (1982). See also Lively, 1992: 174. The Supreme Court has upheld the distress sale policy as constitutional; *Metro Broadcasting v. FCC,* 110 S.Ct. 2997 (1990).

31. Between 1978 and 1989, only thirty-eight such sales were effected. For this and other details, see Holsinger, 1991: 435f.

32. See Report and Order, In the Matter of the Revision of Programming and Commercialization Policies, Ascertainment Requirements, and Program Log Requirements for Commercial Television Stations, 98 F.C.C.2d 1076 (1984).

33. Deregulation of Radio, 84 F.C.C.2d 968 (1981).

34. These are the rules currently in force for radio; see Deregulation of Radio, 104 F.C.C.2d 505 (1986). For television, see Revision of Programming and Commer-

cialization Policies, Ascertainment Requirements, and Program Log Requirements for Commercial Television Stations, 104 F.C.C.2d 358 (1986).

35. For radio, see *Office of Communications, United Church of Christ v. FCC,* 707 F.2d 1413 (D.C.Cir.q 1983); *Office of Communications, United Church of Christ v. FCC,* 779 F.2d 702 (D.C.Cir. 1985). For television, see *Action for Children's Television v. FCC,* 821 F.2d 741 (D.C.Cir. 1987). Details can be found in Franklin & Anderson, 1990: 771f.

36. See *Office of Communications, United Church of Christ v. FCC,* 359 F.2d 994 (D.C.Cir. 1966); Carter, Dee, et al., 1994: 378ff.

37. Compare the events described in *Broadcasting,* 21 March 1988: 43f. and 18 July 1988: 33. CBS termed it a "victory" that it only had to compensate a competitor $187,000.

38. Formulation of Policies and Rules Relating to Broadcast Renewal Applicants, Competing Applicants, and Other Participants to the Comparative Renewal Process and to the Prevention of Abuses of the Renewal Process, 4 F.C.C.Rcd. 4780 (1989); Amendment of Sections 1.420 and 73.3584 of the Commission's Rules Concerning Abuses of the Commission's Processes, 5 F.C.C.Rcd. 3911 (1990); Amendment of Section 73.3525 of the Commission's Rules Regarding Settlement Agreements Among Applicants for Construction Permits, 6 F.C.C.Rcd. 85 (1991).

39. Even more intensive protected status is enjoyed by owners of a local cable franchise. The local licensing authority cannot refuse to renew the franchise unless it proves that the operator of the cable facility persistently violated conditions of the franchise. See 47 U.S.C. 546.

40. The traditional networks are NBC, the American Broadcasting Companies (ABC), and the Columbia Broadcasting System (CBS). But the Fox Network, founded in 1985, has become so large that it is now correct to speak of the big four networks. However, Fox was initially able to get a one-year exemption from the fin–syn rules that limit the other large networks.

41. See, for example, 47 C.F.R. 73.658. In the radio sector, these restrictions have largely been abolished; see Elimination of the Requirement to File Network Affiliation and Transcript Contracts with the Commission: Network Affiliation Agreements, 101 F.C.C.2d 516 (1985). In the television sector, the FCC has already rescinded the obligation that contracts be of limited duration; Review of Rules and Policies Concerning Network Broadcasting by Television Stations: Elimination or Modification of sec. 73.658(c) of Commission's Rules, 4 F.C.C.Rcd. 2755 (1989).

42. Cf. 47 C.F.R. 73.658(k). This is primarily to ensure that broadcast time during this important section of programming is available for other programmers, in order to aid in program diversity.

43. A comprehensive, thorough analysis of the situation of and perspectives for the networks can be found in Setzer & Levy's contribution to the FCC's working paper Broadcast Television in a Multichannel Marketplace, 6 F.C.C.Rcd. 3996 (1991): especially 4073ff., 4084ff., 4099f.

44. In the Matter of Evaluation of the Syndication and Financial Interest Rules, 6 F.C.C.Rcd. 3094 (1991) as modified at 7 F.C.C.Rcd. 345 (1991).

45. *Schurz Communications, Inc. v. FCC,* 982 F.2d 1043(7th 1992).

46. Second Report and Order, 8 F.C.C.Rcd. 3282 (1993); reconsidered at 8 F.C.C.Rcd. 8270 (1993).

47. This could be witnessed quite clearly with the expansion of the American multiple-ownership rules in 1984. This led in 1985 to transactions involving broadcaster ownership rights in the record amount of $30 billion; see *Broadcasting,* 30 December 1985: 35.

48. See, for example, In re Revision of Radio Rules and Policies, 7 F.C.C.Rcd. 2755 (1992).

49. The history of and purposes behind the multiple-ownership rules are dealt with in FCC, Notice of Proposed Rulemaking, Multiple Ownership of AM, FM, and Television Broadcast Stations, 95 F.C.C.2d 360 (1983): and in FCC, Report and Order, Multiple Ownership of AM, FM, and Television Broadcast Stations, 100 F.C.C.2d 17 (1984).

50. In re Revision of Radio Rules and Policies, 7 F.C.C.Rcd. 2755 (1992). See Conrad, *The New York Law Journal,* 24 July 1992: 5ff.

51. In re Revision of Radio Rules and Policies, 7 F.C.C.Rcd. 2755 (1992).

52. See Review of Regulations Governing Television Broadcasting, MM Docket No. 91—221, F.C.C. 92—209 (June, 1992). *Broadcasting,* 23 March 1992: 44. For pressure being exerted by television broadcasters in this direction, see *Television Digest,* 31 August 1992: 8; Conrad, *The New York Law Journal,* 24 July 1992, 5ff.

53. Compare *Broadcasting,* 16 March 1992: 27. The scope and significance of these agreements, however, received differing appraisals. Whereas Congress pressured the FCC to deal with the problem, the FCC pointed out that "only" 6–7% of stations had entered into some form of an LMA; Bureau of National Affairs, Antitrust and Trade Regulation Report, 12 March 1992, vol. 56, no. 1556: 329.

54. In re Revision of Radio Rules and Policies, 7 F.C.C.Rcd. 2755 (1992): confirmed In re Revision of Radio Rules and Policies (released 4 September 1992). At the same time, LMAs must also be registered in public files, which have to be submitted to the FCC together with the annual reports on ownership shares. See also *Television Digest,* 16 March 1992: 2.

55. The FCC decides cross-ownership of UHF television and radio on a case-by-case basis. See Lively, 1992: 156.

56. As the financial interest rules that denied TV networks any ownership rights or financial interest in the programs they broadcast are being phased out, the networks have become more attractive takeover targets of giant entertainment organizations such as Disney and Paramount. They will now be able to use a network to distribute their own shows without ownership constraints—and, of course, they have no commitment whatsoever to continuing the network tradition of responsible broadcast journalism.

57. The rules can be found in In re Amendment of Part 76, Subpart J, Section 76.501 of the Commission's Rules and Regulations to Eliminate the Prohibition on Common Ownership of Cable Television Systems and National Television Networks, 7 F.C.C.Rcd. 6156 (1992).

58. See also Section 6 referring to network–cable and telephone–cable cross-ownership.

59. Compare In re Revision of Radio Rules and Policies, 7 F.C.C.Rcd. 2755 (1992).

60. Compare Bagdikian, 1990. In 1987 alone, more than 1,300 radio stations changed owners. See *Broadcasting,* 6 February 1988: 61.

61. Compare Review of Commission's Regulations and Policies Affecting Investment in the Broadcast Industry, 7 F.C.C.Rcd. 2654 (1992), for further efforts to strengthen investments in the broadcasting sector.

62. Tender Offers and Proxy Contests, 59 R.R.2d 1536 (1986).

63. Compare *United States v. RCA,* 358 U.S. 334, 79 S.Ct. 457 (1959). Antitrust law has, however, found little application in the broadcasting area. Nevertheless, it was used to eliminate the NAB code (see Section 5.4.1).

64. Compare, for example, *CBS, Inc. v. FCC,* 453 U.S. 367, 101 S.Ct. 2813 (1981); *FCC v. WNCN Listeners Guild,* 450 U.S. 582, 101 S.Ct. 1266 (1981).

65. However, it is astonishing that the public trustee concept is being talked about again. For instance, R. Zaragoza, president of the Federal Communications Bar Association, has termed the reform of the comparative licensing procedure the "watershed in the reemergence of the public trustee concept"; *Broadcasting,* 14 May 1990: 31.

66. See *Writers Guild of America West, Inc. v. FCC,* 423 F.Supp. 1064 (C.D. Cal. 1976); reversed *Writers Guild of America West v. ABC,* 609 F.2d 355 (9th Cir. 1979).

67. See Report and Order, In the Matter of Revision of Programming and Commercialization Policies, Ascertainment Requirements and Program Log Requirements for Commercial Television Stations, 98 F.C.C.2d 1076 (1984).

68. The Senate initially wanted to extend this approach to shows dealing with sex and drugs, but it lacked support for such action; *Broadcasting,* 5 June 1989: 27f. For the recently enacted regulations, see Section 5.4.3.

69. The Department of Justice has even extended waivers of the antitrust law to allow networks to cooperate in voluntary joint efforts to control violence; Carter, Dee et al., 1994: 433.

70. Such was proposed by Senator Simon; *Christian Science Monitor,* 26 August 1992, 9. First Amendment concerns raised in the legislature were countered with the argument that the act does not contain any mandatory obligations. But as a result of clever political pressure, a de facto obligation seems to be coming about.

71. See Teeter & Le Duc, 1992: 381ff.; *FCC v. WNCN Listeners Guild,* 450 U.S. 582, 101 S.Ct. 1266 (1981).

72. *Telecommunications Research Action Center v. FCC,* 801 F.2d 501 (D.C.Cir. 1986), *cert. denied,* 107 S.Ct. 3196 (1987).

73. In re Complaint Syracuse Peace Council Against WTVH Syracuse, Memorandum Opinion and Order, 63 R.R.2d 542 (1987); the decision was upheld in *Syracuse Peace Council v. FCC,* 867 F.2d 654 (D.C.Cir. 1989). For the abolishment of the fairness doctrine, see also Le Duc, 1988: 62ff. For the political interplay with Congress, see also Teeter & Le Duc, 1992: 394f.

74. Earlier efforts to do so, however, were vetoed by President Reagan. See Teeter & Le Duc, 1992: 396f.

75. On the relationships between the three main U.S. network pressure groups that attempted to influence programming, see Montgomery, 1989.

76. In addition to the cited work by Rowan, 1984, see Simmons, 1978; Schmidt, 1978: 202f.; Coyne, 1981: 591ff. See also *Brandywine-Main Line Radio, Inc. v. FCC,* 473 F.2d 79 f. (D.C.Cir. 1972) (dissent by Chief Judge Bazelon).

77. Report in the Matter of Inquiry into Section 73.1910 of the Commission's Rules and Regulations Concerning the General Fairness Doctrine Obligations of Broadcast Licensees, 102 F.C.C.2d 145 (1985).

78. The obscenity of the material is determined according to "(a) whether the average person, applying contemporary community standards, would find that the work, taken as a whole, appeals to . . . prurient interests . . . , (b) whether the work depicts or describes, in a patently offensive way, sexual conduct specifically defined by the applicable state law, and (c) whether the work, taken as a whole, lacks serious literary, artistic, political or scientific value." *Miller v. California,* 413 U.S. 15, 93 S.Ct. 2607 (1973).

79. The FCC defines indecent programs as those which "describe, in terms patently offensive as measured by contemporary community standards for broadcast medium, sexual or excretory activities and organs, at times of the day where there is reasonable risk that children may be in the audience." Pacifica Foundation, 56 F.C.C.2d 946 (1975).

80. During 1972 there were more than 2,000 listener complaints; by 1974 that figure had reached over 20,000. Teeter & Le Duc, 1992: 383f. n. 34.

81. *FCC v. Pacifica Foundation,* 438 U.S. 726, 93 S.Ct. 3026 (1978).

82. See Infinity Broadcasting Corp. of Pennsylvania, 3 F.C.C.Rcd. 930 (1987); New Indecency Enforcement Standards to Be Applied to All Broadcast and Amateur Radio Licenses, 62 R.R.2d 1218 (1987).

83. There is a wealth of literature on the mainly First Amendment problems associated with the regulation of indecent speech. See, for example, Gayoso, 1989; Passler, 1990.

84. *Action for Children's Television v. FCC,* 852 F.2d 1332 (D.C.Cir. 1988).

85. The statutory obligation was a part of the budget appropriations law. The rules subsequently enacted by the FCC can be found in Enforcement of Prohibitions Against Broadcast Obscenity and Indecency in 18 U.S.C. 1464, 4 F.C.C.Rcd. 457 (1988).

86. Report of the Commission, In re Enforcement of Prohibition Against Broadcast Indecency in 18 U.S.C. 1464, 5 F.C.C.Rcd. 5297 (1990).

87. These principles were enumerated by the Supreme Court just prior to this case in a decision rendered on "dial-a-porn" telephone services, which was clearly distinguished from the "Pacifica" holding; *Sable Communications v. FCC,* 492 U.S. 115, 109 S.Ct. 2829 (1989). See, for example, *Broadcasting,* 3 April 1989: 57.

88. The FCC defined children as persons under the age of seventeen. Earlier, the FCC had chosen age twelve as the limit.

89. *Action for Children's Television v. FCC,* 932 F.2d 1504 (D.C.Cir. 1990), *cert. denied,* 112 S.Ct. 1281 (1992). *Action for Children's Television v. FCC,* 304 U.S. App. D.C. 126 (D.C.Cir. 1993), *vacated,* 15 F.3d 186. For the reactions to the decision by the Court of Appeals, see, for example, *Broadcasting,* 20 May 1991: 33f. and 27 May 1991: 41f. The D.C. Court of Appeals also held that upon license renewals, the FCC is required to investigate charges of allegedly obscene broadcasts by the licensee; *Monroe Communications Corp. v. FCC,* 900 F.2d 351 (D.C.Cir. 1990).

90. This applies particularly to radio stations. See, for example, *Broadcasting,* 28 August 1989: 27f.; 4 July 1989: 49f.; 30 October 1989: 28f.; 18 February 1991: 60f.; 13 January 1992: 91f.; 2 March 1992: 29f.; 25 May 1992: 56. But in so doing, the FCC largely reacts only to complaints (which were, however, numerous) and does not conduct its own research. Compare *Broadcasting,* 7 January 1991: 100ff. and 27 January 1992: 14f. Following this trend is the termination of the two leased public

access channels in the New York cable systems by the new cable television franchise contract of New York City. Both of these channels showed a number of sexually indecent programs. See also sec. 10 of the Cable Television Consumer Protection and Competition Act of 1992.

91. See Children's Television Programming and Advertising Practices, 75 F.C.C.2d 138 (1979). This multivolume report by a special task force was, however, recounted in official FCC documents only in a (toned-down) short form. See also Tucker & Saffelle, 1982: 657ff.

92. Report and Order, In the Matter of Children's Television Programming and Advertising Practices, 96 F.C.C.2d 634 (1984).

93. In addition to this objective, the Act contained advertising restrictions for children's programming, dealt with in Section 5.5.

94. See In the Matter of Policies and Rules Concerning Children's Television and Commercialization Policies, Ascertainment Requirements and Program Log Requirements for Commercial Television Stations, 6 F.C.C.Rcd. 2111 (1991).

95. With respect to this provision, "children" are persons younger than sixteen years.

96. The adoption of this broad term represented a victory by representatives of the media industry over public interest groups, who in hearings had called for a narrow interpretation that, for example, would have excluded fiction programs. See also *Broadcasting,* 4 February 1991: 22ff.

97. To this end, the FCC offers "permissive guidelines," which are intended to give broadcasters accessible criteria.

98. The subsidized programs may initially be shown only on public television and are available to other broadcasters for noncommercial use after two years.

99. Children's Television Report and Policy Statement, 50 F.C.C.2d 1 (1974).

100. In this particular area, children's programs are taken to mean those that are originally produced for children under the age of twelve and broadcast for such an audience.

101. See, in particular, In the Matter Policies and Rules Concerning Children's Television and Commercialization Policies, Ascertainment Requirements, and Program Log Requirements for Commercial Television Stations, 6 F.C.C.Rcd. 2111 (1991).

102. According to a study by an institute at Indiana University, most broadcasters used only ten advertising minutes per hour for children's programming, thereby conforming to the new standard; Bureau of National Affairs, 1991: 585.

103. But many broadcasters have moved their commercials for these figures to shows coming immediately before or after the program in question since this seems to be more effective. For this reason, they are not even affected by the new rule. See *Broadcasting,* 12 November 1990: 33f., with commentary to this effect by Peggy Charren. It therefore comes as no surprise that both the advertising and broadcasting industries are most satisfied with the FCC; *The New York Times,* 10 April 1991: D7.

104. Policies and Rules Concerning Children's Television Programming; Revision of Programming and Commercialization Policies, Ascertainment Requirements, and Program Log Requirements for Commercial Television Stations, 6 F.C.C.Rcd. 5093 (1991).

105. Policies and Rules Concerning Children's Television Programming; Revision of Programming and Commercialization Policies, Ascertainment Requirements,

and Program Log Requirements for Commercial Television, 6 F.C.C.Rcd. 5529 (1991). See also *Washington Post,* 2 August 1991: D1.

106. This is set forth in the rule itself, which also states that in the cable sector the FCC will even rely exclusively on control by the general public.

107. In the Matter of Implementation of Section 4 (g) of the Cable Television Consumer Protection and Competition Act of 1992, Home Shopping Station Issues, 8 F.C.C.Rcd. 5321 (1993).

108. See the finding on 22 of the Cable Television Consumer Protection and Competition Act of 1992, in House of Representatives, Conference Report 102-862, 14 September 1992: p. 41: "Congress finds and declares that . . . increased numbers of females and minorities in positions of management authority in the cable and broadcast television industries advance the Nation's policy favoring diversity in the expansion of views in the electronic media." A different view is taken in *Lamprecht v. FCC,* 958 F.2d 382 (D.C.Cir. 1992).

109. See *United States v. Southwestern Cable Co.,* 392 U.S. 157, 88 S.Ct. 1994 (1968).

110. *FCC v. Midwest Video Corp.,* 440 U.S. 689, 99 S.Ct. 1435 (1979).

111. See Cable Rate Regulation, 8 F.C.C.Rcd. 5631 (1993). The regulation established three classes of rates: basic, premium, and other. For basic cable, the FCC created a table of benchmarks based on the average September 30, 1992, rates of systems subject to effective competition. Basic rates that exceeded the benchmark had to be reduced by 10% or by the benchmark, whichever is higher. The FCC soon learned about the difficulties in fixing adequate rates. In 1994, the Commission raised the benchmarks in its table of benchmarks, and later it approved new rate rises, outside the existing regulated tiers (see *The New York Times,* 11 November 1994: D1, D17).

112. *Turner Broadcasting Systems, Inc. v. FCC,* 114 S.Ct. 2445 (1994). For earlier must-carry rules, see *Quincy Cable TV, Inc. v. FCC,* 768 F.2d 1434 (D.C.Cir. 1985); *Century Communications Corp. v. FCC,* 835 F.2d 292 (D.C.Cir. 1987).

113. Compare note 112.

114. However, the 1992 act prohibits the award of exclusive franchises.

115. *City of Los Angeles v. Preferred Communications, Inc.,* 476 U.S. 488, 106 S.Ct. 3034 (1986).

116. For the various forms of regulation in these areas, see Carter, Dee, et al., 1994: 487ff.

117. Some of them were challenged in court. See, for example, Common Ownership of Cable Television Systems and National Television Networks, 7 F.C.C.Rcd. 6156 (1992).

118. Compare Video Dialtone Order, 7 F.C.C.Rcd. 5781 (1992).

119. Further Proceedings on Cable-Cross-Ownership Rules, 69 R.R.2d 1613 (1993).

120. Selected comments are to be found in the bibliographies in Williams & Pavlik 1994: 255 ff. and in *Media Studies Journal,* Winter, 1994: 172ff.

121. The $33 billion sale of Telecommunications, Inc., to Bell Atlantic in October 1993, the biggest merger in American history until now, was particularly spectacular: "A combination of the nation's most powerful cable company with one of the nation's most profitable public utilities" (Grossman, 1994: 28).

122. His role is important to the extent that he nominates commissioners. The

influence of policy on the selection process can be seen in commissioner biographies. See Cole & Oettinger, 1978: 16ff; see also Barnett, 1988: 118.

123. See Krasnow et al., 1982: 281f. Whether this statement would also be correct if deregulatory policy were included must be left aside here.

124. For a critical analysis that in particular deals with the myths of the neutral marketplace, see Ingber, 1984: 1ff. See also Fiss, 1987: 781ff.

125. The contrary holding in *Red Lion Broadcasting v. FCC,* 395 U.S. 367, 390, 89 S.Ct. 179 (1969) has been of little consequence.

CHAPTER 2

Great Britain

Although not the oldest in the world, the British broadcasting system was for many decades certainly the most admired. In particular, the British Broadcasting Corporation (BBC) was archetypical for the establishment of broadcasters on all continents. In addition, the duality of public and private commercial broadcasting as set up in Britain in the 1950s was at least taken into consideration as a model in many countries. Furthermore, the lofty journalistic status enjoyed by the BBC, the remarkably diverse, ambitious programming offered by the private broadcasting companies, and a number of innovations, such as the critically acclaimed cultural offerings of Channel 4, are all spoken of in the highest terms. In view of the leading role British broadcasting has played in the media-policy discussions in many countries, a description of the anatomy of the British broadcasting system is merited, with emphasis on the role of broadcasting supervision.

To the extent that this is done from a comparative standpoint, however, particular attention must be paid to the fact that the "duopoly" in Great Britain was for decades structured quite differently than the new duopoly of public and private broadcasting introduced in other Western industrialized countries. Moreover, British broadcasting supervision was able to draw upon a specific political culture and supervisory tradition, which are not to be found elsewhere in the same form. It must also be kept in mind that the British broadcasting system is currently in a state of transformation: The Broadcasting Act of 1990 (Broadcasting Act 1990, London: Her Majesty's Stationery Office) represents a dramatic turning point, whose way was paved by the 1988 white paper "Broadcasting in the '90s" (Home Office, 1988). Not only have the fundamen-

67

tal structural elements of the broadcasting order been remodeled—in particular, the monopoly-like duality of public and private broadcasting is to be eliminated—but the system of broadcasting supervision has also been undergoing extensive change. It is still too early to make substantial observations on the viability of the new broadcasting and supervision system. However, a departure can be noted from the old system, while effectiveness and efficiency of the new order are not yet amenable to analysis.

It continues to be instructive to analyze the experiences thus far with broadcasting supervision. No other European country has had such lengthy experience with supervision in dual broadcasting than Great Britain. British experiences are therefore especially important for lessons on the possibilities for and conditions of successful supervision. Aside from this, it can be expected that many supervisory traditions will continue to have an impact on the new broadcasting order as well and thus also have far-reaching consequences for the new legal framework.

The first three sections of this chapter describe the basic elements of the British broadcasting system. Attention will then turn to a depiction of the former supervisory authority (4) and its fields of activity (5). Subsequently, the relationship of the supervisory authority to the State will be dealt with (6). A concluding section offers a description of the most important elements of the new regulation (8). The latter is limited to several factors likely to be of importance and is not intended as a comprehensive description and analysis.

Before embarking, it should be noted that in contrast to deregulation in the United States, for example, the reform for the 1990s in Britain was only to a limited extent justified with the argument that previous broadcasting supervision had failed (see Home Office, 1988). Primarily it has been the result of the political concern for a fundamental reorganization of the broadcasting system and, as a result, a pursuit of de- and reregulatory objectives.

1. DEVELOPMENT AND STRUCTURE OF THE BRITISH BROADCASTING SYSTEM

1.1. Public Broadcasting: The British Broadcasting Corporation

Great Britain is endowed with a public broadcasting body, the BBC, which, stemming from a paternalistic–elitist tradition (cf. Smith, 1986: 10), was and is characterized by high professional standards and journalistic independence. The BBC, successor to the private British Broadcasting Company founded in 1922, was established in 1927 on the basis of a royal charter and a license from the Postmaster General for the operation of radio broadcasting. From 1936 to 1939 and since 1946, it has broadcast television. Currently the BBC supplies programming to two television channels and five radio networks, as well as to a variety of local radio broadcasters. Its managing body—the Board of Governors—is nominally appointed by the Queen in Council, although the right effectively lies with the Prime Minister (Gellner, 1990: 319). Traditionally this appointment is preceded by an accord between the government and the opposition, as well as—at least prior to the Thatcher era—by contact with the BBC itself (Madge, 1989: 145). The Board itself appoints the Director General. The composition of the Board is not governed by the representation principle; rather, it aims to appoint "remarkable men and women . . . of the highest calibre" (Pilkington Report, 1962: 123)—in other words—drawn from the "great and the good." This does not, however, rule out regional origin and political stance from being taken into account de facto.

The idea of feedback at the BBC from interest groups in society is put into effect institutionally by establishment of a variety of internal advisory bodies, which serve to facilitate the articulation of special interests (Briggs, 1985: 89) and "public opinion" (Madge, 1989: 103). However, these bodies have not been given direct decision-making authority. Participation through bodies is differentiated both from a regional standpoint (there are broadcasting councils in Scotland, Wales, and Northern Ireland) and with respect to specific issues. In addition, public meetings have been held in order to generate feedback from viewers with regard to broadcasting operations (Madge, 1989: 115ff.).

The BBC's legal commitments are not set down in formal statutes but, rather, ensue from the above-mentioned royal charter and the license conditions, as well as from guidelines set by the BBC itself. For instance, the Code of Practice issued by the BBC sets forth principles regarding balance and impartiality and obligates it to provide for information, education, and entertainment in its programming. Though it establishes a framework for BBC operations, the royal charter does not lay a foundation for wide-ranging control powers by the government. On the other hand, the license, which is currently granted by the Home Secretary, provides the government with authority to intervene, including the right to prohibit the transmission of a program. The BBC is also precluded from broadcasting its own views on matters of public policy other than those relating to broadcasting.

Editorial independence is one of the founding principles of the BBC. In spite of its broad authority to intervene, the government has consistently made sparing use of its reserve powers and thereby contributed to the formation of broad political consensus that the independence of broadcasting be respected. This consensus has remained intact for decades. Consequently, the twelve members of the Board and the Director General are not selected according to criteria of political affiliation; to this extent, political permeation of the BBC has also been avoided. External political pressure, too, has been substantially deflated by the board. Only in recent times—in the Thatcher period—has the political pressure become greater (see Section 6) and the BBC's internal powers of resistance weaker. Nevertheless, the BBC has remained worldwide a paragon of independent broadcasting. There is probably no other public broadcasting order anywhere in the world where employees can orient their work according to journalistic criteria with the same degree of independence as at the BBC.

The BBC is financed solely by the license fee, whose level is fixed by Parliament upon proposal by the House Secretary. This provides the political authorities with the possibility of exerting influence on the development of the BBC. In recent times it has been used at least as a means of channeling the BBC's general developmental opportunities and of constraining its role in competition with private broadcasters. Of late, the fee has been tied to the retail price index. Although this indicates that the standard for setting fees has become more objective, it spells a considerable risk for the future of the BBC. In particular, increases in the retail price index are substantially lower than wage and salary increases in the broadcasting sector, not to mention increases in the acquisition or production costs for broadcasting programs. As a result of new competition, dramatic cost increases can be noted throughout the world. In a white paper, the government even announced that the BBC's future growth rate would remain below the retail price index; this was designed to motivate the BBC to look for other sources of financing, such as pay-TV (Home Office, 1988: 9). In autumn 1994, however, the government announced that until 1996 the license fee will be pegged to the retail price index. Suggestions that financing through advertising be implemented in part led to the establishment of the Peacock Committee, which, however, rejected corresponding proposals (Peacock Committee, 1986). The government has accepted this result. Thus, the BBC is not in the running as a competitor of private broadcasters for financing in the form of advertising (cf. Department of National Heritage, 1992: 33f.).

Following up on liberal economists' calls, the white paper announced that the mode of financing the BBC would have to be broadened to include pay-TV. The BBC reacted to this with the introduction of an experimental, nighttime pay-TV program specifically for physicians. It has also begun to develop specialized pay-TV programs on both channels during nighttime hours (see Checkland, 1989: 19ff.). The financial pressure has been mitigated by the recent announcement that the license fee would be tied to the retail price index until 1996. It appears as if the government will accept the licence fee as the best (or least negative) source of BBC funding.

During the course of the discussions surrounding the reform of the British broadcasting order and the expansion of the private sector, the government has often felt compelled to stress the BBC's role as the "cornerstone" of public service in British broadcasting. It remains to be seen whether this is ultimately an indication of the will to maintain the status quo. In any event, such assurances serve to deflate criticism of the radical changes that have already been introduced to British broadcasting recently. Although there is much rhetorical commitment to high-quality programming on the side of the policy makers, programming regulation is being relaxed. The organizational and financial safeguards for the trustee concept of broadcasting are being eroded. Supervision is being weakened. Whatever else happens to the duopoly, the government can repeat that at least the BBC will remain an oasis of public-service quality provision. However, initiatives continue to be made to change the BBC's economic basis, and this is bound to impact upon its role as provider of broadcast programming. For instance, calls are still being made by liberal economists to shift the BBC's mode of financing more strongly toward pay-TV.[1] As mentioned above, though, they seem to be less influential at present. In addition, a basic discussion has begun on the future role of the BBC. In light of the fact that the BBC's charter will expire in 1996, the government has called upon the public to comment on what role the BBC should then play (see Department of National Heritage, 1992); the BBC has also come up with its own proposals (BBC, 1992).

1.2. Commercial Terrestrial Broadcasting

Great Britain's broadcasting system differs from all other Western European broadcasting orders by its early start in the duopoly of public and private broadcasting. Commercial television began in Great Britain as early as 1954. The Television Act of 1954 created the Independent Television Authority (ITA; for details, see Sendall, 1982), a central licensing and supervisory institution for Independent Television (ITV). This was then restructured as the Independent Broadcasting Authority (IBA) by the Broadcasting Act of 1973, which at the same time extended the IBA's supervisory powers to Independent Local Radio (ILR), established in the interim under the Sound Broadcasting Act of 1972. As a result of the Broadcasting Act of 1990, the IBA and the ILR have been replaced by the Independent Television Commission (ITC) and the Radio Authority.

In many respects, the IBA was modeled on the BBC. Although commercial broadcasting was to be introduced, commercial aspects were, as far as possible, not to have a negative impact on the quality of programming. In particular, they were not to lead to American-style programming conduct. In setting up private broadcasting, the IBA took care to avoid the unleashing of unbridled competition of the American kind among braodcasters. The IBA was designed as a very different kind of supervisory body to the U.S. Federal Communications Commission (FCC); it was far more interventionist.

The members of the IBA were appointed by the Home Secretary, who could also revoke their appointment.[2] As with the BBC Board of Governors, members of the IBA were not appointed with regard to group affiliation or other representational considerations but, rather, from the ranks of the "great and the good." Although not employed full-time, the members were supported by an administrative machinery, which, under the leadership of a director general, encompassed some 1,300 employees and several regional offices (IBA, 1990: appendix 2ff.). This was in turn supplemented by a number of advisory bodies composed of representatives of social interests and IBA personnel (Potter, 1989: 92, 96) and at the end numbered an additional 700 persons (IBA, 1989: 34).

The bases for the broadcasters' operations were the franchises that were issued by the IBA for a limited period to the fifteen regional television companies, TV-am for morning television, and roughly seventy-five local radio companies. Each television company essentially was endowed with a regional monopoly; but for the morning television program broadcast by TV-am,[3] a national monopoly was awarded.

The local radio companies also had local monopolies until 1989. For news broadcasts, they had formed a network (Independent Radio News). As a reaction to the delays with the new broadcasting legislation, in 1989 the IBA began to award so-called incremental franchises in addition to the normal franchises (Brown, 1988: 19). In this context, "incremental" meant that stations were licensed for areas that were already being serviced by established ILR stations but also that they were not made subject to the requirements of full- (public) service programming provisions. For instance, they were not obligated to broadcast news, and they were allowed to target their programming at specific communities of interest (IBA, 1990: 24f.). These changed requirements in the face of an unchanged legal framework were expressly justified by the IBA as an anticipation of the coming "light-touch" regulation by the yet-to-be-formed Radio Authority, which had already been announced in the 1988 white paper. As the established ILR stations were in sound financial condition, the IBA sought on the whole to satisfy the widespread desire for more specialized, varied broadcasters. However, it limited the number of new franchises to twenty-four in order to avoid overly confining the maneuvering room of the future radio supervisory authority (Brown, 1988: 19). At the time when these "incremental radios" began their operations, a second news network was launched, which serviced many of these franchise holders (IBA, 1990: 24).

The broadcasting companies did not have possession of their own transmission facilities, nor did they have the right to broadcast their programs over outside transmission facilities. Rather, the IBA itself was the holder of the transmission license.[4] It provided for the transmission of programs produced or put together by the ITV and ILR companies. The IBA's function was, however, not limited here to transmission; from a legal standpoint, it also took on programming responsibility itself. In particular, it had both the right and the duty to approve schedules and monitor the contents of programming, including advertising. At the same time, however, in the franchise contracts with the ITV and ILR companies, it committed itself to transmitting the programs they produced, unless there was a reason for objection. The duties placed on the IBA by the broadcasting laws to ensure high-quality, balanced programming were shifted by the IBA to the program companies with the aid of the franchise contracts and its day-to-day supervisory activities.

The IBA was also selected to take charge of an ambitious project: the creation of an intellectually and culturally challenging channel, named Channel 4, launched in 1982. It was called upon to serve minorities and "appeal to tastes and interests not generally catered for" (Broadcasting Bill 1980, clause 3) by broadcasting cultural and discriminating, high-quality entertainment programs (cf. also Lambert, 1982; a portrait of this channel is sketched by Bock & Zielinski, 1987: 38ff.). This formed the second commercial television channel, which was designed as a supplement to and not as a competitor of the other ITV broadcasts. Channel 4 was owned by the IBA and did not produce its own programming but rather obtained it from the ITV companies, independent producers, and the international market for programs. News broadcasts were drawn from Independent Television News (ITN), the joint news service of the fifteen ITV companies.[5] Because of a relatively high share of commissioned productions from independent producers, Channel 4 became a crucial promoter of this market.

With the replacement of the IBA by the ITC and the Radio Authority in 1991, a new phase a broadcasting regulation has begun in Great Britain. Particular attention has focused on the allocation of regional licenses for Channel 3 (see Section 7.3).

1.3. Cable and Satellite Broadcasting

In Great Britain, both cable and satellite are used in broadcasting. The transmission of programs via cable was regulated[6] by the 1984 Cable and Broadcasting Act. It also applied to satellite broadcasting when programs were picked up by cable networks. To this extent, the conduct of cable operators was subject to control by the Cable Authority, which started operation in 1985. When communications satellites were used, no special licensing and supervisory requirements were placed on them (for details, see Wiedemann, 1989: 166ff.). The granting of franchises to operate channels on a DBS was the province of the IBA.

Representing a departure from the rules then in force for the BBC and the IBA, the Cable and Broadcasting Act was governed by the concept of far-reaching deregulation. It was in this manner that both the government and the majority in Parliament hoped to be able, on the one hand, to meet more quickly industrial-policy objectives in the satellite and cable sector and, on the other, to create a media counterweight to the programming offered by BBC, ITV, and ILR.[7]

1.3.1. Satellite Broadcasting

Following an initial euphoria, which gave way very quickly to much delay and disappointment, satellite broadcasting developed slowly but steadily in Great Britain. It is still effectively a niche market, and until recently it was the subject of fierce competition. Leaving aside the infancy of this technology—particularly, the use of the telecommunications satellite *EUTELSAT F1* for the British "Super Channel"—the satellite broadcasting era can be deemed to have truly commenced with the use of the telecommunications satellite *Astra,* whose sixteen channels make possible the transmission of programs that can also be received directly. The trailblazer in the use of these satellites was Rupert Murdoch with four Sky channels. The broadcasting satellite *Marco Polo* has also been in service since 1990.[8] The IBA awarded franchises for all of its five channels for a fifteen-year term to the British Satellite Broadcasting Consortium (BSB), the first three in 1986, and the remaining two in 1989. Thus, BSB became the rival of Sky Channel. But in 1990, Sky and BSB announced they would merge, concluding the most costly "media battle" in British history (see, e.g., *The Times,* July 1989; *Journal des medias,* 4 September 1989: 12; *Financial Times,* 7 August 1989; *Marketing,* 20 July 1989). It is worth noting that the regulatory body was not informed of the Sky–BSB merger until after it had been agreed. This rendered the authority powerless as its only sanction was to withdraw the BSB license, which would simply have meant that those viewers with BSB equipment could not receive the new service until they bought new equipment, with minimal financial consequences to the new company.

The merger arrangement concluded between the opponents demonstrated just how little respect the actors themselves paid to the competition principle. The market for satellite programming was deemed too small for more competitors. Satellite channels do in fact incur heavy losses (cf. Collins, 1993: 118; Tunstall, 1992: 251). The integration between the two companies and the British newspaper industry, moreover, harbors the risk that intensively interlocked intermedia systems will emerge. For instance, these companies can use their power in the press market plea to support advertising campaigns for satellite channels. The Sky Channel, linked to various newspapers by way of Rupert Murdoch, has clearly made use of this opportunity from the outset.

Satellite television is characterized by specialty and target-group channels. For four of these channels, especially the cinema channels, which are especially attractive for viewers, pay-TV has been selected as the form of financing. The remaining channels are financed with advertising revenues and, at quite high shares, sponsor proceeds (see Burnett, 1989: 18ff.).

1.3.2. Cable Broadcasting

Cable broadcasting has proceeded quite sluggishly in Great Britain and remains far behind expectations with the penetration rate remaining very low. A number of factors were responsible for its weak start, above all the uncertain profit-making prospects and some economic policy decisions by the government regarded as detrimental to investment (Dutton & Blumler, 1988: 287; Cable Authority, 1990: 46). But starting in 1989–1990, the business sector began to develop greater interest in the construction and operation of cable networks. This engendered some bustling franchising activities by the Cable Authority shortly before it was dissolved by the Broadcasting Act of 1990.

Decisive for this development was the strong engagement of North American cable operators and telephone companies in the British cable broadcasting market (Cable Authority, 1989: 6; 1990: 5f.; for figures, cf. *Screen Digest,* August 1990: 192; *New Media Markets,* vol. 8, no. 14, 1ff.). With the virtual saturation of their home markets, they expanded into the similarly English-language, deregulated British market, expecting to gain both access to the European market and experience in cooperation with cable operators and in coordination of telecommunications and television services.[9] In the process, however, the North American companies had to overcome the statutory hurdle that ruled out non-EC citizens as owners of cable facilities.[10] For this purpose, many of them resorted to such diversions as setting themselves up as trusts in the Channel Islands (Cable Authority, 1989: 6).

The profitability of investments in the cable sector is by no means guaranteed (see Kleinworth Benson Securities, 1990: 48ff.). For instance, it is noteworthy that despite the upswing in the cable sector, British Telecom decided in 1990 to pull out of the broadband cable business, since sufficiently quick capital returns were not expected (*Cable and Satellite Europe,* September 1990: 38). Insofar as prospects for success are seen in the cable sector, they are connected to the growing number of satellite programs, which also offer cable households a more attractive choice of programming. It is hoped that this will lead to an increased penetration rate (see, e.g., Cable Authority, 1989: 5; Davey, 1990: 7; Tunstall, 1992: 251). In light of the possibility of direct reception of satellite programming, however, it is in all respects uncertain to what degree this hope will be fulfilled.

For franchising and program supervision in the cable sector,[11] the Cable and Broadcasting Act of 1984 created a new institution in the form of the Cable Authority, a supervisory body with relatively few personnel (Cable Authority, 1990: 46). The cable sector was largely divorced from the public service concept and set up with the objective of far-reaching de- and nonregulation (so-called light-touch regulation). The Cable Authority was not the broadcaster of the programs and thus not itself responsible for them but, rather, was merely a licensing and supervisory body. It was endowed with relatively large powers of discretion with regard to franchising. Franchises were much less detailed than the franchise contracts between the IBA and the ITV companies (Coopers & Lybrand, 1988: annex B). In addition, the commitments placed by law on the franchise holders and specified by the Cable Authority's guidelines were considerably weaker than in the cases of the BBC and the IBA. In particular, the cable companies were basically not obligated to observe impartiality in programming, but, rather, only to respect the limits of decency. Only news and information originating in the United Kingdom were subject to the requirements of objectivity and impartiality. There was no requirement of diversity and balance in the overall programming, merely a "must-carry" rule requiring cable companies to offer BBC and IBA programs as well, but only if capacity allowed. At the same time, however, the companies had to ensure that with regard to religious or politically controversial topics, the views of individual persons, groups, or organizations did not predominate ("undue prominence rule").[12] In monitoring commitments, the Cable Authority as supervisory body lacked—apart from exceptional cases—the possibility of advance control.[13] Even with subsequent program supervision, the Cable Authority dispensed with systematic control (Cable Authority, 1990: 10). Also, in view of its dearth of personnel, it would not have been able to conduct such control in any case. The Cable Authority mainly took action here in response to complaints, though at times on its own initiative as well. On occasion, some companies voluntarily submitted problematic programs for advance review by the Cable Authority as in the case of certain music videos (see, e.g., Cable Authority, 1990: 12).

As a whole, the Cable Authority has not undertaken extensive activities in the area of programming supervision, and it has made little use of formal monitoring instruments—with which it could, for example, demand that a program or advertisement not be broadcast or insist upon inspection of broadcast programs. Above all, it has never resorted to the possibility of revocation of franchise. Formal admonishments as a result of improper conduct in the area of programming have likewise not been documented. On an informal level, however, a number of objections have ensued.

The Cable Authority was charged by statute not merely with the regulation but also with the promotion of cable television (see Cable and Broadcasting Act of 1984, 4 (9)). It came to view this latter task as the focal point of its work.[14] Particularly during the initial phase of extreme reluctance on the part of the cable industry, the Cable Authority assumed more the role of promoter than that of controller of the cable companies. For instance, it took active part in the search for parties interested in cable franchises and for investors and did not impose fees on franchise holders to cover its own costs of operation (Coopers & Lybrand, 1988: No. 30). In order to avoid putting a damper on the start-up chances for cable broadcasting, the Cable Authority consciously practiced restraint in exercising the regulatory competences provided to it under law. With regard to franchising, for example, it sought to discourage applicants for franchises from making promises that they would not be able to keep.

However, in the face of newly emerging competition, with the so-called second wave, or the entry of North American firms in the market, the Cable Authority was presented with the possibility of exercising pressure when necessary on the companies (Cable Authority, 1989: 10). This pressure was nearly exclusively used to ensure the spread of cable and not, for example, to impose programming requirements. Thus, the Cable Authority informally warned franchise holders that were experiencing delays in becoming operational that their franchises might be revoked (Cable Authority, 1989; 10f.). This had such a stimulating effect that in the end only one ultimatum had to be issued (Cable Authority, 1990: 11). Departing from earlier practice, the Cable Authority also hoped to achieve accelerated cable usage by licensing SMATV systems, permitting the occupants of apartment buildings to receive satellite broadcasts even in areas in which franchises for cable networks had already been awarded when the holder of the cable franchise was unable to guarantee provision of service within six months. When new franchises were issued, the Cable Authority raised requirements regarding demonstration of sources of proposed funding in order to prevent the growth of the network from foundering due to financial constraints (Cable Authority, 1989: 10).

The Cable Authority was dissolved in the course of the broadcasting reform of 1990; its functions have now been assumed by the ITC. It should, however, be noted here that the underlying regulatory concept has gained increasing importance in the public discussion in Great Britain. Moreover, in its green paper on the development of radio (Home Office, 1987), the government had considered making the Cable Authority instead of the IBA the supervisory authority for radio. Although these considerations were not further pursued—the Broadcasting Act of 1990 created a new Radio Authority—the government nevertheless made clear that, in the course of deregulating the broadcasting sector, a different supervisory concept was in order than that which had heretofore been typical for the IBA.

2. TRUSTEE CONCEPT FOR BROADCASTING

International acclaim—or, by others, opprobrium—has been expressed for the British concept of committing broadcasting to a trustee role without placing it under state dependence. The trustee concept stemmed, among other things, from the effort to avoid "American conditions" in British broadcasting (for details, see Hearst, 1992: 62ff.). The initial focus was on preventing chaos in the airwaves, which then gradually gave way to ensuring a special public responsibility for broadcasting.[15] The trustee concept was essentially developed by John Reith—later, Lord Reith—the first Director General of the BBC, who firmly believed in the moral and religious values of the British middle class. This concept had as early as 1926 been described by the Crawford Committee of Inquiry, which recommended that "the broadcasting service should be conducted by a public corporation acting as Trustee for the national interest and that its status and duties should correspond with those of a public service" (cited in Annan Report, 1977: 9). Accordingly, under the IBA Act of 1973, the IBA was charged with providing "the television and local sound broadcasting services as a public service for disseminating information, education and entertainment" (IBA Act of 1973, s. 2(2)(a); for the IBA's trustee perspective, see IBA, 1985). In fulfilling public functions, both public and private broadcasters—for the first commercial channel and for Channel 4, as well as for private radio—were made subject to a wealth of substantive, organizational, and financial commitments and corresponding control.

In the area of cable broadcasting, however, the Cable and Broadcasting Act of 1984 substantially relaxed many commitments. The broadcasting reform introduced with the Broadcasting Act of 1990 has led in a broader framework to a loosening of the public service commitments of private broadcasters without, however, fully departing from the public service concept as such.[16] In spite of such relaxations, it can categorically be stated that in none of the other countries studied here has the trustee commitment on private broadcasting been so intensively developed as in Great Britain. The legal requirements placed on the ITV companies have been even more detailed than those in force for the BBC. This detail was based on the assumption that a commercially financed broadcasting system would be especially tempted to sidestep the traditional public service broadcasting commitments. The British concept has thus far been marked by the effort to counteract this evasion intensively with legal and organizational means.

The days of the traditional trustee concept are nevertheless numbered even in Great Britain. As early as 1988 in its white paper, the government largely departed from this concept for private broadcasting (Home Office, 1988: particularly 4ff., 11ff.). Market forces, and not normative requirements, were to determine program offerings. Although there is much rhetorical commitment to high-quality programming expressed by the policy makers, programming regulation is being relaxed. The organizational and financial safeguards oriented to the trustee concept of broadcasting are being eroded. Supervision is being weakened. However, political pressure from the public has resulted in increased concessions once again being made to the trustee concept in the Broadcasting Act of 1990.

3. FORGOING ECONOMIC COMPETITION IN THE DUAL BROADCASTING SYSTEM

The British broadcasting system has thus far differed from others not only in its long BBC tradition. With regard to its "commercial variety" as well, Great Britain has gone its own way. For instance, although commercial television was introduced relatively early, this was accomplished only in the form of a regulated monopoly, which precluded media development from becoming dependent on economic competition. In the radio sector as well, there were local or regional monopolies for many years. The following remarks will concentrate on television. They relate mainly to the state of affairs prior to the introduction of the Broadcasting Act of 1990.

Although the British broadcasting system recognized disseminational and journalistic competition between public and private broadcasting, it declined to do so with regard to economic competition for sources of revenue (Negrine, 1989: 100f., 107f.). The deliberate avoidance of such economic competition was understood as necessary to ensure diversity (Smith, 1986: 2ff.). In contrast to the BBC, which was financed by license fees, private broadcasters were dependent on advertising revenues. They were entitled to all advertising returns in their respective zones of transmission.[17] Competition between the BBC and the ITV companies therefore only existed from a disseminational standpoint and with regard to acquisition of broadcasting rights, the hiring of personnel, and so forth, but not with respect to revenues.

Neither did economic competition occur among the ITV companies themselves. They not only had regional broadcasting monopolies but also revenue monopolies. Only in border zones, between franchise areas that could receive the transmissions of neighboring broadcasters, was there some (limited) competition among various ITV companies. However, this was of minor significance, particularly in light of the fact that the ITV companies had formed a network for the purpose of program distribution[18] and therefore broadcast common core programming. All ITV companies belonged to the Network Program Committee of the Independent Television Association (ITVA), which conducted joint program planning. The five largest ITV companies (Thames, London Weekend Television, ATV, Granada, and Yorkshire) dominated as program suppliers (for programming shares, see IBA, 1986: 32) and thus also as decision makers.[19] As a result of the dissatisfaction on the part of the smaller companies and the pressure exerted by the IBA (see also the veiled remarks in

IBA, 1987a: 5), these arrangements for program supply were later modified in order to provide the smaller companies with improved access opportunities.[20]

Nor did Channel 4 amount to economic competition for the individual ITV companies. Rather than receiving its own advertising revenues, Channel 4 was financed by annual grants from the IBA, which were indirectly paid by the ITV companies.[21] However, the ITV companies had the right to sell advertising time on Channel 4; they thus earned money and retained exclusive control over television advertising in their regions.

In short, although the British broadcasting system opted for a private form of financing for its commercial sector, it made sure that the mechanisms of economic competition did not impose a predominant, let alone overriding, influence on programming. The situation changed to some degree with the introduction of cable and satellite broadcasting, which was financed via advertising. However, as the result of start-up problems, the "new media" long remained an insignificant competitor for the duopoly broadcasters.

With the reform implemented under the Broadcasting Act of 1990, this situation has changed decisively. The ITV structure has been reformed (the previous system was changed and licenses have been awarded for Channel 3), and a fifth television channel is to be created. This may lead to a battle for advertising revenues and pay-TV monies and cause broadcasters to compete with one another from both an economic and, in their respective fields, a journalistic standpoint. However, the first attempt to get a fifth channel started was a dismal failure (mainly due to lack of investor interest). In the radio sector, the competition principle has become the standard. Also, Channel 4 now sells its own advertising, although under complex safeguards.

In the words of the publishing magnate Lord Thompson, the monopolies formerly held by the ITV companies meant that a television license—at least in the lucrative regions—was equivalent to a license to print money.[22] It is true that disseminational competition with the BBC generated a stiff battle for viewer ratings, which—as has been repeatedly stressed by observers (see, e.g., von Hase, 1980: 18f.; Heyn & Weiss, 1980: 145ff.; Lincoln, 1979: 131)—drove both systems to conform to elements of mass appeal. However, this relative monopoly status permitted the ITV companies to survive economically even when they did not orient their programming solely toward the goal of furnishing the advertising industry with a large group of consumers ready and willing to purchase. This situation in turn provided the IBA with the possibility of imposing and enforcing programming conditions even when these collided with the interests of the franchise holders in profit maximization. The monopolization of advertising revenues was even able to serve as justification for skimming off income to fund an otherwise risky project like Channel 4.

4. SUPERVISORY AND CONTROL AUTHORITY: THE IBA

By no means did the monopoly structures in the private broadcasting sector, which were in force until the Broadcasting Act of 1990, alone automatically guarantee a diverse broadcasting program based on the trustee principle. It was impossible to rule out entirely economic or disseminational inefficiency or even abuse of monopoly positions. For this reason, a publicly regulated monopoly was set up in Great Britain. The regulatory authority, the IBA, was endowed with powers that reached far beyond those of other supervisory authorities, such as the three French supervisory bodies or the German state media authorities.

The IBA held the license to transmit broadcast programming. It was obliged, among other acts, to provide the population with programming, other than that of the BBC. In order to be able to fulfill this mandate, it in turn concluded franchise contracts with the ITV program companies. It then used these contracts as instruments to ensure the fulfillment of the commitments imposed upon it by law.[23] Since it was legally responsible for the content of programming, the IBA was entitled to preventive programming control: The television companies were obliged to have their program planning approved in advance by the IBA (see Alvarado & Buscombe, 1978: 87; for details, see also Wiedemann, 1989: 122ff.). The radio companies could be similarly obliged. The IBA advised the program companies upon issues of program planning and was able to issue directives regarding the telecasting or omission of certain programs and on the adherence to or shifting of certain broadcast times. In the franchise contracts, the companies were furthermore obliged, upon request, to make available advance scripts and other broadcasting documents, as well as picture and sound recordings, to provide the IBA with any desired information, and to permit the inspection of their books.

Seen as a whole, however, supervisory practice was quite moderate. Only in exceptional cases did particular programs have to be screened prior to broadcast (see Potter, 1989: 103ff.). On the other hand, it was not uncommon for programs to be changed (see, e.g., IBA, 1990: 13), or portions thereof subsequently edited out, or for programs to be retracted as a result of objections (for criticism, see, e.g., Annan Report, 1977: 190; for intervention by the IBA, see also Lincoln, 1979: 136). Intervention occurred particularly often in the sensitive area of news reporting about Northern Ireland (Negrine, 1989: 123ff.).

The IBA's powers were supplemented with the authority to impose sanctions. For instance, the IBA could refuse to transmit a given program. An even stiffer supervisory measure took the form of the general suspension of the IBA's duty to transmit programming of the ITV companies. This presupposed that the IBA had on three occasions ascertained violations of the franchise and that the possibility of suspension had been communicated in writing. (For the difficulty of using these instruments, see Dix, 1980: 374.) An additional sanction consisted of exercising the power not to renew the franchise when it expired. This threat of such a sanction also had anticipatory effects in the phase of current supervision.

On account of their drastic consequences, however, the "relatively incisive" sanctioning instruments were difficult to implement in practice (see Garnham, 1980: 24f.), particularly since it was by no means certain that another programming company would prove to be more compliant. As with supervisory authorities in other countries, the IBA therefore principally employed its formal sanctioning potential in the form of a threat, in order to stress the importance of supervisory objectives and to remedy problems via informal avenues. Advance discussion of programming and ongoing consultations created starting points for this.[24] In addition, the IBA had developed its own informal sanctioning tools, such as informal, internal disapproval and a public notification. Most important, however were the diverse forms of informal contact. Such communication often took place in special committees (see summary in IBA, 1986: Appendix 4; see also IBA, 1990: Appendix 7)—such as the General Advisory Council and the various advisory committees—but also in an informal fashion directly between the IBA and ITV companies (for details, see Schacht, 1981: 690ff.).

In practice, cooperative relations had developed between the IBA and ITV companies (Dix, 1980: 374) that served as an early warning system and made the use of formal sanctioning instruments largely unnecessary ("agreement by discussion"). For instance, although the IBA had on many occasions expressed reservations about planned programs and was successful in demanding that they not be transmitted, it was never forced to strike the broadcasting of a program by interrupting transmission (cf. Schacht, 1981: 690f.). However, the suggestions, warnings, and objections that were drawn up—and sometimes negotiated—in informal cooperation in all respects appear to owe their effectiveness to the formally strong position of the IBA, which was convincingly able to threaten stiffer measures. The relative inexplicitness of the criteria[25] by which franchises were renewed or not renewed (see Section 5.1 below)—the franchise decisions in 1980 were striking in this regard (see Tunstall, 1983: 209f.; Dix, 1980: 370; Schacht, 1981: 692)—made the franchise holders even more insecure, prompting them to avoid annoying the IBA as far as possible. In any case, the policy of avoiding conflicts by self-restraint was particularly discernible on the part of the ITV companies.

The IBA's strong position was not only to be seen in its relationship to the various franchise holders but also extended beyond the overall responsibility for commercial broadcasting to cover considerable influence on the organization of relations among the ITV companies themselves. For instance, the IBA played a decisive role in the restructuring of the network agreements among the ITV companies in 1989 (IBA, 1989: 7; Spicers Consulting Group, 1989: 4f.).

5. FIELDS OF REGULATION AND SUPERVISION

The IBA Act contained an extensive number of rules dealing with possible supervisory powers given to the government over the IBA (see Section 6.6) and to the IBA over the program companies. These were accompanied by IBA powers set down in the franchise contracts with the companies. The IBA's powers regarding the ITV companies are only treated in passing in the discussion that follows.

5.1. Franchising

Broadcasting franchises in Great Britain were awarded following a comparison of bids submitted in a public tender. The IBA sought to gather data on the bidder by means of a questionnaire dealing with financial, personnel, and organizational strength and the intended programming policy (for the questions asked, see Mahle, 1984: 213). Applicants were also interviewed. In 1980, the IBA began to involve the general public as well, on the one hand by studies on the attitudes of broadcasting consumers and, on the other, by special public meetings (for details, see Mahle, 1984: 209ff.; Dix, 1980: 370). The public was provided with access to the tender documents, insofar as these were not classified as confidential (see Briggs & Spicer, 1986: 72). Copies of the franchise contracts concluded with local radio broadcasters were to be made available to anyone upon request. The ultimate decision about the award of the franchise, however, was made behind closed doors. (For criticism of the decision-making procedure, see Briggs & Spicer, 1986: 13, 136f., 141; but see Annan Report, 1977: 195.) The IBA did not list detailed reasons for decisions reached. In justifying this practice, it pointed out that it was practically impossible to indicate the precise reasons that, following an assessment of initial data and future possibilities, were decisive for the judgment that the particular applicant would be likely to render the optimal services.

In the comparative procedure, the IBA tended to accord particularly great weight to capability—above all, financial capability—and to the objective of ensuring the stability of the ITV network. This orientation strengthened existing franchise holders' chances of a renewal of their franchise (Fraser,

1989: 7). As long as they were able to provide evidence of at least a minimal capability, they had an advantage over applicants that could only deliver promises. However, the IBA has not been deterred from awarding franchises to new applicants on occasion.[26]

Those companies that had a dominant position in the ITV network enjoyed a de facto protected status, given that their departure would have had a marked effect on program offerings. This related especially to the larger companies. Although none of them has ever lost a license outright, the interests of some have been challenged.

Since the existing franchise holders also had well-established systems of contact with the IBA, they knew exactly what the IBA was looking for and were better able to react to this than were outsiders. Pressure for protected status also came from the trade unions. For instance, it was noted as early as the Annan Report of 1977 that the trade unions would demand a guarantee of continued employment of workers in the event of a change of franchise holder (Annan Report, 1977: 193). However, the trade unions have thus far been unsuccessful with this demand, and, in view of the prevailing political conditions in Great Britain, this is not likely to change in the foreseeable future.

The more the existing franchise holders became integrated into other sectors of the media system—for instance, the British Super Channel was originally supported by fourteen of the fifteen ITV companies[27]—the more indispensable they became. Since British ITV enjoyed a good reputation worldwide and had produced a number of high-quality programs, there was also no particular reason to undertake radical changes. In addition the maintenence of the status quo corresponded to a strong tradition in British political culture.

The blanket extension of the franchises of the fifteen ITV companies until 1992 provided for in the Broadcasting Act of 1987 also prolonged their protected status. In justifying this government-initiated measure, it was noted that the ITV companies and the IBA should be given an opportunity to catch their breath, during which time the effects of directly receivable satellite broadcasting on the traditional structures of the British broadcasting system could be studied. However the real reason was undoubtedly to give the government time to consider plans for introducing some dramatic changes without having to face the obstacle to their implementation that freshly reawarded, long-term franchises would have presented. The plans developed in the 1988 white paper for reforming the ITV system and auctioning the franchises (Home Office, 1988: 22–23) made clear that the government was interested in introducing some major innovations in this system. Regarding renewal of the franchises, however, the broadcasting companies were made aware that they should not take steps, perhaps for fear of government plans, to stop investments in the broadcasting sector, let alone to withdraw capital.

5.2. Programming Control

The programming regulations contained in the Broadcasting Act of 1981 confirmed that the British broadcasting system did not aim for external pluralism but rather aspired to broadcasting according to the integration model. The most important programming norms were aimed directly at the IBA. For example, the IBA was responsible for guaranteeing broadcasting's trustee obligations, maintaining a high standard of program quality, and ensuring that programs were balanced and diverse. With regard to substance, a balanced supply of information, education and entertainment had to be imparted. Informational broadcasts or other programs with political or otherwise controversial content had to be impartial. But in order to avoid excessively restricting freedom of opinion in monitoring this requirement and thereby acting as a censor, the IBA pursued a pragmatic policy here. (For conflicts regarding the specification of impartiality in given programs and the problem of personal, position-related programs, see Potter, 1989: 114ff.) More weight was placed on achieving a balance of programming over time than on that within a given broadcast (Potter, 1989: 118).

Neither the IBA nor the program companies themselves were allowed to comment on current political or other controversial topics in the programs. However, after an earlier, even more restrictive rule had been relaxed, members and employees of these institutions and companies were allowed to submit personal statements (Potter, 1989: 112). The law merely prevented an authority or company view from being presented and did not restrict journalistic comment. A further requirement was to ensure that a reasonable amount of programming originated in Great Britain and was produced with British personnel (for details on quotas, see Wiedemann, 1989: 289ff.).

The IBA was subject to stricter obligations with regard to programming issues than the BBC, although the latter voluntarily undertook to observe similar restrictions. As explained above, the detailed regulations were aimed at avoiding specific dangers that the legislature assumed were connected with a commercial broadcasting system. The IBA was given broad discretion in its monitoring activities (Seymour-Ure, 1987: 274). In specifying its requirements, the authority enacted guidelines, such as the code on violence (reproduced in IBA [1985], *Television Program Guidelines,* Appendix 1) and the *Television Program Guidelines* themselves (IBA, 1985, with additions in the *Annual Reports* of 1989 and 1990).

This linking of monitoring policy with economic factors can also be detected with particular clarity in the sector of radio. For instance, the local radio stations were long bound by restrictions (e.g., programming obligations) to an economically unattractive profile. The interest in radio programs therefore remained rather low (Sparks, 1988: 4), and as a result, advertisers saw only a marginal role for radio. IBA made great efforts—in the view of competent observers, with considerable success—to get the program offerings of the ITV companies to meet the desired qualitative objectives and to achieve high ratings. Above all, the stiff criticism contained in the Pilkington Report and the resulting change this triggered in broadcasting law were effective incentives for increased effort on the part of the IBA (Pilkington Report, 1962; for criticism of the report, cf. description in Follath, 1974: 87ff.). For instance, the share of informational broadcasts and of programs in the areas of education, instruction, and so forth, increased.[28]

Moreover, a relatively high in-house production quota was secured, and an increasing number of independent producers were employed. This took place during a phase in which the private broadcasting system in Great Britain was well established and most of the companies were earning sufficient profits as a result of their monopoly positions. The IBA was able to take advantage of this situation as well as the public criticism of programming practice in order to place stronger public service obligations on the ITV companies.

However in 1984, the IBA began to relax programming obligations in order to create better conditions for the program companies in the battle for viewer ratings (Dyson, 1988a: 2f.; Dyson, 1988b: 252f.). It can be seen from the various phases of differing intensity of programming control that the IBA tended to tie the intensity of its supervision to the respective market conditions or, in the words of the Chairman of the IBA: "The [regulatory] approach changes over time, in response to changing public demands and tastes as well as changing economic and mechanical circumstances" (IBA, 1987a: 4).

In the face of this situation, the IBA relaxed restrictions in order to improve profitability prospects (Sparks, 1988: 153; Easton, 1989).[29] In early 1988, for instance, programs that had previously been broadcast simultaneously on FM and AM were allowed to be split with the approval of the Home Secretary, thereby making available more frequencies for the ILR broadcasters for various programs (IBA, 1989: 24). This made it possible for broadcasters to orient their programming more strongly to target groups (see Hughes, 1988: 18; IBA, 1989: 24f.) and to gain a larger listener audience (Easton, 1989; IBA, 1990: 25) and thereby improved financial footing.

5.3. Advertising Restrictions

The statutory programming rules also covered advertising. These were accompanied by special advertising regulations, such as the requirement that programming be separated from advertising spots, limitations on commercial interruption, restrictions on advertising time,[30] and so forth. In accordance with its statutory mandate, the IBA specified statutory duties after conferring with the Home Secretary. It then drew them up in the form of The IBA Code of Advertising Standards and Practice, updating them as necessary. These were accompanied—in accordance with the principle of the greatest possible self-regulation of the broadcasting industry (for the general preference for self-regulation in British media policy, see, e.g., Seymour-Ure, 1987: 274)—by an advertising code that the ITVA (earlier, the Independent Television Companies Association, or ITCA) enacted (ITCA/ITVA, Series of Notes of Guidance, Nos. 1–10, 1981–1989). As a result of these specifications, the advertising rules that were in force in Great Britain were considerably more detailed than those in other broadcasting systems under study here (see Wiedemann, 1989: 120ff.).

The advertising rules restricted the broadcasters but also helped channel competition between them. However, their main goal was to protect broadcasting consumers, particularly against misleading, harmful, or offensive advertising (IBA, 1990: 30; see, e.g., IBA Code, Nos. 8 et seq.), but also against manipulative effects that might result, for example, from a mingling of advertising and program contents (see, e.g., IBA Code, Nos. 5–6). The substantive requirements were supplemented with organizational safeguards. For instance, the principle of separation of advertising from programming was strengthened by splitting decision-making responsibilities: Whereas the acquisition of advertising remained the province of the program companies, the responsibility for programming and thus also the control over the observance of the advertising rules remained with the IBA (see also Pragnell, 1986: 296f.). In addition, organizational barriers between program production and advertising acquisition were sometimes set up within the program companies themselves (Potter, 1989: 193). Moreover, possibilities for mutual consultation and clarification of interests between the IBA and the program companies were explored in special advisory bodies (Pragnell, 1986: 295, 322). Particularly worthy of mention is the Advertising Advisory Committee (AAC), which, in accordance with statute, was composed of representatives from relevant groups, such as consumers and the advertising industry, and of experts.[31] Although it only exercised the role of an advisory body, the AAC had a significant influence on IBA policy, which was made clear by the fact that in most cases the IBA accepted its proposals.

For specialized issues relating to advertising for health-care products, the AAC was augmented by the Medical Advisory Panel. In 1980 the Advertising Liaison Committee (ALC) (see Pragnell, 1986: 322ff.) was created as an advisory body at the highest level, although without any special statutory basis. It was composed mainly of the top representatives from the interest groups involved and headed by the Chairman of the IBA. The ALC took up basic issues and served as a first step in the search for consensus on the principles of advertising regulation (Potter, 1989: 206).

Even when drafting The IBA Code of Advertising Standards and Practice, the IBA also sought to gain the cooperation of the affected interest groups. For instance, a first draft was sent to the AAC and probably also discussed in the

ALC. Following treatment by these bodies and the possible introduction of ammendments, it was then sent to all conceivably affected groups for consultation. Only after the submitted comments had been reviewed was the final version prepared and then presented to the Home Secretary for approval.[32]

The search for cooperation was also continued in the implementation of the rules, particularly with regard to controlling whether a given advertising spot conformed to the guidelines. The treatment of television advertising spots provides a good example of this consensual approach. In the first instance, control occured through self-regulation by the broadcasters themselves. Scripts as well as finished commercials were first reviewed by the ITVA/ITCA Copy Clearance Secretariat (for the formal procedure, see ITCA/ITVA, 1988: Series of Notes of Guidance, No. 2), which proposed changes where necessary or even rejected the spot. A negative decision could be appealed to the Copy Committee of the ITVA, which then gave the case a renewed evaluation.[33] According to information supplied by the ITVA, changes in a submitted spot were only necessary in rare cases, and even more seldom were protests filed against decisions. The ITVA sent a copy of each script to the IBA with its comments, so that the IBA was informed and could intervene before the scripts were produced. On many occasions, the agencies themselves called upon the IBA to review scripts; in this manner, the risk of producing a subsequently worthless film could be reduced (Potter, 1989: 199). In the event that the ITVA and the IBA took different stances, an effort was made to reach agreement by formal and, above all, informal exchange of opinion (ITCA/ITVA, 1985: No. 8, Series of Notes of Guidance, 5). The IBA also checked completed advertising films after the ITCA reviewed them (see also Paulu, 1981: 70).

In view of the imminent switch to "light-touch regulation," the IBA in December 1988 introduced, on what it described as "an experimental basis," a simplified procedure with increased delegation of the control function. This was especially intended to gain early experience with the kind of mechanisms expected in the future, thereby facilitating the transition to the new control authority, the ITC, and its procedures. Scripts for television spots were now reviewed only in particularly sensitive areas[34] or upon voluntary submission by the advertising industry or the ITVA Copy Clearance Secretariat (see IBA, 1989: 29). At the same time, however, the films themselves continued to be controlled in (nearly) all cases prior to broadcast. These changes only led to a quite limited reduction in the number of scripts submitted to the IBA (see IBA, 1990: appendix 3). Producers apparently placed considerable value on the possibility of having a sort of clearance certificate prior to the costly production of spots.

In the radio sector, the kind of large-scale control in place for television was considered unreasonable for a much longer time (Coulson, 1990: 6). In view of the low production costs, there was also no great interest by the advertising industry in advance review of scripts. For local advertising, the IBA dispensed with central advertising control—except in sensitive areas, such as advertising for contraception or alcohol—and instead nominated for each broadcaster an individual who was to be responsible to the IBA and in decentralized fashion to ensure compliance with the advertising rules.

However, the mechanisms in place in Great Britain for the intensive control of advertising were unable to do away with thorny problems related to the enforcement of advertising rules. Qualitative advertising restrictions, such as the prohibition of advertising within programs and against its influence on programming were particularly difficult to implement (Potter, 1989: 109f., which includes an example of the transmission of sponsored events).

Provisions that were hard to put into effect, such as the restriction that advertising only be broadcast during natural breaks in programs, also led to difficulties (Potter, 1989: 194ff. writes that the treatment of natural breaks took place in a permanent gray area). For instance, both the Pilkington Committee and the Annan Committee objected to these commercial interruption opportunities and complained that broadcasters themselves created such "natural" breaks for advertising purposes (see Pilkington Report, 1962: 72–74, paras. 229–236; Annan Report, 1977: 165. Pragnell, 1986: 304f., considered these to be less problematic). The IBA also objected to a recommendation by the Pilkington Committee that advertising be prohibited in children's programs, together with the demand that such programs not be abandoned simply on account of the accompanying loss of advertising revenues (for this demand, see Pilkington Report, 1962: 83; Annan Report, 1977: 165–167; see also Mahle, 1984: 170; Potter, 1989: 201f.).

In contrast, advertising limitations dealing with quantitative restrictions, such as maximum average advertising time per hour, were relatively successful. However, the IBA has also operated flexibly in this regard (cf. Potter, 1989: 193ff., for the shifting of permitted advertising minutes between various hours).

Over the course of time, control of advertising was adjusted to conform with the growing trend toward liberalization (for instance, areas of permissible advertising were expanded; see IBA, 1988: 27; 1989: 28f.; 1990: 9, 11). Whereas the original IBA "philosophy" followed the principle "when in doubt, prohibit," the guiding tenet in the transition phase read, "As long as there is no reason for prohibition, then permit it."

In general, with respect to the enforcement of advertising restrictions, the IBA often became caught up in conflicts of interest. If, for example, the broadcasting of a commercial sparked any public protests, the IBA was called upon as supervisory authority; even though it had itself previously approved the spot. But in a number of such cases, the IBA did not see this previous approval as preventing it subsequently from demanding retraction of the commerical. This normally did not lead to any special problems vis-à-vis the advertising industry; it was entirely consistent with the latter's own interest in avoiding antagonism as far as possible. In cases where justified objections were voiced, commercials were usually voluntarily retracted following informal talks. But even the advertising industry's cooperative stance did not shield the IBA from the fundamental dilemma in which it found itself. While charged with monitoring advertising restrictions, it was at the same time, as a result of its responsibility for the entire private broadcasting system, also supposed to be concerned with the system's economic viability and thus show understanding for the advertising needs of the program companies. (For the consideration paid to the special situation of Channel 4 and TV-am, see Pragnell, 1986: 324ff.) Since it had to balance advertising restraints against other programming requirements, the IBA was often forced to make compromises, which in turn had varying effects at different times, depending on the economic state of private broadcasting.

5.4. Combating Concentration

Regional monopolies characterized the private broadcasting system in Great Britain. In order to alleviate potential misuse resulting from this or risks of intensive multimedia interlocking, efforts were made to keep media concentration and integration in check.[35] As in other countries, these were undertaken by means of tools of competition law on one hand and media-specific precautionary measures on the other. General competition law has developed only sporadically in Great Britain (see Kübler, 1982: 45); nevertheless, precautionary measures, whose observance was essentially the province of the IBA (Veljanovski, 1990: No. 78), were set up for combatting concentration in media. Since the functioning of the British broadcasting system was not based on economic competition, no need was seen to develop an overarching regulatory policy (cf. also Kübler, 1982: 49). The precautionary measures for combatting concentration were mainly sporadic.

For instance, in approving applicants, the IBA often ruled out those whose participation could have led to conflicts of interest: Advertising agents and agencies were generally excluded; record companies, music publishing firms, concert promoters, and so forth were prohibited from receiving radio franchises. Publishers, however, were not absolutely excluded. Nevertheless, participation by publishing companies in television or radio program companies was not allowed to lead to results "contrary to the public interest." (Application of this stipulation has never been resorted to; Kübler, 1982: 47.) The reverse case, of a publishing firm participating in a radio company, was not expressly regulated by law. This was, however, covered by clauses in the franchise contracts relating to diversification into other areas (for details, see Kübler, 1982: 48; see also generally the brief survey in Veljanowski, 1990: No. 78ff.). It should be pointed out that, as a whole, the IBA used the franchise contracts as an additional instrument for setting restrictions and, in so doing, also made an effort to be flexible by taking into account the particular conditions in the various regions.

Integration between broadcasting and the press was not infrequent (see Veljanowski, 1990: No. 65ff., with a listing of various cases of integration). As early as the beginnings of commercial television, the ITA (as predecessor of the IBA) had encouraged publishing companies to invest in the television sector. In particular, the IBA strived to encourage the local press to participate in regional program companies (Annan Report, 1977: 198; see also Royal Commission on the Press, 1977: 4.3.2f.). But on occasion it also sought to achieve reductions in majority holdings by the press in television companies (Annan Report, 1977: 198). All the same, the extent of shareholdings by newspaper publishers in commercial television companies can be deemed "remarkable."[36]

Integration occurred in the radio sector as well. For example, the Independent Broadcasting Authority Act of 1973 granted the respective local newspaper companies a preferential right to acquire shares in Independent Local Radio companies—a policy that was then eliminated by the Broadcasting Act of 1981. Here, the IBA pursued the policy of encouraging newspaper companies to acquire minority holdings, which did not enable them to exercise control over the broadcaster (Royal Commission on the Press, 1977: 4.3.4). In view of the economic difficulties in the radio sector, the IBA's anticoncentration policy, seen as a whole, was quite moderate (Sparks, 1988: 152).

In recent times, publishing companies have discovered an interest in satellite television. They have become heavily involved in providing television services via telecommunications satellites, which are widely removed from national regulation. They have also obtained holdings in BSB (see European Institute for the Media, 1990: Mediafact 3, 6f.).

The IBA was obligated to ensure that no—or at least no problematic— integration occurred among the various regional monopolies (details in Wiedemann, 1989: 243ff., 247ff.). For instance, one programming company was not allowed to participate financially in another; one and the same natural or juridical person was not permitted to hold shares in more than one program company when the holdings in one of these companies exceeded 5% of the voting capital; in overlapping transmission territories, a television program company was not allowed to have dominating influence over a radio program company (or vice versa) or to receive a cable franchise for this region.

Prior to concluding contracts, the IBA had to ensure compliance with the rules regarding participation, and so on. The program companies were obliged in the franchise contracts to provide all important information upon request; and the franchise contracts were required to contain termination clauses for the event that integration be disclosed that would have prevented the granting of a valid franchise, had it been known at the outset.

The precautionary measures against concentration and integration were characterized by a mixture of rigid prohibitions and flexible balancing clauses. The IBA made an effort to make use of its flexibility, and in informal clarifications with the broadcasting industry, it also searched for workable paths to limit concentration while at the same time to enable broadcasters to remain economically viable. It should be noted in passing that the policy of the Cable Authority, which was also entrusted with warding off concentration, was marked by even greater flexibility. It was thus consistent with its function that the Cable Authority intensively supported the growth of cable broadcasting in spite of the many difficulties (see Wiedemann, 1989: 249ff.).

6. SUPERVISION OF BROADCASTING IN THE POLITICAL ARENA

6.1. The Relationship between the IBA and the ITV Companies

The description thus far has shown that the IBA and the monitored ITV companies stood in close association to one another. The IBA's legal status was marked by dominance. But in the light of its practical administration, some observers have instead had the impression that it was a somewhat weak authority.[37]

The IBA applied its legal authority to create a cooperative supervisory association in such a way that the use of its unilateral, regulatory competences could be avoided as much as possible. Nevertheless it also served as a threat, and in extreme cases it did in fact resort to these. The ITV companies were able to benefit from this regulatory relationship because of their regional monopoly positions and the relatively well-functioning programming network between the various companies, which was backed up by the IBA. As a result, basic conflicts between supervisor and supervisee were, despite many short-lived controversies and occasional friction, apparently avoided. Under the basic conditions in effect at that time, a legally strong supervisory authority appeared to be in the interest of the overall association, as long as it in practice applied its supervisory power with care and understanding for the situation of those supervised.

In fact this understanding does not seem to have been lacking. During the initial years of private broadcasting in Great Britain, there was much animosity against commercial broadcasting, which led the IBA's predecessor, the ITA, and the ITV companies to become allies. The ITA moreover strived from the outset to ensure the financial viability of the new media sector and to encourage financially endowed undertakings to invest in it (Potter, 1989: 84ff.). The ITA and the IBA were interested in economic viability not merely for reasons of basic survival[38] but also because its work was nearly exclusively financed by payments from the program companies. The success of the ITV system at the same time also strengthened the political regard for the supervisory authority.

In the course of their common, 35-year history, the IBA and the program companies have developed contact systems that have led to a strong, reciprocal harmony of perspectives and to largely identical interests (Garnham, 1980: 24f.; Dyson & Humphreys, 1985: 367; Dyson, 1988a: 3f.; Gibbons, 1991: 146f.). Strengthening of the social relationships was also aided by the fact that personnel flowed in both directions between the IBA and the ITV companies and that several members of the board had earlier worked in the broadcasting field.[39] There were also many crossovers from the BBC.

The intensive, usually informal, and nonpublic interaction and cooperation has led (as was observed by the Pilkington Committee) to a problematic loss of distance. It was felt that the ITA had too firmly seen itself as "defender" of the ITV companies. According to the Pilkington Report:

> The Authority told us that it saw its relationship with the contractors as one between friends and partners. . . . It is our view that, while there is everything to be said for persuasion, so long as it is effective, the relationship between the Authority and the companies must be that between principal and agents. As the trustee for the public interest, the Authority is answerable. It must, therefore, be master, and must be seen to be master. That the companies are widely regarded as principals and the Authority as their spokesman is unsatisfactory. (Pilkington Report, 1962: 168, para. 572)

Although the amendment in the law prompted by this criticism led to a strengthening of the ITA's legal position, its cooperative administrative style remained unaffected.[40] In addition, the reciprocal consultations have since then increased to some degree.

All the same, the IBA has been successful in its efforts to keep the "entanglement" from reaching the point where its political authority would suffer intensively. Occasional formal supervisory action—which often had a much more severe impact than would have been tolerated by any of the other supervisory authorities under study here—bolstered its integrity. Moreover, the IBA made its independence plainly clear during franchise proceedings. For instance, it strictly refrained from any prior contact with applicants. In the course of the proceedings for awarding franchises, there was considerable tension between the IBA and the ITV companies. Although the relationship was otherwise principally directed toward cooperation, it became periodically strained when the dates for the conclusion of new franchise contracts approached. Nevertheless, the "close camaraderie" (Dyson, 1988a: 10) has never truly suffered. Dyson sees in this a demonstrative example of sectoral corporatism and clientele orientation in broadcasting (Dyson, 1988a: 12).

6.2. Political Neutrality

The IBA has been able to avoid getting drawn into the vortex of political polarization. This is particularly significant for the reason that the IBA was able to have a lasting effect on the programming broadcast. To the outside observer, it may well seem miraculous that to all appearances this power was not abused for political (party) or other unilateral ends.

The mere reference to corresponding statutory duties (s. 4 of the Broadcasting Act of 1981) does not suffice as an explanation. It is also not enough to note that the law prohibited not only the nomination of members of Parliament but also that of persons whose own financial or other interests conflicted with the tasks of management.[41] Rather, the explanation lies in Great Britain's political culture[42] that has created a tradition of political restraint (see Hearst, 1979: 89; Tunstall, 1983: 24), a restraint which has made it possible to establish the control authority as trustee of the public interest. Even the act of creation—the selection of the members—was embedded in a tradition of political restraint. The government simply could not afford to make unilateral use of its right of nomination without risking a political scandal. Yet it was also exempted from having to proceed according to aspects of pluralistic representation. Nevertheless, the nomination at least took into account political orientation and regional origin (for the first twenty years, cf. Potter, 1989: 84ff., 92ff.). In addition, the Annan Committee, which had been presented with a variety of proposals for the procedure for nominating members, recommended that individuals, and not representatives of specific groups, continue to be nominated, although at the same time it expressed the wish that groups and sections of society thus far not taken into consideration be included in the selection process (Annan Report, 1977: 48, paras. 5.23f.).

Independent, experienced, and well-regarded persons were supposed to be selected. Appointments usually went to representatives of the British establishment (cf. also Tunstall, 1983: 212ff., 216), who were drawn from a list of personages the government considered trustworthy and qualified to exercise public office (Hood, 1980: 39f.; Potter, 1989: 91). Numerically speaking, the dominant sector represented here was education (Potter, 1989: 93).

The various selection mechanisms resulted in a relatively elite composition of the Board, which helped to prevent politicizing it (see also Hood, 1980: 38) but at the same time suggested a certain sensitivity to the positions of holders of political, economic, and cultural power. Moreover, it was not possible to rule out the risk that the IBA members would lose touch with user needs in orienting their work. Nevertheless, the degree of independence practiced by the Board was considerable (see also von Hase, 1980: 17).

6.3. Civil Activities

The IBA's legally strong position was one of the reasons the legal system did not enable citizens (broadcast users) to exercise influence directly on programming via legal avenues. For instance, broadcasting was exempt from the jurisdiction of the Obscene Publications Act of 1959,[43] which in other areas—for example in the publishing industry as well as in the field of cable programming—enabled the citizen under certain conditions to take action in court against obscenities in publications. The courts have also been reluctant, if not unwilling, to recognize civil rights of action to enforce statutory program standards (for details on case law, see Wiedemann, 1989: 122ff.).

However, informal opportunities for the citizen to take action do exist. Of world renown is the conservative National Viewers' and Listeners' Association, headed by Mrs. Whitehouse, which seeks to enforce rigid programming standards[44] and which is especially able to count on support from the right wing of the political spectrum. This movement has also found a certain degree of support from the Conservative Party and the British government. In recent times the Consumers' Association, a private body financed by subscribers, has become increasingly involved in asserting what it considers to be the interests of viewers–consumers.

6.4. The Broadcasting
Complaints Commission

Opportunities for citizens to have a say in programming issues are also created by the Broadcasting Complaints Commission (BCC), which was established by the Broadcasting Act of 1981 and has continued after the Broadcasting Act of 1990. The BCC is composed of at least three (formerly five) members, who are nominated by the Home Secretary and are prohibited from pursuing any interests in the broadcasting sector parallel to their activities in the BCC (Seymour-Ure, 1987: 274; Robertson, 1990: 354f.). The BCC deals neither with broadcasting regulation nor with general program supervision but instead with asserted violations of the rights of third parties, particularly the rights to privacy.[45] It is charged with investigating complaints regarding "unjust or unauthorized treatment in broadcast programs" or "unwarranted infringement of privacy in, or in connection with the obtaining of material included in, broadcast programs" (Broadcasting Act of 1981, s. 54(1); for details, cf. Robertson, 1990: 354ff.). It is thus strictly limited to matters involving the violation of subjective civil rights and does not review questions of bad taste or offensiveness, which are the province of the Broadcasting Standards Council (BSC).

Under the Broadcasting Act of 1990, the procedure begins with a written complaint submitted by the injured party or on that party's behalf. If this is deemed admissible, the BCC requests the pertinent documents from the relevant broadcaster (in the area of private terrestrial broadcasting, this was the IBA) and asks it to comment on the complaint. In such cases, the IBA tended to consult first with the program company, which was formally not involved in the proceedings, and then to send the BCC both its own comments as well as those of the program company (Robertson, 1990: 359; IBA Television Program Guidelines, 1985: appendix). A hearing is also often conducted with the broadcaster and the complainant. The BCC is not allowed to pursue complaints when these relate to a matter that is the subject of court proceedings and can more appropriately be dealt with as such (see Robertson, 1990: 358). When the BCC is able to take action, it evaluates the criticized conduct and can demand as sanction that its summary of the judgment be published or broadcast in a certain manner and at a stipulated time. Normally, a copy of the BCC judgment summary is also reproduced in the respective program magazine (Robertson, 1990: 359). Its decisions can also be appealed to the courts.[46]

Initially, the BCC was not very well known to the public. After having made itself more visible through a variety of measures, the Commission found itself faced with a growing number of complaints.[47] Of these, an increasing number were submitted by organizations, interest groups, and firms; this led to increased professionalization and resort to legal counsel in dealing with complaints (Goodwin, 1990: 14). Parallel to this trend, was a mounting criticism of BCC decisions (see, e.g., BCC, 1989; Smith, 1990: 1).

In spite of original plans to integrate the BCC into the BSC (see Home Office, 1988: 35), the merger met with protest, in particular from the BCC (see *Independent,* 19 July 1989: 8). The Broadcasting Act of 1990 left both institutions intact, and they are to continue to work alongside the new external supervisory authority, the ITC.

6.5. The Broadcasting Standards Council

The BSC was established in 1988 and charged in particular with attending to sex and violence in broadcasting and with monitoring the observance of good taste and decency. In its code of practice (BSC, 1989), it enacted detailed, but in many respects weakly formulated, guidelines and attempted to clarify their application through examples. Although it initially operated without a statutory basis, this was provided in the Broadcasting Act of 1990 (Broadcasting Act of 1990, s. 151 et seq.; see also Section 1 for a detailed description of its procedures). The BSC is required to investigate program complaints from citizens that relate to the observance of broadcasting standards and to monitor programming directly. Its authority covers BBC programming as well as that of licensed broadcasters, and the results of its investigations are published.

In justifying this new authority, the government noted, among other things, that problems would arise if the IBA, as the institution then responsible for programming, were at the same time to be entrusted with programming supervision. It was, however, an open secret at that time that the establishment of this new authority also signaled criticism of the IBA's (and the BBC's) course of action. Both the IBA and the BBC had, to all appearances, been too lax or too liberal for the government in its supervision of programming. Subsequently, however, critics complained that the new institution was primarily intended to restrict editorial freedom by intensified control of content. Sharp criticism was directed at the person of the selected Chairman. The mistrust this engendered, as well as the criticism of the new concept, is exemplified by the assertion that the creation of the BSC represented a "substantial step in the direction of the direct censorship of programs" (see, e.g., Sparks, 1988: 148). Although this criticism has waned over time, the BSC has been very careful not to interfere with editorial discretion and to avoid acting like a censor.

The government's desire for increased supervision of programming threatened to collide with another objective it simultaneously pursued in the late 1980s, namely, to deregulate the broadcasting sector more strongly in order to unleash market forces. It is hardly conceivable that in a deregulated environment, a new supervisory authority will be able to undertake more effective programming control than that performed by the IBA under the previous broadcasting structure. But since this programming supervision was supposed to be substantially relaxed with the creation of the ITC, the creation of the BSC could also be considered a sort of insurance for the period of deregulation: The government's confidence in the power of the market did not extend so far as to do away entirely with the desire for a guardian of morals.

6.6. Governmental Supervision of the IBA

The IBA proved to be an independent organization. Despite some attempts by the government to interfere (especially with regard to broadcasts on the Northern Ireland controversy), its high degree of independence was all the more noteworthy in that the government—specifically, the Home Secretary—supervized the IBA. The office of the Home Secretary, in turn, is subject to the control of Parliament.

The starting point for government control of the IBA was similar to that with regard to the BBC: The government issued to the IBA a limited, revocable transmission license. Rights of intervention were set down in statute. The

Home Secretary did not merely appoint the twelve members of the board but was also given authority to fire them at any time. Further, he had the right to issue directives to the IBA regarding the content of programming, advertising, and broadcast times (Broadcasting Act of 1981, s. 29[3]). Theoretically, the IBA could even prohibit given programs. Rights to intervene with respect to programming were, however, only intended to be "reserve powers" (Mahle, 1984: 36; Dix, 1980: 376; Negrine, 1989: 118), and the opportunity to use them was heavily restricted by law. The Home Secretary was also able to exercise influence on financial structure, particularly on the amount of payments that, as a sort of quid pro quo for the exclusive right to use terrestrial frequencies for private broadcasting, had to be made to the Treasury from the revenues of the ITV companies.[48] The government was also able to have a say in the broadcasters' technical capabilities (such as transmission strength) (for details, see Negrine, 1989: 105).

The IBA made no apparent use of the power to issue directives with regard to individual programs.[49] Coming into play here as well was the tradition, initially developed in British society with respect to the BBC, that broadcasting not be permitted to become politicized, let alone be put to the service of the government (Dix, 1980: 376; Lincoln, 1979: 126). If the government were nevertheless to attempt to intervene, such intervention could trigger political outrage or even a scandal. This was all the more foreseeable in light of the fact that the IBA had the right to publish government directives.

As a result, specific program-related directives have not been, or are not known to have been, issued. At the same time, however, general programming directives have been issued, such as the Home Secretary's October 1988 directive prohibiting direct interviews with or statements by persons representing or supporting terrorist organizations connected with Northern Ireland.[50] The IBA (IBA, 1989: 7; IBA, 1990: 13), the BBC, and other organizations (see Stevenson & Smedley, 1990: 69) publicly criticized the unreasonableness of the rule, but to no avail. The legal remedies sought by broadcasting organizations were unsuccessful.[51] Attempts to exercise influence had been made on a number of earlier occasions (Wiedemann, 1989: 94f.; Street, 1982: 89). The Thatcher government not only put more pressure on the BBC (details in Humphreys, 1988: E 73; Dyson, 1988a: 12f.) but also put increased—usually informal—pressure on the IBA, which also led to heated battles between the IBA and the government. For instance, the government's plan to remove radio from the IBA's sphere of responsibility (Home Office, 1987) spurred protest from the IBA,[52] although it also revealed that the IBA had suffered a loss of political pull. Plans to dismantle the IBA disclosed in the 1988 white paper made clear that the era of the IBA had come to an end.[53] The IBA's situation was aggravated further by sharp criticism from the industry, for example, as a result of its blocking takeover bids for Granada and Thames Television (Dyson, 1988a: 28).

7. REFORM OF THE BRITISH BROADCASTING SYSTEM

7.1. The Formative Period

The attempt to reform the British broadcasting system in the late 1980s was, however, only partly motivated by the government's discontent with the IBA. The decisive impulse stemmed rather from an effort to apply the government's broader philosophy of deregulation to the broadcasting sector. In view of technological advances and the triumph of the competition principle in many of the world's broadcasting orders, there were reasons enough to alter the traditional broadcasting system, even in Great Britain. The Conservative government had no doubt that broadcasting had to be further opened to the market. The path for the reform debate was paved by the Peacock Committee (Peacock Committee, 1986). Although it was only intended that this committee take up the issue of BBC financing, it went further and developed concepts for the broadcasting system as a whole. While it rejected financing through advertising for the BBC, it recommended, among other things, that broadcasting license fees be tied to the retail price index (RPI). For the area of commercial broadcasting, the majority on the committee proposed that, in future, franchises be awarded to the highest bidders, as long as special considerations did not justify a departure from this principle.

Defense of public service broadcasting was the thrust of an ensuing report by the Home Affairs Committee (a select committee of the House of Commons) in 1988 (Home Affairs Committee, 1988). The report was, however, unable to deter the government from its plans. Shortly thereafter, in November 1988, the government presented its white paper advocating an extensive opening of the market (Home Office, 1988). Prior to this, in 1987, the government had publised a green paper for radio, which provided for the de- and reregulation of the radio sector (Home Office, 1987). Although based on this document, the government's 1988 white paper primarily concentrated on television. The government stated that the planned creation of competing, commercial television broadcasters would not interfere with the quality of programming but, rather, would lead to a strengthening of the "quality, variety and popularity" of the broadcasting system (Home Office, 1988: 4). The government placed its trust in market forces, which, together with new transmission possibilities and correspondingly diversified program offerings, were to enable the interests of both the advertising industry and viewers to be better satisfied than did the existing system. Quality was to be improved through free consumer choice. To this extent, the government pledged to abolish unnecessary and outdated regulations while it maintained or reformulated important standards on content.

One striking feature of the 1988 white paper was that the proposals were mainly based on the general deregulatory philosophy, yet it did not refer to systematic analyses of broadcasting reality or comparative studies of the experiences in other deregulated media systems. Even the proposed reform of broadcasting supervision was not supported with reference to shortcomings in the existing supervisory concept, let alone failures by the IBA; rather, it was merely noted that at a time when there are a large number of competing supplies, there is no need for restrictive supervision such that "light-touch regulation" is in order similar to that exercised in the past by the Cable Authority (Home Office, 1988: 11, 20).

The publication of the white paper generated an intensive public discussion.[54] Although it stuck firmly to its basic concept, the government found itself compelled to undertake substantial modifications. These culminated in the enactment of the Broadcasting Act of 1990, an expansive body of rules that imposes obligations on private broadcasters in Great Britain. The philosophy of deregulation has been cloaked in the guise of detailed reregulation and, in some cases, even overregulation.

7.2. Important Provisions of the Broadcasting Act of 1990

The Broadcasting Act of 1990 addresses the BBC only in isolated instances (see, e.g., Broadcasting Act of 1990, s. 186–187; see also s. 180), and at first glance, the BBC seems to be spared from its authority. However, changes in related areas will have substantial repercussions for the BBC, in particular with regard to the admission of new broadcasters and their exemption from some earlier obligations. In addition, discussion is taking place about the future of the BBC when its current royal charter expires in 1996 (Department of National Heritage, 1992; BBC, 1992).

The prior ITV system was decisively altered, and the radio sector, which had previously been restructured in part, was deregulated to an even greater extent. A new Radio Authority was created specifically for the licensing and monitoring of radio. The discussion that follows focuses on television and its regulation.[55]

Of particular interest is the fact that the regional monopolies held thus far by the program companies were formally terminated, to allow future introduction of economic competition in the area of private, terrestrial broadcasting. The special IBA construction was also eliminated. The IBA and the Cable Authority were replaced with a newly established licensing and supervisory authority, the ITC, which does not function simultaneously as broadcaster (see Home Office, 1988: 22). The ITC is composed of a chairperson and a deputy chair, as well as eight to ten additional members all appointed by the Secretary of State. The ITC is entrusted with the regulation of television and certain other services.[56]

The current ITV system was formally replaced by Channel 3. The holders of regional licenses granted by the ITC have transmission rights on this channel, as do the holders of licenses for broadcasting supraregional or national programs at specific times of day. Ten-year licenses are to be awarded both for regional and for supraregional/national television services. Channel 3 will be joined by Channel 5, a new commercial broadcaster with a national flair whose reach is to be determined according to the frequencies remaining following consideration of Channels 3 and 4. However, its future is somewhat uncertain, following the ITC's refusal to accept any of the initial bids for it.

The applicants for Channel 3 and Channel 5 licenses must submit a cash bid, which aims at the payment of an annual license fee and thus represents a quid pro quo for the use of the transmission possibility.[57] The amount to be paid on the basis of the cash bid after the first year is dynamic for subsequent years: It changes according to developments in the retail price index. In addition to this fee, there are payments consisting of a percentage of advertising, sponsor or pay-TV revenues, which are fixed in the invitation for bids and in the ensuing license. Thus, in contrast to the other broadcasting orders studied here, the market is also taken as a regulator in the granting of the license. However, the amount of the cash bid is not, in itself, decisive for awarding the license. In the first place, applicants must fulfill certain minimum requirements as to their capabilities and the programming they propose to offer—the so-called quality threshold. But even when these requirements have been fulfilled, it does not mean that the license always goes to the applicant with the highest offer. Under exceptional circumstances, in particular with respect to the high quality of the proposed programming, the license may instead be awarded to a different bidder. The licensees must post a bond as a guarantee that they will meet the quality programming obligations.

This exception clause was a response to the sharp protests about the purely economic criteria by which licenses were originally to be awarded. It reflected doubts voiced by the legislature as to whether the broadcasting order could be structured solely according to financial criteria. Such doubts about pure market control shaped other rules as well. For instance, Parliament expected that despite normative programming duties, commercial broadcasting competitors would ignore important user needs and, in particular, fail to attain the high quality level held to be desirable in the broadcasting system as a whole. As a result, supplementary provision continued to be furnished by the culturally oriented Channel 4. Continuing its previous work,[58] this broadcaster remained operating in order to satisfy programming needs not catered to by commercial broadcasting. Channel 4 also broadcasts commercials and is now directly entitled to the revenues from these; since this may not offer sufficient financing, the commercial operators of Channels 3 and 5 are also required to make a contribution.[59] In this manner Channel 4 receives a "prescribed minimum income" of 14% of the total revenues of all television companies as estimated by the ITC. Should Channel 4's revenues exceed this prescribed minimum income, one-half of the excess is to be paid to the ITC, and, of the remaining amount, at least 50% is to be paid into a reserve fund. If, however, the income estimated by the ITC for the following year is less than the prescribed minimum income, and furthermore, the difference cannot be compensated by resources from the reserve fund, then the ITC finances the amount of this difference by imposing a levy on all Channel 3 license holders.

Thus, the broadcasting system is marked by a division of disseminational tasks, which, when necessary, is coupled with a financial redistribution at the expense of those broadcasters that are better able to generate earnings for the very reason that they ignore certain types of programs. (It should also be mentioned that the Broadcasting Act obligates the ITC to give reasons for its licensing decisions, which can be challenged in court. In this way, as in other European countries, British broadcasting regulation now has a strongly legalistic character.) The conceivable financial burden of the redesigned system on the commercial operators of Channels 3 and 5 may therefore be quite severe. They will have to undertake great efforts to earn revenues. Competition will be mainly focused on advertising revenues. Since the number of competing broadcasters set by the legislature is still relatively small,[60] a diversification strategy with regard to programming is, as analyses in media economics have repeatedly demonstrated (see the older standard works by Noll, Peck, & McGowan, 1973; Owen, Beebe, & Manning, 1974; see also, e.g., Bates, 1987), extremely unlikely; instead, it is likely that an attempt will be made to reach the largest possible share of the mass audience and to make the program scheduling particularly attractive for the advertising industry. The probable results for programming in the long run are by no means unpredictable.

The Broadcasting Act of 1990 attempts in part to counter the likely negative effects on programming, which the legislature itself apparently expected, by setting, among other things, requirements dealing with the amount and quality of informational, regional, religious, and children's programs. In addition, the act calls for a "proper proportion" of programs of European origin and for a 25% share of broadcast time for programs by independent producers. (For programs of European origin, the act refers to article 4 of the EC television directive. For independent productions, however, a higher quota—namely, 25%—is set than that in article 5 of the directive.) The corresponding norms provide for programming-related duties that are to an extent counter to the market in that they were likely motivated by, on the one hand, issues of programming policy and, on the other, policy concerns regarding economic structure.

The program requirements are contained in the licenses for Channels 3 and 4. The ITC is required to publish in its annual report the deficits of Channel

3 licensees that have failed to comply with these requirements. In its first performance assessment, which has been elaborated in a procedure involving viewer consultative councils, ITC staff, and the licensees, the ITC reported that almost all minimum amounts in the four statutorily required categories had been met.

The period under consideration was a transition period, so that it is doubtful whether it can be taken as an indicator for future programming. It is hardly amazing that, in face of an environment not yet stable, the Channel 3 network programs shared one characteristic: safety first. As the ITC stated, "The overall feel of the network schedule was cautious and predictable" (ITC, 1994: 2). Some tendencies noticed by the ITC, however, may herald a change: Drama series and documentaries seemed drawn to crime stories to a greater degree than in the past. With respect to adult education, there was a significant reduction in quantity, combined with a shift of emphasis in content toward leisure and related themes. Religious broadcasting was removed from its Sunday early evening slot and concentrated into two hours on Sunday morning. Current affairs programs remained in peak time, but the ITC also remarked on this that "the pressure to deliver ratings was never far away and occasionally showed in a loss of nerve in the choice of subject and treatment" (ITC, 1994: 4). With respect to news programs, the ITC—supported by political pressure— needed to formally remind the television companies of their license obligations in order to ensure that "News at Ten" remained in its slot.

While these developments may reflect a greater influence of economic considerations on programming, children's programs thus far seem not to have come under pressure to the same degree. However, these remarks represent very preliminary conclusions. The impact of the evolution of the new broadcasting order remains to be assessed.

In contrast to the situation for Channel 4, the Broadcasting Act of 1990 does not apply the term "public service" to commercial broadcasters. Nevertheless, commercial broadcasters hold a number of duties that in some respects clearly follow from the tradition of public service. Particularly worthy of mention here are the general conditions attached to the license, which apply to all broadcasters jointly. Some of these aim at positive guarantees, as with the requirement that news be presented with "due accuracy and impartiality" (Gibbons, 1991: 105ff.). A code enacted by the ITC specifies programming obligations in detail, especially the duty of impartiality. These obligations are accompanied by precautions designed to avert dangers, such as prohibitions dealing with programs that "offend against good taste or "decency" or that "incite to crime or lead to disorder or are offensive to public feeling." Attempts at the use of techniques for subliminal manipulation are similarly forbidden. The "persons providing the service" are not permitted to take stances on "matters of political or industrial controversy or relating to current public policy." Furthermore, the ITC is to enact detailed rules on the portrayal of violence and on calls for donations, as well as for the protection of children and young persons. The monitoring of such program standards is the task not only of the ITC but also of the BSC, insofar as the portrayal of violence and sexual conduct and the observance of standards of taste and decency are concerned.

An inconsistency in the rules, and one that is frequently cited by critics, becomes clear here (cf. Stevenson & Smedley, 1989: 64ff., for the criticism voiced on the white paper). The concept of deregulation and thus the confidence placed in market forces have been renounced in favor of the protection of certain political and moral values. Despite claims to the contrary, consumer sovereignty is not trusted, and far-reaching procedural and substantive control of programming is made possible. These rules also allow opportunities for influence by the government: It can dictate not only that governmental announcements be broadcast; it may also force broadcasters to refrain from including in the programs any matter or classes of matter specified in the notice. This picks up on the former right of intervention, which has recently been resorted to, particularly, as mentioned above, in relation to news coverage of events in Northern Ireland. Furthermore, intensified substantive control of programming has been achieved in that the jurisdiction of the Obscene Publications Act of 1959 now extends to the formerly exempted field of broadcasting as well. On the whole, it is evident that "light-touch economic regulation" has been supplemented with "heavy-handed moral reregulation" (see, e.g., the Scottish Film Council's remarks on the white paper, cited in Stevenson & Smedley, 1989: 65).

Advertising restrictions are to be set out in detail in an ITC code that is to take into account international obligations, such as at the EC level (see Gibbons, 1991: 90ff.). In the Broadcasting Act itself, political advertising as well as advertising relating to "industrial disputes" is forbidden. The advertising regulations in the Broadcasting Act may be supplemented, if necessary, with rules by the Secretary of State, once these have been passed by both the lower and the upper house of Parliament. In modified form, the Broadcasting Act retains for Channel 3 the previous regionalization concept for the ITV companies, continuing to permit the interexchange of programs in order to facilitate a unified, national pallet of programming (networking arrangements) (Tunstall, 1992: 248). ITN continues to act as Channel 3's provider of national and international news bulletins. The holders of regional licenses are obliged to broadcast a certain amount of regional programming. In order to restrict concentration effects, the Director General of Fair Trading may be called in, primarily in an advisory capacity.

The legislature has addressed concentration processes in other respects as well, which will probably have stronger repercussions in the future as a result of de- and reregulation (see Broadcasting Act of 1990, schedule 2). The multiple ownership restriction places limitations on the number of possible licenses. At the same time, however, it is still possible to obtain, for example, two regional licenses for Channel 3 and in addition, though under certain restrictions, to own shares in companies that hold other licenses. Hotly contested cross-ownership between broadcasting and the press is permitted, although limited in a highly complex regulation, in each direction. The ITC is provided with opportunities to control changes in the composition of a license holder, albeit in a very restricted way.[61] Seen as a whole, the rules to combat concentration are more detailed, and though they are more widely applicable than before, they have to be deemed as somewhat weaker in substance.

The ITC, which is called upon to monitor license commitments, has lost the role of broadcaster previously assigned to the IBA and thereby the legal authority to conduct comprehensive control of programming. As supervisory authority, it may enact general rules (in particular, in the form of codes), ensure that norms are complied with at the licensing stage, and thereafter perform ongoing supervision. The set of formal supervisory sanctions is graduated. For instance, the mildest measure the Act provides for in the event of a violation of license conditions is a correction or apology by the broadcaster, as well as the prohibition of continued violations. Not only may fines be imposed; the license term itself may be reduced. Finally, the license may be revoked following prior warning and failure to remedy the infraction within a given period. In all cases, the affected party must be given a "reasonable opportunity of making representations" before the imposition of sanctions (Broadcasting Act, 1990).

There are a number of additional ways to supplement the set of formal sanctioning instruments with informal communication. The ITC has found incentives to continue the IBA's tradition of seeking to resolve infractions via informal channels. In view of the fact that the act employs many vague legal terms that are in need of interpretation, a particularly interesting alternative is the search for consensus. Despite the relatively long license periods, the license holders are undoubtedly interested in maintaining a good relationship with the ITC, especially in order to preserve the chances for renewal.

7.3. The 1991 Licensing Procedure

The awarding of Channel 3 licenses in 1991 garnered great attention. The ITC made an effort not to put the focus on the government's intention of acquiring revenue and tried instead to award licenses only after careful examination of expected capabilities.

Forty applications were received for the sixteen licenses up for bids.[62] All but four of these licenses went to the incumbent franchise holders. Three former license holders that were not exposed to competing offers received their licenses with minimal payment pledges. Of the remaining thirty-seven applicants, thirteen failed to surmount the quality threshold. The other twenty-four were analyzed comparatively on the basis of, above all, the information supplied on type of programming and on economic viability as well as the financial bid they submitted. Only five of the thirteen available licenses then went to the highest bidder. Two former license holders (TV-am and Thames TV) lost their licenses to applicants that had submitted higher bids.

Clearly, the amount of the bid represents a not insignificant yet relatively lower-ranking factor in licensing decisions. Apparently, the ITC did not have a great deal of confidence in the market as a medium of license allocation. Instead, it put greater weight on prognostic estimates as to whether the applicant's economic and journalistic bases were adequate to ensure a viable broadcaster. If, in view of expected revenues, the financial bid appeared too high to be able to generate sufficient earnings for quality programming, the license was refused. In making their bids, the applicants had to act under a high degree of uncertainty. They had no knowledge of competitors' bids, nor could they safely predict future income and expenses. Indeed, in view of the transition from the previous monopoly structures to competitive structures, prognoses about future developments were especially difficult to make. The ITC was also faced with this uncertainty. This did not mean that in doubtful cases it followed the estimates of the applicants; rather, it attempted to come up with its own appraisals. Since the criteria were operationalized only to a limited extent, and thus comprehensible, many observers felt that the result was more of a lottery than an auction. For this reason, the ITC encountered more criticism than approval[63]; especially prominent was the criticism of Margaret Thatcher (see *Independent,* 23 October 1991), whose unyielding pressure had brought about the new rules in the first place.

The application procedure saw a collision of diverse philosophies on the organizational structure of future broadcasting in Great Britain. Although British broadcasters had thus far produced a majority of programs themselves, the British government had long worked on support for independent producers. It was thus aiming at the decades-long practice in the United States where broadcasters or networks mainly purchase programs or commission them— that is, produce virtually nothing themselves. In the most spectacular case of nonrenewal of a license, Thames TV's license for the London area was awarded to a competitor, Carlton TV, which had been particularly insistent on the new publisher–broadcaster model and was therefore able to calculate its bid with, among other things, a minimal amount of personnel. Thames, which had a great deal of production experience and a large amount of programs in stock, subsequently decided to continue operations as a program production company. In this way, a powerful company—one that is also tied multimedially and internationally with other enterprises—has moved into the production sector. This rearrangement may likely have a considerable influence on the production landscape in Great Britain. In 1993, for example, the largest contribution of independent production to the network programming came from Thames.

7.4. Overall Appraisal

An overall assessment reveals that the special British path of a dual broadcasting system begun in the 1950s has been abandoned, although in its invitation to apply for regional Channel 3 licenses (cf. Hearst, 1992: 72), the ITC has tried to preserve some of its spirit. However, the structural elements that had previously marked this system have been eliminated. There will no longer be either private regional monopolies for broadcasting or protection against competition for advertising revenues. Although the number of possible competitors is limited for terrestrial broadcasting, additional competition can be expected from satellite and cable broadcasters. There has, however, been greater continuity than originally expected. Channel 3 is still composed of fifteen regional TV companies that will continue to operate as producer–contractors. The franchise areas remain the same as before. The network still exists, albeit run by a new Network Control Centre that will centralize scheduling (and weaken the grip on the network of a handful of megapowerful TV companies). The ITC is a slimmed-down IBA with a large degree of personnel continuity, although of course it is now a regulator pure and simple (and it has privatized its transmission service).

Whereas in the previous broadcasting system, the focus was on the effort to prevent economic motives from predominating in programming conduct, economic stimuli are now being strengthened. As a result of the financial levies, broadcasters are under more pressure to orient their programming toward profitability objectives. It is hard to imagine that the programming commitments, which are formulated in the tradition of public service broadcasting, will be able to form a counterweight. The ITC, which is called upon to supervise, is in a structurally weak position in relation to the broadcasters—at least weaker than that of the IBA in relation to the program companies. In particular, it is unclear how the ITC as external authority hopes to ensure that commitments in licensing and supervision will be observed when they conflict with the broadcasters' economic and disseminational interests. It remains to be seen whether the ITC can be more effective than external supervisory authorities in other deregulated broadcasting orders, such as the FCC in the United States. With respect to traditional supervisory fields, for example, advertising rules, quality commitments, and norms for battling concentration, the Broadcasting Act might turn out to be more of a symbolic or ritual set of norms.

The Broadcasting Act also goes to remarkable lengths to cover foreign satellite channels, even those that do not have licenses awarded in Great

Britain. However, the control possibilities opened by the act are only limited (see Collins, 1993: 121f.). The act's means of combatting further concentration in the media sector are likewise weakened by loopholes. Critics fear that the additional rules are likely to set off a new wave of concentration (see Murdock, 1992: 231ff.). The heavy-handed efforts to increase the number of independent producers in program production and to restrict broadcasters to the role of publisher will not necessarily lead to greater diversification in the media industry. Rather, it cannot be ruled out that the production sector will also be characterized by increased concentration. The refusal to grant a broadcasting license to Thames TV and its subsequent restructuring as a production company might be the start signal for the development of powerful production giants, which could in turn conceivably be linked with other larger enterprises (cf. Murdock, 1992: 231f.).

The Broadcasting Act is marked by an inherent contradiction between deregulation from an economic–institutional standpoint and overregulation with regard to certain programming issues, particularly with regard to ethical and moral affairs. In the latter case, Parliament does not rely solely on the ITC but rather puts the BSC to work as an additional authority, thereby preserving the government's powers to intervene. In addition, citizens may take action by turning to the BCC when their right to privacy has been infringed upon. Although neither the BSC nor the BCC possesses a more effective set of sanctions than the ITC, it is not entirely inconceivable that the increased responsibility for programming control will trigger anticipatory self-censorship activity by the broadcasters themselves. This will, in turn, lead to the rigidity in programming conduct that the government intended with the creation of the BSC. As a number of authorities are entrusted with supervisory powers, the effectiveness of ethical and moral programming obligations might be substantially greater than that of other programming commitments, such as the duty of impartiality, the observing of program quotas, and even ensuring that various user interests are satisfied.

The Broadcasting Act of 1990 refers with a surprising lack of self-consciousness to the "high quality" to which it aspires in the palette of its programming. But this target goal is neither specified in detail nor put into practice in a manner that can be easily administered for supervisory purposes. The ITC's "invitation to apply" tried to step in here. Moreover, Channel 4 stands ready as a deficiency guarantor in the provision of high-quality programming. At the same time, however, it is uncertain whether it will be able to maintain the high quality of its previous program offerings under the new conditions. Competition has become greater, and Channel 4 is now itself responsible for advertising revenues. Although it enjoys a sort of financing guarantee, it is dependent on the ITC for the estimate of its financial needs and the receipt of compensation payments by Channel 3 license holders. The ITC is caught in a conflict of interests. At the moment, it is impossible to predict how it will decide when the financial earnings interests of the Channel 3 and 5 license holders collide with Channel 4's financing interests.

The Broadcasting Act of 1990 displays yet another contradiction. The government implemented its white paper in order to do away with "unnecessary regulatory barriers." Broadcasting norms were supposed to become "lighter, more flexible and more efficiently administered" (Home Office, 1988: 6). Yet a glance at the Broadcasting Act of 1990 reveals a highly complex set of rules, whose implementation will undoubtedly prove to be most difficult, particularly in light of the statutory technique adopted, with the many cross-references and clarification attempts. The quality of the provisions in the Broadcasting Act of 1990 remains behind that of continental European and North American broadcasting laws. Moreover, despite the many statutory definitions, the normative requirements are by no means unambiguous and strict. On the contrary, license holders and/or the supervisory authority must work with considerable normative latitude, a latitude that can probably be only partially reduced with codes and general, abstract directions.

In view of the lack of clarity regarding rights and duties, it seems likely that the supervisory authority and the license holders will seek to develop cooperative ties and to resolve problems primarily via informal avenues. Since ITC personnel is to a great extent composed of former employees and functionaries of the IBA, there are a number of possibilities for making use of communication networks already in existence.

In interaction with the ITC, the Channel 3 and 5 license holders will probably be able to respond successfully to criticism of program quality and violation of program quotas in such a way that they will not be the target of accusations. Certainly they can argue that they render considerable financial compensation to the state and at the same time contribute to the financing of the cultural Channel 4, which, for its part, is required to service niches in programming. As a result of the many structural reforms, it may be expected that the programming conduct of Channel 3, as well as that of Channel 5, will differ considerably from that of the former ITV program companies. By no means can it be ruled out that adjustments will be made to conform to the conduct of typical commercial broadcasters in other countries, such as in the United States or Italy. Indirect repercussions for the BBC's programming conduct are also likely.

With its new Broadcasting Act, Great Britain has joined the trend toward market broadcasting, but it has also made an effort to maintain a tighter rein on programming substance than have other dual Northern and Western European broadcasting orders. The philosophy of deregulation has allied itself with the tradition of public service broadcasting. It remains to be seen whether the British broadcasting concept will survive the new multichannel environment that will evolve in the near future as digital compressing will multiply the number of available transmission facilities. The British model will have to be adjusted further in the course of converging technologies and the establishment of the emerging Information superhighway.

NOTES

1. For example, Peacock at the Edinburgh Television Festival. See the Independent, 27 August 1990.

2. For the problem of the selection of personnel, cf. Tunstall, 1983: 211f.

3. Following the 1991 franchise round (see Section 7.3), TV-am lost its license to Sunrise TV.

4. With regard to the following remarks, see the Independent Broadcasting Authority Act of 1973.

5. The share of programming supplied by independent producers amounted (in broadcast hours) to 54% of the commissioned program transmissions in 1989–1990, which in turn made up 58% of total transmission time (Channel Four, 1990: 14).

6. The Cable Broadcasting Act of 1984 was repealed by the Broadcasting Act 1990.

7. For the promotion of cable and satellite in British politics, see Dyson & Humphreys, 1985: 363ff.; see also Dyson & Humphreys, 1988.

8. For the complicated history surrounding this satellite, see Goodfriend, 1988: 161ff.

9. Burnett, 1989: 22ff.; Davey, 1990: 7; Tunstall, 1992: 252. For the spread of telecommunications services in cable networks, see also Cable Authority, 1989: 5. For details concerning the legal framework, see Cable Authority, 1990: 32.

10. For details, see Part II for schedule 2 to the Broadcasting Act of 1990. Pursuant to s. 1(2) of Part II, the prohibitions for non-EC citizens do not apply to cable services or to non-British satellite services, thus lifting the earlier restriction. As a result, the former construction as Channel Island trusts is no longer necessary.

11. On the other hand, the establishment and operation of the cable network from a technical standpoint are regulated separately. A license issued by the Department of Industry and Trade and the Office of Telecommunications is essential. For details, see Wiedemann, 1989: 194f.

12. For details, see Cable Authority, 1988b: 9f. (4).

13. However, all sponsoring of news programs or reports on current affairs, for instance, had to be agreed upon in advance with the Cable Authority; see Cable Authority, 1988a: Code of Practice on program Sponsorship 6iii.

14. See J. Davey, Director General of the Cable Authority, in an interview with the author: "If we don't promote, we have nothing to regulate."

15. For the concept of public service and public responsibility, see Gibbons, 1991: 32ff.

16. For instance, the public service obligation is expressly placed on Channel 4 (see s. 25[2] of the Broadcasting Act of 1990). But for the other commercial channels as well, there are considerable programming commitments that in many respects continue the tradition of the trustee concept.

17. For financing, including the issue of equalization of financing among several companies and the duty of paying a levy to the state, see Mahle, 1984: 188ff.

18. Programs were typically produced by the individual ITV companies for showing across the ITV network. The network was therefore a criterion affecting producer decisions.

19. Efforts were made with the support of the IBA to involve the other companies more strongly in program provision; see Wiedemann, 1989: 113; Spicers Consulting Group, 1989: 4.

20. However, these changes often continued to be viewed as insufficient. For details on this problematic area, see Spicers Consulting Group, 1989: 4f.

21. Subsequent to the government's plans as announced in its white paper (Home Office, 1988: 24ff.), Channel 4 financing was changed by the Broadcasting Act of 1990.

22. The situation was different for the smaller television stations and for radio; see Tunstall, 1986: 131.

23. For the contents of these contracts, see Mahle, 1984: 203ff.; Wiedemann, 1989: 121ff.

24. For the close cooperation between the IBA and ITV companies regarding the control of programming, see also Potter, 1989: 105ff.

25. See also Section 5.1. This was also on occasion criticised politically; see, for example, Social and Liberal Democrats, 1988: 4.4.1. For more on the 1980 decisions, see Tunstall, 1983; 209f.; Dix, 1980: 370; Schacht, 1981: 692.

26. For example, in 1968, a company for the west of England and Wales (TWW) was replaced by Harlech Television. In 1980 two companies lost franchises for the south of England and southwest England respectively, and two larger companies, Central and Yorkshire, were obliged to cater more specifically to subregions of their areas.

27. In 1988 the ITV companies canceled their participation. The Super Channel had not proved to be particularly successful, especially outside Great Britain.

28. For data on program shares in the IBA Annual Reports, see IBA, 1990: 22f.

29. The policy of relaxing restrictions was facilitated by the strengthening of private radio as sought for in the political arena. This found expression in the government's 1987 green paper; see Home Office, 1987.

30. The advertising time restrictions were comparably strict. For instance, a maximum of seven minutes of advertising per broadcast hour daily average was permitted, during prime time (6–11 PM: a maximum of 7.5 minutes per broadcast hour). See IBA, 1988: 27.

31. See ITCA & ITVA, 1985: Series of Notes of Guidance No. 8, 4; IBA, 1990: 31. The members of the AAC have been published in the IBA annual reports, for example, IBA, 1990: Appendix 7.

32. For the comprehensive consultations prior to the new code for radio advertising by the future Radio Authority, see Coulson, 1990: 6.

33. The Copy Committee dealt with an average of eight cases in each of its monthly meetings.

34. These covered roughly 30% of the scripts.

35. Although the 1988 white paper adhered to this position in principle, it did not consider relaxations to be inconceivable; see Home Office, 1988: 31–32.

36. At least, this is the position taken by Hood, 1980: 87. For detailed data, see Mahle, 1984: 261ff.

37. For instance, Dyson, 1988a: 31, also reiterates the appraisal of the broadcasting companies. See also Tunstall, 1983: 208, who asserts that although the IBA is apparently located in every niche of British broadcasting, it has thus far not proved that it reigns supreme.

38. For the relevance of this aspect as well, see, for example, Dyson, 1988a: 3.

39. For IBA members, see IBA, 1989: Factfile, 5f.; further examples of this situation of "revolving doors" include the chairman (see the portrait with resume in *Television Week,* 6–12 April 1989, 13) and the director (portrait and resume in *Broadcasting,* 3 March 1989) of the BSC.

40. This, at least, seems to be the conclusion reached by Garnham, 1980: 25.

41. A personal association with the interests of the advertising industry, the program companies, the manufacturers of transmission facilities, or the competition (the BBC) was forbidden by law.

42. This is all the more significant as there is no written constitution with basic civil rights. However, the European Convention on Human Rights does function as a substitute, at least to some degree. See Bailey, Harris, & Jones, 1991: 1f., 749ff., 819ff.

43. The Broadcasting Act of 1990 has changed this; see s. 162 et seq. This is the subject of severe public criticism.

44. For instance, the planned broadcast of a documentary on Andy Warhol had generated such heated public controversy that a member of the organization took the case to court; see Potter, 1989: 124ff., generally, 147ff.

45. Since neither English nor Scottish law recognizes a right to privacy, the existence of the BCC will often be the only remedy for breach of privacy.

46. See, for example, the case of the C4 Hard News, in Smith, 1990; Goodwin, 1990.

47. Whereas some 152 complaints were lodged in 1987–1988, the number skyrocketed to 550 in 1989–1990; see BCC, 1990: 1.

48. There were at times intensive discussions regarding the manner in which this levy to the state was to be calculated. Debate particularly centered around the question of whether the levy was to be determined on the basis of advertising revenues or with regard to profits. For the amount, see Wiedemann, 1989: 150.

49. For examples of general program prohibitions, see Mahle, 1984: 103f.

50. This directive was the direct outcome of the highly controversial broadcast of the program "Death on the Rock" about the shooting of alleged IRA terrorists in Gibraltar, a program that the government had unsuccessfully sought to ban. See also Home Office, 1988: 36. The same directive was issued to the BBC. See Bailey et al., 1991: 309ff.

51. *R. v. Secretary of State for the Home Office, ex parte Brind* (1991) 1 A.C. 696, House of Lords.

52. See, for example, the statement, "Future of UK Independent Radio," IBA, 1987, and the further development of this by the IBA, "Independent Television in the 1990s," London, April 1988.

53. One of the IBA's chief worries was the question of whether it would be possible to shift most of its personnel to the new ITC; see *Broadcast,* 11 November

1988, 3. This has since been favorably resolved. See, for example, Members of the Shadow ITC, in IBA, 1990: 36.

54. For an evaluation of the written statements, see Stevenson & Smedley, 1989.

55. For details on the particular rules in the Broadcasting Act of 1990, see Reville, 1991, and Gibbons, 1991. For a critical appraisal, compare Miller & Norris, 1989.

56. The latter involve services offered by using the spare capacity within the signals carrying a television broadcasting service, as well as closely specified cable service offerings.

57. This concept applies in other sectors as well, such as for the granting of licenses in the cable sector, for so-called local delivery services, and for national radio.

58. For the considerations behind the creation of a special Channel 4, see Negrine, 1989: 102, 111ff.

59. At the moment Channel 4 is very successful and does not need any financial help.

60. Also to be taken into account, however, is other programming broadcast by cable and satellite. But in view of the dominant role that terrestrial broadcasting will continue to occupy for many years, it is to be expected that with respect to the competition among programming broadcast terrestrially, diversification activities will be characterized only insignificantly by respect for cable and satellite programming.

61. The ITC could only veto changes in the composition of license holders until the end of 1993, although it retains the power to ensure that the obligations in the license are observed. License holders have been open to takeovers since then. Mergers between several licensees were already set in motion (ITC, 1994: 2).

62. The following analysis is based on Murdock, 1992: 226ff.; see also Alvarado, Locksley, Paskin, 1992: 329ff.

63. TV broadcasters have made no bones about the claim to run commercial TV in a way that is fundamentally different from former practice, by acting more as a publisher–contractor (rather like Channel 4 but far more commercially oriented) than as a producer–contractor (like the traditional ITV companies). This could turn out to be the pattern of the future, and it could amount to a major change indeed.

CHAPTER 3

Germany

The German media order offers a particularly rich source of experience in broadcasting supervision. Although regulation and supervision of private broadcasting began relatively recently, the fact that each of Germany's states (until 1990, there were eleven; in 1995, there are sixteen) has its own supervisory concept (fifteen have supervisory authorities) has enabled a variety of experience to be gained. Moreover, the German broadcasting order is characterized by the intensive interaction of law, not simply on the part of the courts but also by parliaments. For this reason, it provides a particularly good subject for studying the extent to which it is possible to influence the substance of the broadcasting order with the aid of law and its implementation.

1. DEVELOPMENT AND STRUCTURE OF THE GERMAN BROADCASTING SYSTEM

1.1. Historical Development

Broadcasting commenced in Germany in 1923 under the supremacy of the German Reichspost (Postal Authority). It was conceived of as entertainment broadcasting, which not only was to develop as free as possible from political conflicts but also was to avoid intervening in the political process—apart from the broadcasting of official government information (Ross, 1986:

58). In the early 1930s, however, it was given a growing role in the battle of political opinion as the government's mouthpiece. From 1933 onward its centralized form of organization offered the National Socialists a good basis for political subjugation and for abuse for propaganda purposes (see BVerfGE 12, 205ff., 208ff.; Lerg, 1980; Diller, 1980: 16ff.). Therefore, following Germany's defeat in World War II, the tradition of state broadcasting was not only deemed inappropriate but also explicitly ruled out as a starting point for embedding broadcasting in a democratic framework in the Western zones of occupation. Instead, it was the British BBC that served as a role model, exemplifying impartial broadcasting committed to the common good. From the BBC Germany derived the requirement of independence and the concept of a broadcasting committment to the public service idea.

In light of the fresh memories of the abuse of broadcasting by the National Socialists, however, an effort was made to ensure even greater independence from the state with institutional and legal approaches than was the case with the BBC. Broadcasting was decentralized into various broadcasting authorities. In addition, an attempt was made to integrate the most important proponents of social interests as guardians of diversity and independence in programming (for details on development, see Bausch, 1980: 46ff.) and thus to create a structure for broadcasting that exemplified the pluralism theory. Broadcasting was financed by license fees, at first exclusively and then later (beginning in the mid-1950s) joined by advertising revenues. Attempts by the federal government and newspaper publishers to gain access to broadcasting were thwarted by the Bundesverfassungsgericht (Federal Constitutional Court) in 1961 (BVerfGE 12, 205; Bausch, 1980: 430ff.). In the late 1970s, publishers as well as other interested private parties began once again to call for the licensing of private broadcasting. Following stormy debates of media policy (Montag, 1978; Humphreys, 1990: 195ff.; Hesse, 1990: 20ff.), a compromise was reached in the 1980s with the launching of so-called pilot cable projects. From an official standpoint, these served as a means of gaining experience with a broadcasting system of the future; in reality, however, they represented the first step toward a dual broadcasting system and thus the beginning of private broadcasting (Hoffmann-Riem, 1991e: 53ff.).

However, in the German Democratic Republic, formed from the Soviet zone of occupation, once state broadcasting was introduced it eventually became a tool in the realization of the socialist order and thereby a tool for political control of citizens (Riedel, 1977; Gerber, 1990; Schuster, 1990: 32f.). It was managed centrally and subjected in every aspect to the political imperatives of the East German Communist Party (Sozialistische Einheitspartei Deutschlands, or SED). Although citizens were prevented for decades from receiving Western programs, this began to be tolerated in the 1980s and to some degree even supported by the spread of cable networks in order to encourage citizens to remain in areas in which these programs could otherwise not be received. Following the political overthrow in 1990, decentralization was initiated and broadcasting freed from the system of state patronage. The accession of the German Democratic Republic to the Federal Republic of Germany on 3 October 1990 marked the beginning of the adaptation of the East German broadcasting system to its counterpart in West Germany (Hoffmann-Riem, 1991d).

The German system is currently characterized by the coexistence of non-profit-oriented public broadcasting and private broadcasting, which is nearly entirely dependent on advertising revenues. Pay-TV, though growing, is in its infancy, and thus far it has been offered solely by private companies. Both public and private broadcasting are regulated by special broadcasting laws in each of Germany's states (Bundesland); these are accompanied by interstate treaties between the states, in particular, the Interstate Treaty on Broadcasting in Unified Germany (ITB; Staatsvertrag über den Rundfunk im vereinten Deutschland or Rundfunkstaatsvertrag) of August 31, 1991, which replaced the interstate treaty of 1987. Apart from the area of foreign broadcasting, the Federation (Bund) itself has no jurisdiction in the broadcasting sector. Each state has enacted its own broadcasting laws, and since these often deviate from one another, the state of the law is not uniform (Ricker, 1985).

The states have established public broadcasters (*Landesrundfunkanstalten*), some of which are responsible for several states. They are independent of the government and financed by fees as well as advertising revenues. Together they form a network called the Working Group of Public Broadcasting Organizations of the Federal Republic of Germany (Arbeitsgemeinschaft der öffentlich-rechtlichen Rundfunkanstalten der Bundesrepublik Deutschland, or ARD). The public broadcasting organizations broadcast both radio and television. In the television sector, they are responsible for the nationally broadcast first channel (ARD) and for the so-called third channel, which varies from state to state and tends to offer programming that is culturally more challenging. Finally, there is the second channel, Zweites Deutsches Fernsehen (ZDF; Second German Television), a television institution established by the states jointly. The public broadcasting organizations cooperate with foreign German–language broadcasters in the television program 3 Sat. A German-French cultural channel, ARTE, has also been established.

In the area of private broadcasting, radio and television programming are offered by a number of separate, though often economically linked broadcasters. There are several television broadcasters that provide nationwide transmission of "full programming,"[1] namely, the particularly successful broadcasters RTL and SAT 1 as well as some smaller operators.[2] Regional TV broadcasters, as in Berlin–Brandenburg and Hamburg, are an exception. Radio is more strongly regionalized and localized; however, the provision of ready-made programming by regionally or nationally operating suppliers is on the rise.[3]

Public broadcasting is under an obligation to offer so-called internally pluralistic "integrated programming." In other words, the programs it broadcasts must present all social interests in a balanced fashion and, in particular, are forbidden from pursuing any one-sided interest or viewpoint. A given palette of programming must be diverse. For the purpose of monitoring the legal requirements, the broadcasting laws have provided two special organs of the broadcasters, namely, the Broadcasting Council (Rundfunkrat) and the Administrative Council (Verwaltungsrat). The pluralistically composed Broadcasting Council's main tasks are to ensure monitoring independence and diversity in programming; that of the Administrative Council is to ensure that the broadcaster's administration and financial management comply with the regulations. Since supervision is thus accomplished internally, that is, by an organ of the broadcasting authority itself, there is no need for an additional, external supervisory body. Nevertheless, each state government makes sure that its broadcasters observe the relevant laws, even though its supervisory powers are somewhat restricted (so-called limited legal supervision). State governments are prohibited from intervening in the shaping of programming itself (for details, see Berendes, 1973; Hesse, 1990: 130ff.; Herrmann, 1994: 358ff.).

Although some of the broadcasting or media laws (e.g., in Hamburg, Bremen, and North Rhine–Westphalia) borrow from the internally pluralistic broadcasting model with respect to private broadcasting, they are generally oriented toward the so-called externally pluralistic broadcasting model in this regard. Broadcasters may finance their operations from the market (advertising, sponsoring, pay-TV), and as long as there are enough—normally, more than three—competing broadcasters, they are not required to offer balanced programming. Laws specifically regulating private broadcasting are likewise oriented toward the idea of broadcasting as a public service, thereby placing broadcasters under statutory programming duties, though these are less stringent than for public broadcasting (for details, see Hesse, 1990: 159ff.).

Whereas public broadcasting often has its own transmission facilities (e.g., for the first television channel and for the ARD radio stations in the western sections of the then Federal Republic), all other parties are reliant on the Bundespost/Telekom[4] as operator of most transmission networks. It is responsible for ferreting out exploitable frequencies within the scope of international agreements, developing and expanding the cable network, and stationing telecommunications and broadcasting satellites. It is, however, forbidden from influencing the way these are used for programming. The initial decision on the use of frequencies by public or private broadcasters is today the responsibility of, in some states, the state media authorities and, in others, the state governments. However, in 1991 the Federal Constitutional

Court ruled that allocation decisions by a state government were unconstitutional when the criteria for allocation were not set by statute (BVerfGE 83, 322ff.; for details, see Scherer, 1987; Eberle, 1989). Frequencies allocated to private broadcasting are awarded to private broadcasters by the respective state media authority. Since the Bundespost/Telekom, the state legislature (or sometimes the state government), and the state media authority are forced to work together, the allocation of transmission possibilities is a highly political act.

In view of the scarcity of terrestrial transmission possibilities, the construction and expansion of the cable network have assumed high political status. The media policy debates in the early 1980s were mainly directed at the development of the cable network. The Bundespost pushed forward its cabling efforts with high internal subsidies, and approximately 50% of all households in the western states are now cabled; in the former GDR, the number is less than 20% (numbers as of September 1994; see *TV Courier/Dokumentation,* 34(22-D), 15 November 1994: 10).

German telecommunications and broadcasting satellites are operated by Telekom. The German television satellite *TV SAT2,* which used the new D2 MAC norm (originally legally stipulated but now no longer the case; Müller-Römer, 1994: A 152f.), did not gain wide acceptance. Very few television sets were equipped to receive its signals, making it disfavored among television broadcasters. It was discontinued at the end of 1994. Television programs are transmitted mainly via telecommunications satellites, such as *EUTELSAT* and Luxembourg's *Astra,* which are extremely successful in Europe (for details, see Müller-Römer, 1992: A 128ff.; Müller-Römer, 1994: A 149ff.). These signals can be received with ordinary equipment. Satellite reception is widespread in the territory of the former GDR because of the lack of cable.

The legal requirements placed on the broadcasting of radio and television do not differ with respect to the type of transmission technology used. The statutory requirements are, however, relaxed when programs that are monitored elsewhere or can also be received with normal reception equipment are merely relayed by cable operators (for details, see Hesse, 1990: 209ff.).

In the mid-1990s Telekom is planning by means of digital compression to substantially increase its broadcasting capacity and thus reduce the frequency scarcity that has existed until now. There have also been various field tests with interactive televison as well as the integration of telephone and cable networks. Telekom and the German communications industry wish to take an active part in setting up the information superhighway, not least for world market and political reasons. Political discussion about the media concerning a suitable reaction to the reduction of frequency scarcity and the convergence of the various services has already begun (cf. Hoffmann-Riem & Vesting, 1995; Engel, 1995; Kull, 1993). However, it is the subject of much less public concern than the political discussion on the media that took place in the 1980s.

1.2. Broadcasting under the German Constitutional Order

1.2.1. The State of the Law

In none of the other countries studied here has law had such a marked influence on the development of broadcasting as in Germany (for details, see Hoffmann-Riem, 1994b). A decisive role is now played in this regard by the case law of the Bundesverfassungsgericht (Federal Constitutional Court), which has been called upon to decide fundamental disputes regarding the broadcasting order; in so doing, it has, much like a legislature, established basic guidelines, thereby thwarting political conflicts in their infancy. The court's point of departure is the basic right of freedom of communication (article 5 of the Basic Law), which places a mandate on the legislature to detail the structure of the broadcasting order by statute. These laws must ensure the free, individual and public formation of opinion and do this in a broad sense—that is, one not limited to mere news coverage or the conveyance of political opinions but, rather, encompassing every conveyance of information and opinion, including entertainment programs (BVerfGE 12, 260; 31, 326; 35, 222f.; 57, 319).

The fundamental assumption underlying the court's reasoning is that in all of its areas broadcasting can have an effect on the individual and collective orientations of citizens and may thus come to be of importance in all areas of life. Freedom of broadcasting serves the free formation of opinion, that is, the freedom of recipients to be able to inform themselves comprehensively. Since this freedom is endangered, it is in need of special protection. In the first place, it must be ensured that the individual is free from government predominance and influence. However, other types of misuse that may one-sidedly influence public opinion, particularly by private advocates of certain interests, must also be prevented. A situation must therefore be averted where powerful holders of opinion who are in possession of considerable financial resources and transmission frequencies are able to predominate in the formation of public opinion (BVerfGE 57, 323). In addition to such "negative barriers," positive guarantees are also needed in order to guarantee that the diversity of existing opinion is articulated as broadly and thoroughly as possible and that comprehensive information is made available. Broadcasting cannot be left to the free play of forces—especially market forces—since, in the view of the court, this free play is unable to prevent the accumulation and abuse of power in the broadcasting sector.

In justifying broadcasting regulation, especially the necessity of statutory guarantees for diversity and independence in the formation of public opinion, the court does not merely emphasize the scarcity of frequencies and the great financial expense that continue to be necessary for successful broadcasting. Even if this special situation were someday to vanish, statutory precautions would still be necessary to protect independence and diversity of opinion: The likelihood of a concentration of power over opinion and the risks of abuse of this by one-sided interests are simply too great (BVerfGE 57, 322f.). The court views broadcasting as an instrument of power, and it does not feel that the economic market alone is capable of limiting this power. As a result, the legislature is called upon to provide for diverse communication structures that are also able to satisfy minority interests.

Nor does the court believe that private broadcasting can fully satisfy such communication interests on its own. Nevertheless, it has ruled private broadcasting to be constitutional, though only within the framework of the dual broadcasting order. The court presumes that the best way to ensure diversity of opinion and independence is by the aid of fee-financed public broadcasting, since this is not at the mercy of the advertising industry and thus subject to the dictates of viewer ratings. As long as public broadcasting provides for comprehensive, diverse types of communication, the statutory requirements on private broadcasting may be relaxed—but not abandoned—in comparison to those for public broadcasting (BVerfGE 73, 157f.; 74, 325f.; 90, 78f.). Therefore, in order to be able to make private broadcasting legally possible, guarantees for the existence of public broadcasting are first necessary. In other words, public broadcasting is to enjoy the protection of its existing stock of transmission possibilities and its mandate, as well as the guarantee of an adequate financial basis (BVerfGE 90, 60ff.); furthermore, it must be able to share in future developments, such as those in transmission technology (BVerfGE 74, 326; 83, 299). Although private broadcasting, in contrast, does not enjoy a corresponding constitutional guarantee, its ability to function must be ensured once it is permitted.

Even though the diagnoses and prognoses underlying the court's reasoning are extremely controversial (cf., instead of others, Engel, 1994: 185ff.; Mestmäcker, 1988; Engels, 1986), they nevertheless form the basis for German media laws, which have sought to implement the court's requirements. According to German constitutional law, the requirements placed on broadcasting are to be regulated in detail by statute (the principle of rule of law), and burdensome state measures may be attacked in court. The result is a broadcasting sector that is pervaded with laws and legal requirements. In practice, however, the legislatures have run up against a number of legal, political, and economic limits to their ability to enact sufficiently detailed statutes, which has presented the supervisory authorities with substantial problems in operationalizing and implementing these statutes. They necessarily entail broad discretion, thereby creating a considerable risk of subsequent court battles. In a variety of cases, frustrated applicants for licenses or addressees of supervisory action have initiated litigation, though usually without success.

However, in spite of the many controversies, the limited specification of statutes, and the economic interests affected, the number of court disputes has been surprisingly low in the area of licensing, although it has increased of late. The low number of disputes is due to the fact that applicants have no claim to a license but, rather, only that the licensing decision be free of error (however, state laws vary considerably here; cf. Hesse, 1990: 173; Wagner, 1990a: 184ff.). The state media authorities are endowed with a considerable amount of decision-making discretion (Wagner, 1990a: 187ff.; Hesse, 1990: 174), and the way this is administered is subject to only limited judicial review (OVG Lüneburg, DVBl. 1986, 1112, 1114; VerwGH Mannheim, NJW 1990, 340, 341; Wagner, 1990a: 168ff., 187ff.).

It should also be noted that license applicants sometimes tend to avoid seeking legal recourse when they are competing for other frequencies, in order not to worsen their chances in such future procedures. There also used to be relatively little court action taken with regard to ongoing supervisory action, although this too has increased of late. In 1993 several broadcasters challenged actions by the supervisory authorities to implement advertising restrictions, trying to force lawmakers to soften the requirement. As a result of the restraint practiced by the supervisory authorities in other areas, the number of court actions is still relatively small (cf. Rödding, 1989: 651; *Kabelrat der Anstalt für Kabelkommunikation Berlin,* 1989: 164f.; Hellstern & Reese, 1989: 50ff.). The authorities have refrained as far as possible from formal supervisory measures and instead searched for informal solutions (see Section 4.6). Since it has not yet been definitively resolved to what extent an individual may appeal for judicial relief from informal supervisory actions (for the problematic, see BayVerwGH, NVwZ 1987, 435, 437), restraint is probably born of prudence as well. Above all, broadcasters are interested in long-term cooperation with supervisory authorities—as are, to an equal extent, the supervisory authorities—and they generally tend to be unwilling to risk losing this positive relationship by recklessly instituting litigation.

The fact that broadcasting supervision is so heavily regulated leads to a situation where the relationship between the supervisory authority and license applicants and/or broadcasters is often marked by legal argumentation and reference to procedural rights. Legal arguments are particularly used by a party to support its position, whether in the applicant's documents or in the supervisory decisions by the state media authority. In practice, however, interaction is by no means primarily characterized by reference to legal norms. Rather, informal structures of interaction have emerged in the decision-making process, and within the framework of these, a variety of informal steps have come to be essential.

1.2.2. Federalist Fragmentation

Since legislation and supervision of broadcasting fall within the competence of the states, the broadcasting order is federally fragmented. This feature also distinguishes the German broadcasting order from those in other countries under study here. At first glance, the federalist regulatory technique may seem somewhat anachronistic, since broadcasting, of course, does not stop at borders and since the broadcasting sector is marked by widespread, even trans- and internationally interlocked systems. The adherence to federalism signals the decision that broadcasting is a cultural commodity: Culture is, in legal terms, a matter for the states. Federalist diversity is considered to be a means of ensuring at once local and regional identity and programming diversity. Particularly in view of the temptation of political influence (see Section 4.2), federalism allows one-sided influence to be combated from many sides.

However, federalism does not preclude a certain amount of consolidation. The interstate broadcaster ZDF and the ARD network are examples of this, as are the Interstate Treaties on Broadcasting. Furthermore, the case law of the Bundesverfassungsgericht has resulted in a uniformity of basic principles. The EC's television directive[5] has generated further consolidation. Coordination between state media authorities has also been stimulated by the fact that private broadcasting tends to orient itself around markets and not state borders. As a result, important forces in the economy have pressed for consolidation.

The state media authorities have already sought to coordinate their activities by creating the Direktorenkonferenz (Conference of Directors) as well as a number of joint committees (Wagner, 1990a: 145ff.; DLM, 1993: 267ff.).[6] Coordination is intended to help prevent the emergence of differing standards for evaluating similar situations and thus different supervisory measures (for coordination, cf. Ladeur, 1989: 717ff.; Rödding, 1989: 649; Benda, 1989: 5ff.; *Kabelrat der Anstalt für Kabelkommunikation Berlin,* 1989: 101ff.). Alongside formally assured intrafederal coordination, one also finds a variety of informal coordination processes. Since the beginning of 1993, the media authorities have experienced increasing difficulties with coordination; considerable conflicts have developed (cf. BayVerwGH, ZUM 1993, 296ff.; VG München, ZUM 1993, 294ff.; BVerfGE ZUM 1994, 639ff.; see also Haeckel, 1993: 21ff.).

The discrepancies in state media laws have led to wide and varied experience with broadcasting supervision within a relatively short period of time. The state media authorities were often able to apply different supervisory strategies and, if they so desired, to learn from one another, for example, through exchange of information in the Conference of Directors (see also Ring, 1988: 159ff.).

Federalist fragmentation has often been criticized by private broadcasters, since it prevents them from gaining a clear overview and hampers actions in larger markets. For instance, broadcasters that wish to disseminate their programming terrestrially in more than one state must acquire a license from each state concerned and observe the statutory requirements of that state. Moreover, the various state media authorities often have different types of requirements (e.g., the inclusion of so-called regional window programming in overall programming). But in spite of public complaints about this situation, some broadcasters nevertheless stand to benefit from it. For instance, they have sometimes been able to play one state against another, such as by threatening to locate in the state that most closely conforms to its expectations. This threat was, however, not effective insofar as the broadcasters were not dependent on terrestrial frequencies (see Section 3.7).

In addition, federalist diversity created a certain amount of protection against foreign competitors. Although German law does not permit foreigners to be excluded from the licensing process, the jungle of broadcasting law and the wide variety of responsible actors have for some time worked to discourage, or at least delay, foreign, multinational media firms from venturing into the German media market. Legal and administrative fragmentation served to give German license applicants, who were better experienced with German law and German officials, a practical advantage over foreign competitors. Above all, the initially somewhat hesitant attempts by multinational entrepreneurs, such as Berlusconi or Murdoch, to gain a foothold in the German broadcasting market were launched too late to be able to win any of the particularly coveted terrestrial television frequencies.[7] This has relieved the supervisory authorities of having to take special steps to shield the domestic media industry against foreign broadcasting competition. In the meantime, however, foreign media companies, including U.S. multimedia enterprises, have bought shares in TV companies and in this way have entered the German market.[8]

2. BODIES AND PROCEDURES OF BROADCASTING SUPERVISION

2.1. Supervision of Public Broadcasting

As mentioned above, the supervision of public broadcasting is incumbent upon internal organs, namely, the Broadcasting Council and the Administrative Council. This internally pluralistic organizational structure, which is typical for German broadcasting law, is not to be found in the other broadcasting orders under study here (however, similar structures are to be found in other countries, such as in Austria, see Steinmaurer, 1994: B 160ff.; for the Swiss, see Saxer, 1994: B 196ff.). So-called socially relevant groups, which are usually enumerated by statute, delegate representatives to serve in the Broadcasting Council; some delegates are elected by Parliament. These representatives then take on responsibility for the functioning of the broadcasting authority. The fact that they are recruited from a variety of spheres of interest is designed to facilitate a reciprocal balance of interests and thereby reduce the risk of one-sided use of broadcasting. Legal responsibility for programming, however, rests with the director general (*Intendant*), the monocratic head of the authority.

Nevertheless, the Broadcasting Council, which is responsible for programming matters, possesses advisory and supervisory powers. It may focus on both individual programs as well as the general structure of programming. Although it normally restricts itself to criticism of programs following broadcast, in some organizations, it passes judgment on sections of programs prior to broadcast in exceptional cases. The Broadcasting Council also takes part in programming-related decisions on personnel, organization, and budgetary matters and to this end is endowed with considerable powers of intervention (for details, see Fritz, 1977; Kabbert, 1987; see also Cromme, 1985: 351ff.). Originally conceived of as guarantors of independence from the state and other power-holders, the Broadcasting and Administrative Councils have, however, in reality emerged as agents of political influence (for reports by insiders, see Menningen, 1981: 185ff.; Vogel, 1983: 1ff.; Plog, 1981: 55ff; see also Fritz, 1977; Schneider, 1981: 120ff.; Ronneberger, 1986: 301ff.). The members of these bodies usually form coalitions along party lines, meeting in so-called cliques (*Freundeskreis*) to agree upon joint conduct. A normally small, remnant group of those without party ties usually constitutes the "gray clique." The clique aligned with the respective state government functions at the same time as the transformer of executive interests. These bodies have often misused their right to take part in the appointment of managing personnel for political (party) ends. If the governing party has been in office for decades, as in Bavaria, its influence on the selection of personnel will be most certain to have been used as fully as possible. On the other hand, in regions with changing majorities in the government, as in the northern states, a proportional system has emerged.

In spite of political affinities, public broadcasting is on the whole marked by political balance and a relatively high degree of journalistic independence. From a comparative standpoint with respect to independence from the federal government, the German public broadcasting authorities more closely resemble the BBC than, for instance, the Italian RAI or the French broadcasters Antenne 2 (now France 2) and FR 3 (now France 3). Proponents of political interests have a relatively insignificant amount of influence on daily programming work. Because of the restriction of the Broadcasting Council's control over programming to exceptional cases, journalists are left with enough room to orient their work according to their own professional and ethical values. At the same time, however, even control measures that are limited to exceptional cases have an effect on day-to-day work, since they help to define the zone of "permissible" activities and thereby prompt or even provoke anticipatory compliance.

The state government's power to monitor the lawfulness of broadcasting activities in the event internal control bodies fail to function properly has been placed under so many legal limitations that, in practice, it has only been exercised in extremely few cases. However, the government's supervisory competence helps to promote informal respect for government interests. It generally tends to exercise its influence in informal interactions not easily grasped by third parties. In addition, although unable to vote, a government representative usually takes part in the meetings of the Broadcasting Councils.

2.2. Supervision of Private Broadcasting

2.2.1. Status of the State Media Authorities

The supervision of private broadcasting is strictly separated from that of public broadcasting. The supervisory bodies are autonomous juridical persons given independence vis-à-vis the government and financed by a portion of the fees collected from viewers and listeners (see Hoffmann-Riem, 1994a: 68 ff, 72ff.). Their names vary from state to state.[9] They are normally composed of two organs: The monocratic executive organ, usually termed the director, is responsible for day-to-day administration. He is joined by a collegial board, designated as the media council (*Medienrat*), state broadcasting committee (*Landesrundfunkausschuss*), executive board (*Vorstand*) or broadcasting commission (*Rundfunkkommission*). In most states, this is a larger organ that—following the example of the Broadcasting Council in public broadcasting—follows the model of pluralistic representation and is responsible for basic policy issues (see Wagner, 1990a: 113ff.). There are smaller executive boards within some of the media authorities (Hamburg, Berlin–Brandenburg, Sachsen, Mecklenburg–Vorpommern).

It seems that the smaller organs are able to deal with day-to-day supervisory work more intensively than the large, pluralistic ones, thereby forming a substantial counterbalance to the respective director. For instance, the former executive board in Baden–Württemberg carved out a remarkably autonomous position in relation to the director (and to the government),[10] and this despite basic political concordance between the majority of the board, the state government, and the director. In the smaller bodies, unanimous decisions have been able to be reached in most cases, and the formation of close-knit cliques according to political leanings has for the most part been avoided. In the larger (pluralistically composed) bodies, large voting majorities are likewise to be found, although these consistently display a more diverse voting pattern. Their size, and thus their complexity, has led to efforts to form cliques (cf., e.g., Reese, 1989: 357). However, a group of politically neutral "grays" has emerged only rarely.

In large as well as small bodies, subcommittees are established to carry out the various functions, although these may usually only make recommendations. The role of the subcommittees has become increasingly important over time, and while this has resulted in a certain dilution of the decision-making power of the main bodies, it has also promoted their decision-making capabilities. Both the main bodies and the subcommittees have grown considerably dependent on the aid of assistants at the managing office, who are responsible to the director. An office's personnel and organizational makeup are thus important for the way in which functions are carried out. To this extent, substantial variations among the different state media authorities are evident (for instance, on the use of financial resources by the state media authorities, cf. Wöste, 1990: 81ff.).

A particularly thorny problem in the daily execution of tasks results from the differing levels of professional expertise possessed by Board members. Because of the recruitment procedure—delegation by groups—it is not certain that all members will have the requisite experience in the media sector. For the delegating groups, media competence is only one of several selection criteria, but the extent of professional expertise may be decisive for the power to influence. Insofar as media experience was present among the initially appointed members, it usually stemmed from earlier experience in public broadcasting, in associations, or directly in media companies.

For many board members, difficulties are especially caused by the frequent argumentation involving legal terminology, a result of the statutory pervasion of the broadcasting order. This gives lawyers (including the Director and his staff, who usually have some legal training) an elevated position of power and leaves those members without a legal background with a feeling of relative helplessness. As a result, nonlawyers have also sought to cloak their arguments in legal terms—without, however, always being able to meet the professional standard of their legally trained colleagues. When legal experts state that a given law does not permit a certain solution, nonlawyers can usually only look on helplessly. Even when there is reason to suspect that legal arguments are only being used to mask other, especially politically motivated decisions, an influential advantage still exists in favor of the legal "pros."

As a result of the undefined, value-neutral terms set forth in the state media laws, the organizational options, and the many underlying media-policy controversies, there has been a considerable amount of legal uncertainty. This is confounded by operational uncertainty. For instance, the media laws surrounding the licensing procedure require that broadcasters also be assessed for their economic, journalistic, organizational, and other capabilities. However, there is a dearth of sufficiently reliable numbers on, for example, the required costs of local radio or anticipated returns. Particularly in the initial phase of licensing, the applicants had an advantage, since their figures—for example, profit or loss prognoses—were difficult to review (cf., e.g., Hoffmann-Riem & Ziethen, 1989: 249f.). Over time, however, the state media authorities have had a number of studies prepared for ascertaining cost and revenue structures (Teichert & Steinborn, 1990; Rinke Treuhand GmbH, 1989; Seufert, 1988; see also Schröder, 1990), thereby improving somewhat the decision-making situation.

There are factual uncertainties in other areas as well, as in the evaluation of the organizational structures of license applicants or broadcasters, particularly economic integration at the horizontal and vertical (supplier) levels (on horizontal integration, cf. Röper, 1990: 755ff.; 1992: 2ff.; 1993: 56ff.; 1994: 125ff.; on verticle integration, cf. Haeckel, 1988: 191ff.; Schurig, 1988: 50ff.; Wöste, 1991: 561ff.).[11] Even the full-time employees of the managing office have difficulty sorting out such arrangements. The ever more ingenious methods of interlocking, using figureheads, and so forth, have made it nearly impossible to gain an up-to-date overview. Volunteer board members are especially overtaxed. Although they sometimes make an effort to fill in the gaps in information with their own research, this is ultimately a venture with little prospect of success.

Any comment on the status of the state media authorities would be incomplete without reference to the special structure in Bavaria. In the German constitutional and social order, Bavaria nearly always claims a special position. In the broadcasting sector, this special status is the result of a unique political conflict: In the early 1970s, the political majority—controlled in Bavaria by the Christian-Socialist Union (CSU)—since the beginnings of the FRG sought to extend its influence over the Bavarian Public Broadcasting Authority. This was opposed by a popular referendum, whose initiators at the same time desired to prevent governmental influence from extending to private broadcasting, should that be established at some point in the future. In order to thwart the referendum, a provision was adopted in the Bavarian constitution, making it mandatory for broadcasting to be operated "under public responsibility and as a public corporation" (for details on the dispute, see Hesse, 1990: 21f.; Schmitt-Glaeser, 1979: 59ff.). This seemed to preclude private broadcasting. Nevertheless, it was subsequently permitted, notably in a construction distantly related to that of the former British IBA. The object of this rule was to implement organizationally the legal requirement of public responsibility while facilitating private investment by individual broadcasters.

For the "operation of public broadcasting," the Bavarian State Authority for New Media (*Bayerische Landeszentrale für Neue Medien,* or BLM) was set up. Legally speaking, the BLM is considered a broadcaster, but it actually serves as an umbrella under which private broadcasters are allowed to operate according to entrepreneurial principles. The private broadcasters licensed by it are subject to its programming authority. The Bavarian constitutional court has held that the BLM's control possibilities are sufficient to ensure programming responsibility (BayVerfGE 39, 96ff.). However, its statutory powers are considerably weaker than those held by the British IBA. Although the BLM may demand that programs be submitted prior to broadcast (see Lerche, 1984: 255f.), this procedure is not used to conduct comprehensive programming control (see BLM, 1989a: 22f.; 1990: 23; 1991: 24; BayVerfGH, ZUM 1992: 378, 380ff.). Like other state media authorities, the BLM is restricted to reviewing conformity to statutory limits, and its measures are subject to judicial review. There are many indications that, apart from the different legal construction, its status does not essentially differ from that of other state media authorities (see also Hesse, 1990: 205). At least in supervisory practice, there are no substantial differences to be seen, particularly none that would indicate that private broadcasting in Bavaria approximates public broadcasting.

2.2.2. Functions of the State Media Authorities

The operation of private broadcasting may commence only after a license has been issued by the responsible state media authority. For the mere retransmission of programs, notification is sometimes sufficient, as long as the broadcaster is subject to control at the place of origin. Some states, however, require a special permit (see Hesse, 1990: 210ff.). Retransmitted programs are also subject to constant supervision, which is, however, not as stringent as supervision of licensed broadcasting (see Hesse, 1990: 213ff.).

Applicants for a broadcaster license must fulfill certain minimum prerequisites, such as those with respect to their economic and journalistic capabilities (for details on this and the following, see Hesse, 1990: 183ff.). Demonstration is required that the statutory requirements will be fulfilled. Licensing is not limited to German legal subjects; however, to ensure liberty from the state, its representatives, and to a large degree also those of the communities, are excluded. The license is granted in a comparative procedure, which enables the state media authorities to select the "best" applicant and formulate stipulations. This is the most important point of departure for control of the broadcasting order, particularly for preventive control of diversity (Wagner, 1990a: 177). Normally, the state media authority calls upon competing applicants first to reach agreement with one another—for example, to agree on the formation of a joint broadcaster association or on a split frequency, that is, use of the same frequency by several broadcasters at different, prearranged times of day. If they are unable to do so, or to agree only in part, a preferential decision has to be made (see Wagner, 1990a: 180ff.), which has very often favored the financially stronger applicant (cf., e.g., Wagner, 1990a: 182; Hoffmann-Riem & Ziethen, 1989: 247ff.). Nevertheless, the state media authority is required to take into account a variety of criteria that vary from state to state but mainly aim at ensuring programming diversity (see Section 3.1). In this regard, the proposed palette of programs or the composition of the company (e.g., the degree of variety among the participating interests, with the inclusion of special cultural associations) may be just as important for the decision as precautions to ensure pluralistically composed programming councils or guarantees that the autonomy of journalists will be protected against pressure from the side of capital (see Wagner, 1990a: 187ff.).

In North Rhine–Westphalia and Hamburg, efforts were made to find ways to restrain as much as possible the influence of owners on programming conduct in the area of local radio. These efforts were motivated by criticism of the effects that commercial forms of financing were having on programming. A particularly prominent role has come to be played by the "twin-pillar model" (*Zwei-Säulen-Modell*) set forth in North Rhine–Westphalia's State Broadcasting Act (see Stock, 1987; Hesse, 1990: 206ff.; Prodoehl, 1987: 235ff.; Schröder & Sill, 1993; see also BVerfGE 83, 325ff.). On the one hand, the model envisions an operating company responsible for providing technical facilities for production and broadcast and acquiring advertising contracts, thereby ensuring financing. These companies are generally owned by newspaper publishers and, for a minority, by the communities. On the other hand, broadcasting itself is undertaken by a pluralistically composed association independent of the operating company. This association makes the basic decisions as to programming and hires the top-level journalistic personnel, particularly the chief editor.

With this structure, an attempt is made to square the circle: Although broadcasting should look for financing on the market, the operating company's responsibility for obtaining funding is not to have any effect on programming decisions. It goes without saying that conflicts are inevitable.[12] The North Rhine–Westphalian State Media Authority (LfR) has special powers of intervention at its disposal in order to ensure the survival of broadcasters; above all, it must pay attention to the economic capabilities of broadcasters in staking out transmission territories. This type of broadcasting was launched in 1990 (for initial experiences, see Hirsch, 1991; Schröder & Sill, 1993: 173ff.). As early as 1987, broadcasters in Hamburg were licensed according to a similar model; however, it has since been abandoned as the result of negative experiences (cf. HAM, 1992: 5f., 9). For this reason the new Hamburg media law of 1994 favors local radio that is completely noncommercial.

In addition to granting licenses, the state media authorities are responsible for the day-to-day supervision of the conduct of broadcasters (see Wagner, 1990a: 219ff.; Holgersson, 1993: 157ff.). With the exception of Bavaria's BLM, they are not empowered to request advance review of programs, but they may lodge objections to them. Supervision is limited to monitoring conformity with statutory requirements. The supervisory and sanctioning tools range from a mere notice of statutory violation to an objection and beyond to an order to refrain from certain violations in the future (see Wagner, 1990a: 230ff.; Holgersson, 1993: 160ff.). Statutory violations can lead to revocation of the license—although usually only after prior, repeated objection—and, in some cases, temporary suspension of the license or prohibition of specific program offerings is also provided for. Subsequent directives designed as an additional specification of the license are also permissible.

3. FIELDS OF REGULATION AND SUPERVISION

3.1. Guaranteeing Diversity

Emanating from the requirements stipulated by the Bundesverfassungsgericht, the main objective in the statutory structuring of the broadcasting order is to guarantee diversity of information and communication, which is an indispensable element for the citizen's free formation of opinion. The distinction between, on the one hand, imperative control of conduct and, on the other, the provision of a structure ensuring (as far as possible) diversity, that is, structural control, is clearly reflected in the statutory precautions (see Chapter 8).

3.1.1. Structurally Based Control of Diversity

Structurally based precautions to ensure diversity include the Bundesverfassungsgericht's landmark decisions on externally and internally pluralistic broadcasting models and the implementation of these, particularly in the form of organizational and financial requirements. Since the rules for private broadcasting have borrowed heavily from the public broadcasting model, the corresponding structural precautions represent a characteristic feature typical of the German broadcasting order.

The statutory requirements are rather weak as far as anticoncentration rules are concerned (for details see Section 3.2). Those laws that set up private broadcasting solely according to the externally pluralistic, market model are somewhat reserved. They operate—as does the Rundfunkstaatsvertrag (Interstate Treaty on Broadcasting) for nationwide broadcasting—under the refutable presumption that externally pluralistic diversity ceases when there are several (usually, three) competing broadcasters. But in the event that adequate program diversity cannot be achieved in this manner, the law calls for additional precautions, particularly internally pluralistic guarantees. Broadcasters may guarantee diversity by way of a pluralistic composition of companies in which different persons or enterprises hold shares (*Anbietergemeinschaften*) or the establishment of programming councils, or they may be obligated to offer internally balanced "integrated programming."[13]

The Bundesverfassungsgericht has generally deemed it a structural guarantee of diversity when the broadcaster is an *Anbietergemeinschaft* (BVerfGE 74, 330; Ziethen, 1989: 82ff.). A broadcaster composed of different enterprises may in fact offer certain guarantees against one-sidedness. On the other hand, if diversity of communication is to be positively ensured, there must be adequate possibilities for the owner to intervene in programming. However, this may collide with the requirements of the specific programming work of journalists, especially their journalistic autonomy. In addition, it can be seen that the *Anbietergemeinschaften* formed in Germany are in all respects characterized by similarity among the participating interests, and the supervisory authorities by no means place strict requirements on the breadth of plurality of a given broadcaster. For instance, in the start-up phase, a number of newspaper publishers set up *Anbietergemeinschaften,* and while the state media authorities took the often divergent orientations of the participating newspapers (or the competitive conduct of the newspaper market) as a sign of plurality, they overlooked that in many respects newspaper publishers have identical interests, which may collide with the interests of other social groups. (For the problematic of assuring plurality through *Anbietergemeinschaften* composed of newspaper publishers, cf., e.g., Hoffmann-Riem & Ziethen, 1989: 266f.) In addition, cooperation among publishers, or media giants, in broadcasting *Anbietergemeinschaften* furthers cartel-like coordination, which can have effects in their other fields of activity (e.g., in the area of daily newspapers). Dysfunctional side effects on competitive behavior in other media sectors are therefore likely. Cooperation in *Anbietergemeinschaften* promotes coordination in the media field and may thus per saldo lead to a loss of plurality in the media order as a whole.

On occasion, programming councils have been created within private broadcasters as an intraorganizational guarantee of programming diversity. However, in all cases they have only weak advisory capacity. In some instances, they have been set up voluntarily by the broadcasters in order to gain an advantage over competitors in the comparative licensing procedure; in others, they have been expressly demanded by the state media authorities. In conformity with the principle of the greatest possible self-regulation, the composition of the programming councils and the recruitment of their personnel have normally been left to the broadcaster itself.[14] Since broadcasters are interested in a smooth relationship between programming council and broadcasting operations, the programming councils hardly reflect the entire spectrum of socially relevant forces. In view of their restricted powers, they have only limited effects on broadcaster behavior. They often serve only to give an illusion of pluralism.

A further safeguard to ensure diversity is the inclusion of representatives of the cultural field in programming (for the statutory rules, see Ziethen, 1989: 85ff.). But in practice, it has become evident that commercially oriented broadcasters, as anticipated, are reluctant to cooperate meaningfully with forces that have special cultural demands but are unwilling to take on economic responsibility. However, as long as cooperation with such forces improves the chance for receiving a license, applicants sometimes make great efforts to obtain the cooperation of such representatives and highlight this in their applications. But once the license is granted, active engagement in this cooperative direction is no longer evident, and the state media authorities are powerless to change this.

Some state laws make use of a concept that was discussed during the 1970s in Germany as well as in other countries: precautions for so-called internal media freedom for the press and in part also for public broadcasting. The goal here is to provide editorial autonomy via institutional protection from other functional units such as the advertising department or the publisher. Underlying such a concept is the hope that with their autonomy sufficiently ensured, journalists will have, as well as use, increased possibilities to heed their professional obligations, thereby contributing to plurality or otherwise reflecting and representing the interests of the general public on behalf of their professional ethics. (For details on this concept, see Hoffmann-Riem, 1972; 1979; Kübler, 1973. For criticism, see, e.g., Weber, 1973; Lerche, 1974.) Measures to ensure internal freedom of the press and subsequently also freedom of broadcasting were introduced sporadically in the FRG, but they met with considerable resistance, particularly from newspaper publishers. Accordingly, publishers protested when some state legislatures sought to implement this concept for private broadcasting. The state laws in Hamburg, North Rhine–Westphalia, and Bremen do, however, cover related precautions within the framework of comparative licensing procedures. (The Bundesverfassungsgericht has condoned this as lawful; see BVerfGE 83, 318ff.) Some broadcasters have thus created provisions (albeit weak ones) designed to provide their journalists with a say in certain decisions.

Still another guarantee of diversity might consist of licensing diversely aligned broadcasters, thereby strengthening programming competition among them. In practice, however, little use has been made of this strategy. Particularly in television, but in radio as well, the broadcasting scene is characterized by broadcasters who tend to take a relatively nonaligned, "middle road," who back off from assuming any sort of political stance, and who more or less restrict political news coverage. (For observations in the television sector, see, e.g., Gellner, 1990: 273; Faul, 1989: 33f., 45f. See generally, Krüger, 1991: 308). In contrast to Great Britain, German media laws do not impose the obligation of neutrality on broadcasters. In the FRG, radio broadcasters with an "alternative" profile were licensed only in rare cases, as with Radio Z (in Nuremberg) and Radio Dreyeckland (in Freiburg). Although they are widespread in other countries, Germany decided against permitting the emergence of community radio stations (cf. Kleinsteuber, 1988a; 1988b), which result from local initiatives and seek to break out of the commercial setting. The mere existence of a licensing procedure, however, thwarted such a model from the outset. In television, an attempt was made to ensure a degree of substantive plurality by requiring that a relatively leftist–liberal political commentary program, Spiegel TV, be made a mandatory programming component (of the broadcaster RTLus). Spiegel TV then ran up against stiff opposition from conservative circles. Nevertheless, it has proved to be so popular that the controversy waned, and its program slot today is assured even without corresponding requirements.

3.1.2. Conduct-Based Control of Diversity

Insofar as the structures of the broadcasting order are unable to ensure sufficient diversity, it may become necessary to impose conduct-based control, which is aimed at the protection of vulnerable values (cf. Blumler, 1992). This control may in particular be used to enforce programming obligations, which are sometimes designated in laws with the terms "programming principles," "commitments," or "guidelines." Statutory programming commitments symbolize that the law has assumed responsibility for regulation, but they alone cannot ensure that the objectives, such as diversity, are ever achieved. When broadcasters object to the required programming conduct for journalistic or economic reasons, their opposition must be expected. In view of the fundamental constitutional priority that accords journalistic autonomy to broadcasters from external intervention, that body in charge of the supervision of programming conduct is, for legal reasons, placed in a difficult situation.

In light of the constitutional protection afforded freedom of broadcasting, as well as the specific legal prohibition on interference, broadcasters have good prospects for success should they mount a legal battle on supervisory measures in program matters in court, especially on account of the vagueness of statutory standards, for example. The state media legislation in Germany has been unable to implement and guarantee certain target values, such as balance and diversity. Legislatures have generally fallen back on vague formulas that are predominantly expressed as appeals for desired programming, and since these are often in need of interpretation, they do not amount to enforceable programming norms. If the supervisory authority nevertheless attempts to enforce these, it runs the risk of sitting as a judge over opinion.

The danger is particularly great that borderline or unconventional viewpoints are more likely to be suppressed than mainstream ones. If broadcasters expect intensive supervision, the threat of self-conformity looms with its resultant constriction in plurality brought about by intimidation. However, this problem does not arise when broadcasters seek to sidestep the plurality standard for other motives—for example, for the objective of reaching broader audiences or for the financial reasons of avoiding large expenditures—by cutting informational programs or refraining from politically controversial broadcasts. In this case, the reference to intimidation through programming supervision may simply serve as yet an additional justification for programming restraint in fact motivated by other reasons. It is according to this pattern that television and most radio broadcasters seem to work.

The state media authorities have recognized such dilemmas and proceeded cautiously in controlling diversity. They quickly grasped that promises of diversity by broadcasters as early as in the licensing phase offer little assurance. This became particularly evident to the extent that the laws called for such information on the programming schedule, indicating that diverse programming was to be expected. However, programming schedules hardly permit substantive directions to be determined; rather, at best, they ensure that certain topics or program categories will be given coverage. But despite the aspirations of authorities, programming schedules that were submitted were generally deficient in this regard (cf., e.g., the remarks by Hoffmann-Riem & Ziethen, 1989: 257ff.; see also as an illustration the application submitted by Radio Schleswig-Holstein, reproduced in *Media Perspektiven,* 1986(9): 587ff.). Moreover, following commencement of broadcasting operations, the proposed programming schedules were largely equivalent to scratch paper. Especially when a presumption is made in favor of the broadcaster's good intentions—as the state media authorities certainly have done—it must be expected that changes will become necessary as operations progress.

The state media authorities have normally requested broadcasters to submit supplementary information, even though they regularly receive overly general, noncommittal responses; detailed information is at best provided for specific projects. In addition, doubts often crop up about whether broadcasters will keep promises they have made. Even in cases where these were set out in the approval notifications for the license, as in Hamburg and Berlin, both sides are well aware that this initial agreement to comply is no guarantee of successful legal implementation. All the same, some state media authorities have sought, within the framework of informal measures, to "collect" on the promises. There is no indication that sanctions have been employed which go beyond this, however. Nevertheless, there have been attempts to ensure increased attention to the elements of "full programming" (i.e., serving to inform, educate, advise, and entertain) called for in the laws.

According to the concept of plurality, diversity is assessed with reference to the overall palette of programming offered to viewers, which means that media authorities can on no occasion undertake supervisory measures against individual broadcasters for lack of diversity.[15] At the same time, even in such models the laws provide for programming obligations, such as the duty to give reasonable, fair consideration to the positions of persons or groups significantly affected and the duty to report in an objective, comprehensive manner that has been introduced for informational programs. There are also obligations regarding conscientious research and prior review of allegations of fact. These programming commitments have led to state media authority intervention in only exceptional cases.[16]

3.2. Combating Concentration

In the interest of diversity of communication, the Bundesverfassungsgericht has called for measures to combat a possible emergence of monopoly power over opinion (see BVerfGE 73, 158ff., 172ff.). In so doing, it allowed the legislature to employ the general law against monopolies. At the same time, it was made clear that, in addition, special precautions were needed in broadcasting law to prevent the emergence of multimedia power over opinion when this threatens to jeopardize diversity of opinion (BVerfGE, 73: 176). With these stipulations in mind, most laws of broadcasting first borrow from general law banning monopolies by requiring that a broadcaster furnish proof of compliance with the basic provisions relating to supervision of mergers. To these are added independent, broadcasting-oriented provisions, for example, against multiple ownership. These offer generous terms with respect to the number of possible licenses. For instance, the ITB permits nationwide broadcasters to operate one full-programming channel and one special interest programming channel focusing on information (see note 1) each in radio and television, that is, up to four channels (for details on the rules, see Wagner, 1990b: 173ff.; Holznagel, 1991: 269ff.; Engel, 1993b: 557ff.).

The rules are applicable not only to multiple ownership of broadcasters but also to some other forms of influence on programming, for example, through a dominant position in the supply of programs or by way of contract and articles of incorporation. No one single owner is permitted to hold 50% or more of the shares or voting capital of a national television broadcaster.

Restrictions regarding cross-ownership between broadcasting and press companies are rare and only come into play in some instances when so-called double monopolies threaten to emerge at the local or regional level (see Wagner, 1990b: 177f.; Porter & Hasselbach, 1991: 121ff.). On occasion, these are tolerated as long as broadcasters observe certain precautions to promote some diversity. Taken as a whole, the rules are weak. For instance, market-dominant press companies are not prevented from operating in other transmission territories or from acquiring holdings in broadcasters transmitting programs state- or nationwide.

German media law offers a certain degree of protection against monopolistic concentration to the extent that the license is essentially inalienable (see Wagner, 1990b: 179f.). Difficulties are, however, presented by the definition of an improper transfer, particularly in the case of mere changes in stockholder relations or shareholding figures. The state media authorities are required to be notified of such changes. But in practice, notification duties are complied with only to a limited extent. The legal power of the state media authority to clarify complex ownership situations is restricted. Nevertheless, the 1991 Interstate Treaty on Broadcasting requires that changes in the ownership of national broadcasters be reported, any violation of which can be punished with a fine. Changes may be undertaken only once the responsible state media institution has given its consent.

Just a few years after the introduction of private broadcasting, national television broadcasters became closely tied to one another through ownership structures (see the annual analyses by Röper in *Media Perspektiven,* e.g., Röper, 1994: 125ff.; see Heinrich, 1994: 297 ff; Europäisches Medieninstitut, 1994). The initially dominant influence by press companies has expanded through the activities of such multimedia giants as Bertelsmann, Berlusconi, and Luxembourg's CLT—and particularly as a result of the enormous influence of the film dealer Leo Kirch. The degree of concentration has become higher than in any other country under study here.

Each state's legislature—and, as a result, the respective supervisory body—is caught in a dilemma in combating concentration. The resort to experienced, financially endowed media companies may in fact serve as a sort of reassurance of the successful operation of private broadcasting. It is no coincidence that in other countries as well licensing authorities have encouraged press companies to enter the private broadcasting business. In the FRG, this was not particularly necessary, since it was mainly the large press companies that pushed for the introduction of, and their own participation in, private broadcasting (see Bausch, 1980; Montag, 1978). The statutory clauses requiring that broadcasters possess sufficient "capabilities" have been interpreted by the German state media authorities to make experience and financial reserves in the publishing sector an indication of capability. Mergers in which there was evidence of participation by publishers were often given preference in the allocation of frequencies; in this manner, such broadcasters received attractive frequencies in the radio sector in many states. They were similarly successful in the allocation of television frequencies (especially SAT 1). This resulted in intensive, multimedia integration at the very outset of private broadcasting in Germany, which has increased ever since (see, e.g., Röper, 1989: 533ff.; 1990: 755ff.; 1992: 2ff.). Although several states (such as Bavaria and Baden–Württemberg) had originally gone to great lengths to take countermeasures, at least in the radio sector, they have since widely been forced to capitulate to the general trend (for Bavaria, see, e.g., Gustedt, 1992: 5ff.).[17]

When the state media authorities attempt to monitor compliance with the restrictions, they encounter many difficulties, associated with the reliable determination of an infraction. For instance, there are informational problems due to the lack of expert knowledge on the part of members of the control bodies, particularly when it comes to unscrambling complex ownership relations. Moreover, these relations often change so quickly that it is nearly impossible to come up with current data. A survey of the ownership relations in the Kirch Group, for example, is apparently nowhere to be found. The rapid fluctuations in ownership relations have caused some members of state media authorities to feel like the proverbial hare in the race, ever trying to play catch-up with the tortoise. In 1992–1993, the state media institutions stepped up their efforts to lessen the degree of concentration, but visible results have yet to be seen. At the end of 1994 the Conference of Directors conceived the idea of a joint committee "to ensure diversity" (Gemeinsame Stelle Vielfaltssicherung); its main task is to guarantee, in keeping with media law, that there is control of concentration nationwide.

Nevertheless, attempts have been made to place limits on further expansion of the influence exercised by Kirch and UFA (a Bertelsmann subsidiary), and these have sometimes been successful. However, this effort has also led to disputes—occasionally before the courts—between the state media authorities.[18] It became apparent that these bodies came to see themselves as advocates of "their" companies, that is those headquartered within their jurisdictions (see Section 3.7).[19] In addition, the state media authorities had considerable difficulty obtaining the necessary information. For this reason, they have sought to obtain from the companies detailed information regarding cooperative relationships and contractual agreements. But these companies were willing to cooperate only in part and have defended their actions by claiming that these records are confidential business data. On many occasions, they delayed complying with requests or responded in some other obstructive fashion. As a result, the state media authorities have publicly been labeled "paper tigers," and the steps they have taken to combat concentration have been criticized as being primarily of a symbolic nature. This can hardly be denied. Yet it must be acknowledged that the state media authorities have at least been able to enforce some restrictions and to some degree limit the further expansion of the power wielded by Kirch and Bertelsmann.

As can be observed in other broadcasting orders, the attempts to check excessive consolidation in Germany have largely come too late. Although some corrections in the laws have been achieved, as with the cap on shareholding and introduction of contractual powers to influence company policy, these have done nothing to alter the high concentration of the media industry. Ironically, the Bundesverfassungsgericht itself has contributed to this situation. With the objective of ensuring plurality in ownership structures, it has, as mentioned above, encouraged the creation of *Anbietergemeinschaften* intended to limit the shares that a given company can hold in any one broadcaster. This has prompted media companies to acquire holdings in as many companies as possible. As a result of the cooperation seen in (often several) *Anbietergemeinschaften,* it is unlikely that shareholders will be interested in stiff competition among those broadcasters in which they hold shares. While these broadcasters do compete for ratings, their actions at the level of program acquisition and distribution are very often closely aligned.

Thorny problems of supervision have also cropped up on the level of program supply and networking.[20] Program suppliers are subject neither to licensing obligations nor to direct broadcasting supervision. These obligations are, therefore, encountered only indirectly when objections are lodged against a broadcaster because of program supply, particularly when supply relationships result in exercising substantial influence on the broadcaster, or when the local orientation of programming as required in the license is threatened by supraregional program supply. However, in 1991 under the Interstate Treaty on Broadcasting, supply of programming for national broadcasting became equivalent to ownership of national broadcasters. This has resulted in greater powers of investigation and intervention for the state media institutions. Because of the frequent shortages in the availability of programs and the difficulty to fund in-house productions, broadcasters are often forced for economic and journalistic reasons to intensify cooperation with other broadcasters and with special supply companies.

It becomes all the more difficult to take action against media concentration in light of the fact that the economic rationale behind integration and cooperation is often quite convincing. In a media system that allows broadcasters to orient their operations according to economic criteria, supervisory measures can be easily carried out only as long as they do not endanger the broadcasters' economic viability. Insofar as the ability of local broadcasters to survive was in jeopardy (as with radio in a number of states), the state media authorities were forced to employ rather broad standards in interpreting their statutory discretion. However, this interpretation also served to benefit some broadcasters whose economic capabilities would have basically remained intact even in the face of anticoncentration measures. In view of the difficulties associated with untangling complex revenue, and particularly profit-making, situations, problems would be raised if revenue factors were to be made the decisive criterion for licensing decisions.

Even though prevailing opinion has deemed the current tools for combatting concentration unsatisfactory, it should nevertheless be kept in mind that if there were no rules concerning monopoly and broadcasting law, the degree of concentration would in all likelihood be much greater. Yet, it should be noted that the set of available instruments has only a limited application: monopoly law covers only a small number of relevant situations (cf. Brinkmann, 1983: 677, 680; Lerche, 1984: 43f.; Hendriks, 1984: 926f.; Hoffmann-Riem, 1982: 269; Wagner, 1990b: 173; and see Chapter 12). Safeguards specifically aimed at broadcasting go further, but they have too many loopholes and are too vague in wording and application to be effective (cf. Holznagel, 1991: 268ff; a more optimistic assessment is made by Wagner, 1990b: 182). But above all, the law does not appear sufficiently able to keep abreast of the constantly changing economic relationships.

3.3. Advertising Restrictions

The supervisory problems associated with the statutory advertising restrictions appear to be significantly smaller. The Interstate Treaty on Broadcasting and the state laws, much like the EC's television directive, contain a proliferation of restrictions, such as those on the permissible scope of advertising, the separation of advertising from programming, the number of commercial interruptions, and the labeling of advertising spots as such. On repeated occasions the state media authorities have had reason to object to violations of these rules. Since the licensing of private broadcasting, newer and newer forms of advertising have been developed. The "Joint Guidelines for Advertising, on Enforcing the Separation of Advertising and Programming, and for Sponsoring in Television and Radio," drafted by the Conference of Directors of the State Media Authorities, seek to specify these restrictions.[21]

Broadcasters have often side-stepped the requirements regarding advertising in blocks and the prohibitions on interruption by breaking up programs into several segments, by forms of concealed advertising (game shows), and so on (for details, see LfR, 1988: 31 ff; 1990: 24ff; for the way particular forms of advertising are treated, see also Hochstein, 1991: 700ff.).

In view of broadcasting's enormous financial requirements and (despite remarkable increases) the still limited advertising "pie," it is obvious that broadcasters go to great lengths to develop and make full use of advertising potential. In this regard, the extensive restrictions on advertising time (in general 20% of the daily transmission time; see para. 27 Rundfunkstaatsvertrag 1991) have not proved to be much of a hindrance. Greater problems have been posed by restrictions on the type of advertising. Broadcasters have shown considerable imagination in this regard in exploring the utmost limits of tolerance. This is a relatively safe venture, since the state media authorities tend first to reprimand advertising violations informally, thereby providing an opportunity for broadcasters to amend the situation "voluntarily." Although this practice is sensible, it works as a sort of call to test without risk how much the state media authorities are willing to accept in certain advertising. It seems sometimes broadcasters test advertising forms in which they ultimately have no interest. In this manner, the state media authorities are able to lodge an objection—and thus score a success—in their advertising control efforts that, ultimately, does not bother the advertising industry and broadcasters. In most cases, however, the state media authority works to come to an arrangement with the broadcaster and thus to rule out the risk of subsequent court conflicts. It appears that the type of interaction on the limits of permissible advertising has led to a gradual displacement of these limits, that is, to a weakening of the requirements (for this problematic issue, cf. Gebel, 1989: 354ff.).

Since the laws work with soft legal terms and authorizations to make exceptions, broadcasters can, within the framework of informal communication networks, push for broader limits. Since advertising revenues form the main source of income and some broadcasters are experiencing problems of financial survival, this tolerance comports with the concern shared by the state media authorities for giving private broadcasting a chance to get started and prosper (Holgersson, 1993: 162f.). Useful from an argumentative standpoint here is the reference to public broadcasting, whose pioneering role in advertising limits[22] has consistently been used by private broadcasters as justification. Since the state media authorities are not responsible for public broadcasting, they can treat its advertising conduct as a field beyond their control. That private broadcasting should not be treated more stringently than public broadcasting is a premise frequently put forth in Germany—in contrast to, for example, appraisals in Great Britain, where it was at least formerly assumed that a commercial broadcasting system engendered particular dangers warranting stricter precautions than those in place for the BBC.

The state media authorities were also required to make exceptions in a number of cases. For instance, the authority in Rhineland–Palatinate permitted for sports broadcasts exceptions from the prohibition of commercial interruption, even when this was not authorized by statute. This practice was subsequently legalized through the granting of legal authority to make exceptions. The power to make exceptions has since been replaced by specialized rules in the 1991 Interstate Treaty on Broadcasting.

3.4. Teleshopping

In the area of advertising, the state media authorities are faced with the particularly thorny problem that private broadcasting tends to develop according to its own, inherent laws, and they find themselves caught in a conflict between the expectations of a critical public and the wishes of the broadcasting industry. This conflict took on critical—and symbolic—dimensions with the emergence of plans for tele- and radioshopping. Teleshopping differs from advertising to the extent that the aim is not simply the conveyance of an advertising message but also the direct stimulation of the audience's willingness to purchase or order. In the process, ingenious mixtures of entertainment programming and merchandising are employed. German media laws initially provided no rules for this. Thus, private broadcasters saw an opportunity to test the limits of acceptability. The first broadcaster to experiment with this, Eureka, had a license in Schleswig–Holstein. Despite many protests regarding the permissibility of teleshopping, the state media authority took no action, thereby signaling that the practice was legally irreproachable. The many undefined terms in the broadcasting laws have also made it possible for other media authorities to treat teleshopping as permissible without there being any special authorization for this in broadcasting statute. Doubts about the propriety of classification as advertising were not considered (see Hoffmann-Riem, 1988; Woldt, 1988; but see Bultmann & Rahn, 1988; Ory, 1988). Accordingly, the licensing of teleshopping was reduced to compliance with advertising restrictions in broadcasting law. Only one authority (in Hamburg) voiced opposition initially, although it did not wish to prevent its director from being able to acquiesce to nationwide consensus, that is, with respect to the federal association. In any event, SAT 1 commenced with its teleshopping broadcasts even though the executive board of Hamburg's authority had previously expressed reservations to this (cf. the parallels to the introduction of teleshopping in France, Chapter 4, Section 3.2). The Interstate Treaty on Broadcasting has now determined that teleshopping is to be treated like advertising.

The inclusion of teleshopping under the category of advertising is of course problematic, not least because the restrictions on advertising time are unsuitable for teleshopping. If teleshopping were to prove a success, of course the restrictions would not be upheld in the long run. Indeed, pressure has begun to increase or rescind the advertising time restrictions or—as has now been provided for in the EC's television directive and in the Interstate Treaty on Broadcasting—to enact more generous time restrictions for teleshopping.

The classification of teleshopping as advertising highlights the primarily economic orientation in the administration of supervisory powers. One problem associated with teleshopping has to do with the fact that because of the ingenious mingling of entertainment programming and merchandising, viewers have hardly any chance of clearly recognizing the advertising segments as such. A similar issue is raised by the ever more popular game shows. If the related problems of lack of transparency and even manipulation of recipients are overlooked by defining such broadcasting forms as advertising—as has occurred with the supervisory bodies—then one loses the very protection that the laws seek to afford by way of the requirements of labeling and block advertising and the prohibitions against program interruptions. However, even the new requirement to identify game shows as advertising for the duration of the broadcast has done little to change this (for ways to avoid the labeling obligation, see Schneider, 1992: 23). In the United States, where there are no comparable rules for advertising, this has not been such a great problem. By contrast, the practice by the German state media authorities has led to a situation where concern for the broadcasting industry is given preference over the protection of specific objectives of broadcasting law. For example, the rationale behind the requirement of clearly separated advertising broadcasts is unable to be upheld when broadcasting schedule reports simply classify the mixed programming–advertising product as pure advertising.

Moreover, it has been overlooked that the statutory restrictions on advertising are in substance designed for traditional forms of advertising and not for game shows or teleshopping as merchandising forms. It is not inconceivable that these restrictions will have to be relaxed still further if teleshopping is prevented from expanding into new areas by advertising rules. This may lead to a further displacement of the limits on permissible advertising.

3.5. Protection of Minors

Freedom of communication encounters an express constitutional boundary in the form of protection of minors. As a result, provisions for the protection of minors specifically designed for broadcasting concerns apply uniformly in all states under the Interstate Treaty on Broadcasting. Compliance with these is monitored by the state media authorities, which, however, sometimes resort to preliminary rulings, for example, the indexing of movies by way of what is known as the voluntary self-control of the film industry (Freiwillige Selbstkontrolle der Filmwirtschaft, or FSK). In addition, there are prohibitions (though not relating solely to the protection of juveniles) regarding incitement to racial hatred, glorification of violence, or the dissemination of pornography. For broadcasts capable of threatening the physical, intellectual, or moral welfare of juveniles, a graduated set of broadcasting time restrictions are in force. The state media authorities are permitted to make exceptions to the time restraints. Such applications have been submitted in great number and are nearly always approved (cf. the statistics in Grams & Hege, 1992: A 108). Difficulties are raised mainly by films that were formerly classified as inappropriate for juveniles but, as a result of changed outlooks, are no longer considered a threat.

The Conference of Directors of the State Media Authorities has drafted "Guidelines for Ensuring the Protection of Juveniles," which have been approved by the state media authorities (see DLM, 1990: 341ff.). These aim only at so-called negative protection; they do not set up obligations to broadcast programs specifically for juveniles or even programs with particular pedagogical value. But even in the absence of such stipulated obligations, several broadcasters have televised a great number of programs especially for juveniles (Schmidt, 1990). Some television broadcasters in fact have made great efforts to set the groundwork as early as possible for long-term "channel allegiance." There are, however, no indications that these broadcasts have also fulfilled qualitative educational standards.

Of late, the number of special programs for juveniles has continued to decline. After the advertising guidelines were sharpened in 1993, particularly with regard to the limitations on advertising in children's programming (for details on the rules, see Hoffmann-Riem & Engels, 1995), some broadcasters announced that they would reduce the number of children's programs even further. However, efforts are also being made to dodge the new rules by rechristening as family broadcasts those programs that were formerly labeled juvenile programming (see *Frankfurter Rundschau,* 12 February 1993: 15). This tactic is just the reverse of that used for many years in the United States: There, some programs originally intended for the family generally or for adults were termed juvenile programming in order to be able to fulfill juvenile program quotas more easily.

After having played such a large role prior to the introduction of private broadcasting, the issue of protection of juveniles faded into the background in the public discussion of German media policy. But in the early 1990s, it flared up once again. Triggered by the increasingly stiff competition among private television broadcasters, but also between private and public broadcasters, there has been an across-the-board increase in depictions of violence in television programming (Groebel, 1993), and there is considerable concern for the effect this will have on juveniles (see Theunert, 1992; Glogauer, 1991). Particular public attention has focused on reality TV. State media authorities have felt compelled to step in, and political parties have called for measures to curb this type of programming (see, e.g., *Frankfurter Rundschau,* February 12 1993: 15). Although critics had predicted the increase in violent programs as an inevitable result of greater media competition, politicians refused to make this fundamental structural problem an issue. Instead, they made appeals to parental responsibility, called for the use of related educational measures, and demanded that broadcasters practice self-restraint, adhere to professional ethics, and create institutions of voluntary self-regulation. In other words, the proposed corrective measures were a reaction to symptoms but not to the causes rooted in the media structure. Since it had long been evident, particularly abroad, that the proposed solutions were at best of only limited usefulness, and since it was unclear why the case should be any different in Germany, the public debate bore the visible markings of symbolic policy. Certainly the public discussions were frowned upon by some advertisers; they exerted pressure upon the broadcasters to provide a program environment for their advertisements that was less open to criticism.

A few TV broadcasters promised that they would attempt to reduce levels of violence on the screen. In order to achieve this aim, the private broadcasters formed Voluntary Self-Control Television (Freiwillige Selbstkontrolle Fernsehen, or FSF; cf. *Funk-Korrespondenz,* November 26, 1993: 33ff.; see the critical analysis by Rossen, 1994: 224ff.). Its functions are to examine programs according to criteria relevant to the protection of children and young people and to make recommendations. At the same time the Interstate Treaty on Broadcasting has been amended. Broadcasters are now duty-bound to appoint a representative for the protection of children and young people. Thus, it would appear that the development toward self-regulation continues.

3.6. Origin Quotas

In contrast to the situation in, for example, France and Great Britain, the legislatures and supervisory bodies in Germany did not place great weight on safeguards to ensure in-house productions, programs of German or European origin, or works by independent producers. This has changed as a result of the EC television directive. Insofar as this issue is taken up in statute, it usually occurs in vague language demanding interpretation. For instance, the Interstate Treaty on Broadcasting requires that full-programming channels provide "a significant share of in-house and commissioned productions, including joint productions within the German-language and European region." The state media authorities have not launched any special activities to specify these requirements or to foster works by independent producers. The protection of national cultural identity was considerably less relevant in the German discussion than in France, for example. In this regard, recent German history may well have cautioned restraint, although there were also leading economic reasons. For instance, the state media authorities largely followed the argument of private broadcasters that in the start-up phase, with its inevitable high losses, it is unreasonable to expect broadcasters to fulfill production quotas and similar requirements. However, the two large television broadcasters (RTL and SAT 1) began to record profits in the early 1990s. This has not led to stricter requirements, although these broadcasters have decided to increase the amount of original quality programming. But this decision is mainly a reaction to the high costs of purchasing films as well as to viewer preferences.

3.7. Siting Policies

Ever since the early days of licensing, the state media authorities have been strongly interested in prompting private broadcasters to locate in their respective states. In addition, each state government sought to exercise influence in an informal fashion on broadcasters and its state media authority in order to secure the location of desired broadcasters in its state. Since Germany does not have a center of cultural policy, a number of cities are competing for the title of "media metropolis" (particularly Hamburg, Munich, Berlin, Stuttgart, and Cologne), and smaller regional centers are taking great pains to encourage part of the prestigious, revenue-generating media industry to locate in their areas. In Germany, much media policy has developed into "siting policy," that is, economic structural policy (cf., e.g., Hoffmann-Riem, 1989c). Prior to granting the license, state media authorities often demand assurances on headquartering and the location of production centers at the place of the license, as well as, on occasion, promises of a certain volume of production at the place of license (for the original requirements imposed on SAT 1 in Berlin, cf., e.g., Hellstern, 1989b: 169f.). The laws in part stipulate that such assurances are positive criteria for the selection decisions under the licensing procedure. However, the promises delivered by broadcasters have often been unable to be fulfilled; some of these have even led to political and court conflicts. The legal enforceability of these assurances is certainly dubious. The state media authorities have since become somewhat more cautious, particularly since broadcasters, once licensed, acquire a legally protected and de facto politically protected status, which substantially improves their bargaining position.

The competition among the sixteen states to have media companies locate in their respective jurisdictions was bothersome. For media companies it was expensive to conform to some of the special conditions found in the laws of the various states or expressed by the various state media authorities. But when—as is usually the case—companies were interested in obtaining licenses for the terrestrial broadcasting of their programming, they had to pay attention to such special desires. Still, the competition among states offered media companies a number of advantages. For instance, they could ask for licenses where, from their point of view, broadcasting requirements were especially relaxed. By the same token, if broadcasting laws were to be tightened up, companies could threaten to relocate in another state. The other states must respect the decision reached by the respective state media authority of the state in which the license has been applied for.

The competition among the states for media company location has meant that some states have tried to structure their media laws loosely. It was obvious that the state media authorities that took part in such policies tended to be more generous when it came to specific licensing decisions. Since 1993, however, this has led to considerable controversies among the states, culminating in a court battle between the state media authorities of Berlin and Munich (see Section 1.2.2). It is as yet unclear to what extent the state media authorities will use this case as the occasion for improving the coordination of media supervision among the various authorities or whether these difficulties will instead destroy the possibilities for harmonious cooperation.

4. SUPERVISION OF BROADCASTING IN THE POLITICAL ARENA

As I hope the foregoing remarks have made clear, the licensing and supervision of broadcasting in Germany are intimately associated with its political system. The following observations seek to offer insight into some additional dimensions of the relationship between the supervision of broadcasting and political interests.

4.1. Aftereffects of Controversies of Media Policy

The observation that political controversies that arise as a law is created tend to reappear in new guise when that law is implemented certainly applies to broadcasting. Following the decision in favor of the dual broadcasting system and thus for the introduction of private broadcasting in the 1980s, critics initially attempted, although only for a brief period, to retain old media-policy positions as far as possible under the changed framework. However, the situation calmed down remarkably quickly, and representatives of those groups skeptical of private broadcasting soon declared themselves willing to work constructively on broadcasting supervision and, in the process, to refrain from lengthy debates or even conflicts on the principles of the dual broadcasting order. The rule allowing representatives of social groups to be included in the supervision of broadcasting in this way proved to be a means of reducing opposition to the introduction of private broadcasting; even critics were integrated into the supervisory role.

In any event, the wide majority of political and social forces came to accept private broadcasting. Moreover, with regard to the situation within the state media authorities, it became apparent that the collegial bodies within these authorities were composed nearly exclusively of groups or political forces that were positively disposed, or at least neutral, to private broadcasting. Top-level personnel as well as the group representatives participating in the collegial organs all seemed to perceive their mandate as promoting, not hindering, private broadcasting.[23] Even with the politically unavoidable inclusion of trade union representatives and—to the extent that political parties with seats in the parliaments were taken into account—the participation of the Green Party, few fundamental opponents of private broadcasting gained access to the organs. However, "opponents" of private broadcasting did not always exercise all of their possibilities for action: For instance, the Greens in Hamburg originally decided for reasons of principle to forgo their right to nominate a member of the executive board of the media authority. Pursuant to the statutory stipulations, the vacant seat fell to the conservative Christlich Demokratische Union (Christian Democratic Union, or CDU) for nomination.

Even the representatives of religious societies (especially the Protestant Church), who formerly were opposed to private commercial broadcasting and even to this day have an ambivalent relationship to it, have come to terms with it. This was certainly aided by the fact that the churches were interested in direct participation, especially in the area of radio. Another critical group consisted of some trade union representatives, who initially saw themselves as a futile minority but have since changed their minds. The overriding factor in the boards is the high identification with the interest in the success of private broadcasting. Particularly in the initial phase, board members saw themselves called upon to exercise their duties actively and with dedication. Supervision of broadcasting at that time did not amount to routine, day-to-day business; rather, it was an active part of the process of reforming the broadcasting order. Furthermore, public attention to the matter was still relatively strong. This demanded an enormous investment of time and knowledge on the part of board representatives as well as the director and the director's staff. The particularly large demands placed on their time and skills during the early days of private broadcasting were exceptional considering that board representatives served voluntarily in these positions. Their sacrifice to supervise media activities was made at the cost of their other occupations. Although this may well have been easy for some members, such as retirees or full-time functionaries of associations, others were extremely busy, upper-echelon group or organization representatives. Endurance limits, as well as initial signs of fatigue, soon began to show. Nevertheless, these board members still continued to be motivated to exercise their functions as actively as possible, particularly since they felt that their job was important in light of the earlier, stormy controversies and that they were obligated to account to their respective groups or organization.

Such frenetic activity during the start-up phase was probably also attributable to people's desire to make names for themselves as well as to set up future avenues of influence. In the early stages, it was still basically unresolved which specific possibilities for informal influence would emerge. Thus, it seemed advisable to stake out zones of influence as early as possible through active participation and to campaign for leading roles.

After completion of the start-up phase, routine operations gradually began to set in. Those board members with other occupations were often unable to keep up the initial supervisory intensity. Furthermore, their political interest in the substance of decisions began to wane once the landmark structural rulings had been made in the form of permits and the less exciting job of monitoring compliance with license conditions and legal norms stood on the agenda. It also became apparent that the possibilities for exercising influence (e.g., on concentration developments or broadcasters' programming behavior) were in fact considerably more limited than many had assumed. This realization put a damper as well on possible steps to resume earlier media-policy controversies. Those who were positively disposed to private broadcasting and who relied on the power of self-regulation of course had no particular reason to push for active exercise of the supervisory role.

With the completion of the start-up phase, internal power in the supervisory bodies shifted increasingly to the directors and their staffs. This was in part necessitated by the fact that the fifteen media authorities coordinated their activities. However, such coordination could not be effectively accomplished by the various collegial organs. Although a conference for the heads of the boards had been set up, the main coordinating bodies were the conference of directors as well as the variety of commissions and working groups in which directors or representatives of their staffs coordinated activities within all of Germany (Schuler-Harms, 1995). In a federal system, such coordination is more a task for a professional staff than by representatives.

4.2. Regard for Political Parties and Governments

Whereas the supervision of German public broadcasting was subject to intensive, and greatly criticized, attempts to influence broadcasting operations on the part of political parties and government alike—indeed, use was also made of internal supervisory bodies for this purpose—the supervision of private broadcasting has not developed in completely the same way. As noted above (see Section 2.2.1), some supervisory authorities have formed political cliques, and party affinities certainly can be observed in decision-making conduct. However, formation along party lines has not taken place in all state media authorities, and on important issues, majorities are often formed across party borders without necessitating the usual, protracted coordination processes.

The limited importance of party orientation is probably most likely related to the fact that the powers of supervisory bodies are substantially smaller and also less effective than in public broadcasting. (This is also the estimation on which the Bundesverfassungsgericht operates; see BVerfGE 73, 170f.) Missing in particular is the right for supervision typical within public broadcasting on decisions regarding financing, budget matters, personnel, and organizational issues. Furthermore, the possibilities for supervised broadcasters to take practical or even legal steps against supervisory measures are more effective than those between collegial supervisory bodies and the director general or other employees in public broadcasting. All of this serves to reduce the incentive of political parties to transform supervisory bodies into their own instruments, although they have on several occasions attempted to gain influence.

The governments in particular have seen to it that the state media authorities do not act in a policy-free zone.[24] For instance, when appointing leading positions, particularly that of the director, attention is always paid so that representatives of the respective political party making up the government are selected. It is especially evident that many directors of the state media authorities were formerly charged with responsibility for media law and policy in their given state government and above all were in charge of drafting the respective state media law. Although the switch in roles does not necessarily mean an automatic continuation of prior governmental policy, the political proximity to the government very often sets the pattern for day-to-day work. Since each state government is in charge of the legal supervision of its state media authority, it also has (restricted) formal powers to monitor activities and, if need be, to object to these (cf. Ring, 1993). As a result, formal intervention may be augmented with informal methods of determining whether certain conduct by a state media authority runs the risk of requiring state supervisory measures. Insofar as the government participates in preliminary decisions on licensing (e.g., in dedicating transmission means to specific broadcasting purposes) or when government representatives take part in the meetings of the boards, there are further occasions for mutual respect. In this regard, a number of state governments have shown particular interest in decisions regarding sites (see Section 3.7).

4.3. Supervision of Broadcasting and Separation of Powers

It has already been pointed out that the broadcasting order has been made intensively subject to laws and regulations. This means that, on the one hand, parliaments or governments (as lawmakers) and, on the other, the courts are able to exercise influence on the supervision of broadcasting. Broadcasting supervision is primarily conceived of as legal supervision limited to monitoring broadcasters' compliance with laws and subject to judicial review. At the same time, however, the supervisory authorities also make "option" decisions, as in the course of comparative licensing decisions or with regard to decisions on economic and journalistic capabilities, but also pursuant to day-to-day supervision to the extent that this follows the principle of discretion. The numerous provisions in the laws open to interpretation require option decisions as well. Even though these are, in principle, subject to judicial review, there still remain a substantial number of compelling prerogatives in favor of the state media authority (cf. OVG Lüneburg, DVBl. 1986, 1114; VerwGH Mannheim, NJW 1990, 341; Wagner, 1990b: 168ff., 187ff.; Ziethen, 1989: 59ff.).

The fundamental possibility of judicial review forces supervisory authorities to proceed with great caution and provides broadcasters with the power to threaten: They may threaten to resort to the courts, thereby forcing the state media authority, at least when it is not entirely sure of its position, to give in. Threatening court action is, however, only a last resort, which, if possible, is avoided, particularly by successful broadcasters interested in long-term cooperation with supervisory authorities (see Section 1.2.1).

Despite the pervasion of laws and regulations, most members of supervisory authorities, as well as such other actors as private broadcasters, do not view supervision as simply being limited to fitting a given fact pattern under the statute. Considerations of propriety and lawfulness are closely related to one another.

Precisely for this reason, programming-related licensing and supervisory decisions are especially problematic. If powers are used extensively, then intimidation can be expected, which may also become a risk for diversity of communication. Potential victims of programming supervision tend not to be broadcasters that have been successful with mass-appeal programs; these normally incline toward relatively colorless, politically neutral programming and thus give only rare cause for triggering supervisory action. Rather, it is the outsiders—such as the so-called alternative radio stations—that might be exposed to intensive supervision in programming matters. They are most likely to provoke complaints from viewers or listeners. Supervision in the wide majority of state media authorities is dependent on such external initiatives (cf. Thaenert, 1990: 37).

4.4. Priority for Limited Self-Regulation

Respect for broadcasters' programming autonomy is a principle common to private broadcasting orders: Journalistic responsibility lies with the broadcasters themselves. Although the basic legal situation is somewhat different in Bavaria (see Section 2.2.1), it does nothing to alter this fundamental broadcasting reality. The principle of broadcaster self-responsibility corresponds not merely to the normative program of broadcasting laws but also to the fundamental elements of private enterprise and journalistic autonomy. The legislatures ensured this for the phase of licensing with the additional requirement that, in allocating scarce transmission frequencies, the state media authorities first leave the solution of conflicts to the competitors themselves. For instance, most laws pay credit to the creation of voluntary *Anbietergemeinschaften* (again, companies in which provider associations hold shares) among license competitors (see Section 3.1.1). If several equally situated broadcasters compete for a license, some state media laws expressly provide that the state media authority is to be bound by an "agreement" or "understanding" reached among them. Sometimes, the authority is expressly called upon to play a mediating role in the process, but even in the absence of such a statutory duty, it can take on moderating functions. In some cases, the laws provide for a proportional allocation of transmission opportunities in the event of competition (frequency splitting), whereby an agreement on the type of distribution is, again, to have priority over a decision by the state media authority.

Expressed in the priority accorded to an understanding among license applicants is the principle of self-regulation, which is essential to a market economy order. The Parliament, which is responsible for the broadcasting order, relies on measures of self-regulation, but, with regard to the its regulatory responsibility, it also establishes a basic legal framework for broadcasting. It thereby gives de facto preference to those license applicants that can develop well in a market-oriented manner. When an agreement or even the formation of *Anbietergemeinschaften* creates advantages, those license applicants that can undertake such ventures on account of their interests receive a de facto privilege. Thus, the relative homogeneity of interests has often made it possible for newspaper publishers to participate in *Anbietergemeinschaften* for radio or television. However, this has offered no guarantee against subsequent conflicts of interest, such as between large publishers and smaller ones. The history of SAT 1 provides enough examples (especially fights between the Springer Publishing Company and Kirch). Unconventional parties interested in a license, or even political outsiders, have little chance of reaching an understanding with others, let alone forming an association with them. The priority given to these agreements thus does not function neutrally.

Agreement among interested parties hence relieves the pressure on the regulators to reach decisions. Examples of successful agreements—sometimes arranged by state media authorities—are numerous (cf. the references by Hellstern, 1989b: 160ff.; Hoffmann-Riem & Ziethen, 1989: 277; Reese, 1989: 363ff.). On occasion, competitors have reached an agreement even without an accompanying demand for such, thus circumscribing the supervisory authority's decision-making discretion more significantly than the latter would desire. Such an impression was at least raised by the agreement reached among competitors in allocating a satellite frequency, the so-called North Track (see the Protocol on the Meeting of the Expert Committee for Programming Issues of the Niedersächsische Landesrundfunkausschuss of 1 April 1987).

The assistance provided by the state media authority in the self-regulation of the broadcasting industry has also resulted in aid to license applicants in the application process. Yet in practice, one sees subtle differences between the two modes of assistance.[25] The reluctance to participate actively in forming *Anbietergemeinschaften* might be motivated by the consideration of avoiding as far as possible any self-commitments or even of having to serve as guarantor for subsequent success. Thus, applicants experienced in the application process or with a large staff of advisers at their disposal have an advantage over newcomers, particularly when the latter have been formed for the express purpose of submitting an application. This was demonstrated particularly clearly by the start-up of broadcasting in the eastern German states (e.g., reports in *Funk-Korrespondenz,* 15 April 1992: 6f.; epd/Kifu, 6 May 1992: 13f.).

Because of their decision-making power, the state media authorities quickly fell into the role of the entity responsible for the subsequent functioning of the broadcasting order they helped to shape. However, the state media authorities have undertaken different strategies on the thorny question of the extent to which they should also assume responsibility for the economic and journalistic success of broadcasters. In accordance with the paternalistic tradition in German government, broadcasting law also provides points of departure for such coresponsibility. For instance, laws that require that broadcasters evidence a minimum financial capability show that broadcasting is not intended to be left to market forces alone but, rather, must also be economically functional for a new broadcasting order to come into being. If, as in North Rhine–Westphalia, the state authority for broadcasting is required by law also to assume responsibility for ensuring that the transmission territory for local programming enables "economically sound local broadcasting," this stipulation of course only relates to the transmission territory; however, it also has an indirect effect on determining the financial viability of the licensed broadcaster. Should such a broadcaster flounder, it is a likely indication that earlier erroneous decisions were made in determining the transmission territory.

The great effect that the media authorities' regard for the success of the broadcasting order can have was particularly clear in the case of the licensing of the first radio broadcaster in Hamburg. Prior to this, sound radio broadcasters had been licensed in the adjoining states of Lower Saxony and Schleswig–Holstein, which primarily aimed at the advertising market in the Hamburg metropolitan area. The Hamburg media authority saw itself under enormous pressure to succeed and sought to ensure that the first licensed radio station in Hamburg could be competitive on the intrafederal level, thus offering proof of the solid footing of Hamburg as a media center and of the survivability of private broadcasting in general. It accorded the first licensee considerable start-up opportunities compared against other licensees, providing it with a distinct head start over other competitors. The effort to succeed in intrafederal competition soon became a handicap for the rest of the radio order in Hamburg (see also Hoffmann-Riem, 1990b: 16). Other strategies have sometimes been tried out elsewhere; in Baden-Württemberg, for example, initially frequencies were awarded as a package in order to provide new broadcasters with reasonable, equal start-up chances (see Hellstern, 1989a: 31).

4.5. Successive Specification of the Requirements for Broadcasting

The many initial uncertainties on the part of applicants and the state media authorities have led to a number of efforts to avoid setting forth the requirements for licensing in absolute categories but, instead, to specify these successively and especially to develop them in interactive processes. By incorporating promises as a legally binding component of the licensing decision, the state media authorities seek to alleviate subsequent judicial control of broadcasting operations. The fears of applicants that they might fall into disfavor as against other competitors should they refuse to acquiesce to the wishes of the state media authorities are used to obtain pledges. On some occasions, promises in the form of unilateral requirements have even been made that the state supervisory authority could hardly have imposed in a lawful manner. Notable here are some location pledges by television broadcasters, which were called for, and received, as a sort of counterperformance for the awarding of the coveted terrestrial television frequencies. Only in the licensing phase does the state media authorities have this sort of bargaining power with regard to such an aggressive strategy of assurances. Once the license is granted, it is very often observed that the readiness to fulfill the promises has waned.

Because of the initial uncertainty on the interpretation of the licensing and supervisory criteria and on the developmental possibilities for private broadcasting, the decision-making and supervisory criteria from the outset were not legally specified in a generalized form, as for example, in the nature of administrative guidelines or regulations. At the same time, however, newsletters, press releases, and certain regulations offered general, abstract orientational assistance, which was then given more specific form in the course of discussions. In the interim, the state media authorities have issued an increasing number of guidelines and regulations. Examples of such specifications include guidelines drafted by the Conference of Directors and issued by the state media authorities in the area of advertising rules and the protection of juveniles.

4.6. Priority for Informal Supervision

As early as in the licensing phase, the state media authorities began to use opportunities for informal cooperation in order to encourage applications likely to receive approval. The interactive relationships have largely been maintained after licensing and even further developed into an extended interactive network. As in other administrative areas, informal administrative action (cf. Bohne, 1981, 1984; Bauer, 1987) became one of the most important modes of supervisory activity. The authorities strived to keep their formal set of supervisory and sanctioning instruments in reserve and to employ them only as a threat, thereby hoping to prompt broadcasters to comply voluntarily with the legal requirements. Supervisory bodies employed clarifying talks with the director, references to statutory violations and calls to remedy these, informal threats of sanction, and so on more often than formal warnings or license revocations. They also relied on flexible forms of supervision, such as setting transition deadlines in order to facilitate the broadcasters' compliance with statutory requirements.

This informal procedure made it possible to clarify uncertainties interactively, to react flexibly, to maintain cooperative relationships as long as possible, and to avoid court conflicts and the associated risks. A formal set of sanctioning instruments in the form of administrative requirements and prohibitions or even the use of compulsory tools and fines can rapidly lose their effectiveness when overused; moreover, they may lead to still further statutory penetration and formalization of relationships, which most participants consider to be an impediment to the fulfillment of duties. Broadcasters also welcome the preference for the "policy of raised eyebrows" and usually react to informal references. However, they are at the same time encouraged by the practice of informal reactions to test the limits of the state media authority's tolerance. Particularly in the area of advertising restrictions, German broadcasters have launched a number of "experiments." It has become possible over time to weaken the advertising rules and, above all, to get permission to use new forms of advertising.[26] Generally, we can observed that in the history of their use, the normative requirements have often been relaxed rather than stiffened by way of reinterpretation in the interest of broadcasters.

Measured against the requirements of the rule of law, informal agreements also meet with reservations, regardless of their practical advantages. The public normally has no access to these agreements, and it is usually impossible to engage the courts in a review of the solutions thus reached.[27] This poses certain risks that in such informal bargaining situations, those who hold the power and are also in control elsewhere in society will prevail and that the law's potential as a power corrective will go unused. Especially in the media sector, one must give serious consideration to the objection that powerful media undertakings (at the moment, particularly publishing companies and multimedia firms) have better access to political power-holders. They have apparently made use of this not only in the phase of legislation but also during important stages in the licensing procedure. It cannot be stated with certainty that—regardless of its measure of independence—a licensing and supervisory authority will be insensitive to power constellations. The many interpersonal ties the directors maintain and the relationships among many members of supervisory bodies and representatives of government policy and the media industry make such fears all the more valid. Licensing and supervisory practices have thus far not presented contrary indications.

In spite of such risks, the possibility of flexibility and informality has been an important tool in fulfilling supervisory functions (see also Grothe & Schulz, 1993: 72 ff; Holgersson, 1993: 102). Since a market economy order basically accords priority to the self-interest of those supervised and since the supervisory authority has only limited insight into the workings of broadcasters (cf., e.g., Thaenert, 1990: 47ff.), the supervisory authority must exhaust the possibilities for fostering interest in cooperation on the part of broadcasters. German law offers no basis for forbidding the options of cooperative interaction and informal supervision that might arise from the above-mentioned risks. At the same time, however, the search for ways to build up procedural and substantive assurances from broadcasters is most certainly tied to the constitutionally anchored principle of the rule of law. To this extent, the constitutional requirement of equal treatment particularly demands that smaller, less powerful broadcasters not be denied the same enjoyment of cooperative relationships and "gentler" supervisory practice accorded to the large, established broadcasters.

4.7. Citizens' Activities and Professional Criticism

The risks associated with informal supervisory practice have thus far been reduced to only a limited extent by control by citizens' groups. External third parties have no rights of participation in the supervisory process. In the course of licensing, competitors provide input to the procedure only by way of the limited possibility of their own applications and may, where justified, take court action in the course of so-called third-party or competitor complaints.

Some state media authorities have involved the public on their own, thereby creating an additional threat and sanctioning tool. Broadcasters normally do not appreciate it when they become the subject of public discussion, let alone when they are publicly stamped as offenders. However, the state media authorities practice restraint with regard to calling in the public. As a result, the press releases, notifications, and so forth are usually formulated in guarded terminology, and the public generally does not have access to the meetings of the supervisory bodies or committees. There is no tradition of institutionalized civil activities in the broadcasting sector in Germany, so even from this side, there is no strong pressure for increased public access. Now that the debates of media policy have waned, broadcasting is no longer a major target of public criticism and corresponding activities.

Media critics[28] have attacked supervisory activities as well, although this criticism has not generated much favorable reaction from the general public. During the phase of pilot cable projects, there was a greater interest among researchers, stimulated not least by the provision of substantial research funds. Since the establishment of the dual broadcasting system, these funds have become more scarce, and most state media authorities have opted to forgo using independent media research to accompany media developments.[29]

Even though public interest has ebbed somewhat, the development of broadcasting, especially commercial broadcasting, continues to be criticized. In 1993, the German President set up a commission of broadcasting experts. In their report, submitted in 1994 (Groebel et al., 1994), they also made proposals to increase the role of the public as a forum for criticism. These proposals include setting up a media council, a kind of "royal commission" that would continuously observe the development of broadcasting. A further suggestion was the creation of a *Stiftung Medientest.* This independent foundation would analyze programs from a consumer's perspective and help recipients find their way through the maze of programming.

NOTES

1. Pro 7, VOX, and RTL 2. Other nationwide channels focus on sporting events (Deutsches Sport Fernsehen, or DSF), news (n-tv), and music videos (VIVA TV). The last three channels are called *Spartenprogramme.*

2. By "full programming" (*Vollprogramme*) is meant broadcasters that offer a full palette of broadcasts, primarily those serving to inform, educate, counsel, and entertain. Some laws prescribe that such programming be offered for at least five hours a day. In Germany, this type of programming is differentiated from the *Spartenprogramme,* or special interest broadcasts, that basically offer only one type of programming or address a special target group (*Zielgruppenprogramme*).

3. For details, see Wöste, 1991: 561ff.; see also Jens, 1991: 570ff., for supraregional participation by the press in private broadcasting.

4. With the postal reforms of 1989 and 1994, the Federal Postal Authority (Bundespost) was split into three separate entities; one of these, Telekom, is now responsible for broadcast transmission. There are only a few privately run cable networks in Germany.

5. "Council Directive on the Coordination of Certain Provisions Laid Down by Law, Regulation or Administrative Action in Member States Concerning the Pursuit of Broadcasting Activities," Commission of the EC, EC-Gazette No. L 298/23ff. (1989).

6. "Gemeinsame Stellen der DLM," "Jugendschutz," "Programm," "Werbung," and "Vielfaltssicherung"; cf. Schuler-Harms, 1995.

7. For instance, the former broadcaster Tele 5, which was held in part by the Berlusconi firm Fininvest, was first allowed to share in terrestrial frequencies after SAT 1 and RTL. But, the foreign ownership of RTL had important consequences: The fact that the main shareholder, the Luxembourg company CLT, was so experienced in the field probably worked in favor of this broadcaster. Foreign holdings in other television broadcasters as well have since begun to rise; see DLM, 1990: 160ff.; Hans-Bredow-Institut (1994: A 301ff.). For example, the new pay-TV channel Premiere is held by the Bertelsmann subsidiary UFA and the Kirch Group's Teleclub GmbH, together with a German subsidiary of the French pay-TV channel Canal Plus.

8. For example, Time-Warner has a share in various local television companies, and Murdoch has a share in VOX.

9. LfK—Landesanstalt für Kommunikation Baden–Württemberg (State Authority for Communication, Baden–Württemberg); BLM—Bayerische Landeszentrale für Neue Medien (Bavarian State Authority for New Media); the former Anstalt für Kabelkommunikation Berlin (Authority for Cable Communication in Berlin); MABB—Medienanstalt Berlin–Brandenburg (Media Authority Berlin–Brandenburg); Bremische Landesmedienanstalt (Bremen State Media Authority); HAM—Hamburgische Anstalt für Neue Medien (Hamburg Authority for New Media); LPR—Hessische Landesanstalt für privaten Rundfunk (Hessen State Authority for Private Broadcasting); NLM—Niedersächsische Landesmedienanstalt für Privaten Rundfunk (Lower Saxony State Media Authority); LfR—Landesanstalt für Rundfunk Nordrhein–Westfalen (State Authority for Broadcasting in North Rhine–Westphalia); LPR—Landeszentrale für private Rundfunkveranstalter/Rheinland–Pfalz (State Office for Private Broadcasters, Rheinland–Pfalz); LAR—Landesanstalt für das Rundfunkwesen Saar-

land (State Authority for the Broadcasting System Saarland); ULR—Unabhängige Landesanstalt für das Rundfunkwesen/Schleswig–Holstein (Independent State Authority for the Broadcasting System, Schleswig–Holstein); LRZ—Landesrundfunkzentrale Mecklenburg–Vorpommern (State Broadcasting Office, Mecklenburg–Vorpommern); SLM—Sächsische Landesanstalt für Privaten Rundfunk und Neue Medien (Saxony State Authority for Private Broadcasting and New Media); LRA—Landesrundfunkausschuss für Sachsen-Anhalt (Saxony-Anhalt State Broadcasting Committee); and TLR—Thüringer Landesanstalt für Privaten Rundfunk (Thuringian State Authority for Private Broadcasting).

10. Cf. Hellstern, 1989a: 11ff. In accordance with the 1992 amendment to Baden–Württemberg's State Media Act, the executive board was to begin to exercise primarily executive functions; its director would henceforth be appointed as a full-time official, thus corresponding to directors in other states. As a result, the acting director resigned.

11. In order to maintain control over this situation, the 1991 Interstate Treaty on Broadcasting requires the periodic commissioning and publication of reports by an independent institute (para 21[6] *Rundfunkstaatsvertrag*).

12. In fact, advertising experts and broadcasting employees responsible for financial matters have consistently emphasized that successful advertising acquisition depends on especially close coordination between marketing and programming, specifically between the advertising department and the editorial office. See, for example, Teichert & Steinborn, 1990: 79.

13. By "integrated programming" is meant ensuring that the programs take into account in a balanced fashion all social interests (see Section 1.1). Journalistic diversity is attained not only through organizational requirements but also with substantive programming commitments, such as duties of neutrality, balance and fairness, and due care. Compare Lerche, 1979: 49f.; Hoffmann-Riem, 1989a: Nos. 174ff.

14. At the same time, however, the state media authority lays down certain requirements, as in the form of guidelines by the LfK; see Hellstern, 1989a: 48. In Baden–Württemberg, two models of programming councils have emerged: in one, the members are appointed by the broadcaster itself; in the other, the members are selected from a circle of socially relevant groups. In some cases, the broadcasters themselves have a seat. Compare LfK, 1989: 22ff.

15. For national broadcasters, this is possible only when the state media institutions establish by a three-quarters majority that external pluralism is not present; it is extremely unlikely that this will ever occur.

16. For instance, in a regional program of the broadcaster RTL, the ULR lodged an objection to a prejudicial headline regarding a public personality in Schleswig–Holstein that, because of abbreviation, gave an incorrect impression. See ULR, 1990: 21.

17. In 1994 ever-increasing media concentration resulted in heated discussions about the future of concentration control in broadcasting. A new form of regulation is imminent (cf. Europäisches Medieninstitut, 1994).

18. The media authorities have fallen out over how to combat media concentration. On the court battles between the authorities in Berlin–Brandenburg and Bavaria, see VG München, ZUM 1993, 294ff.; BayVerwGH, ZUM 1993, 296ff.; BVerfGE, ZUM 1994, 639ff.

19. Compare *Süddeutsche Zeitung,* 22 November 1994.

20. For this problematic, see Schurig, 1988; Wöste, 1989: 9ff.; 1991: 561ff.; Ring, 1992: 39f.; Kübler, 1992: 43ff. Bavaria and Baden–Württemberg especially had to grapple with these problems as regard their concepts for local and regional radio; cf. LfK, 1989: 92ff.; BLM, 1989a: 25f.; 1990: 27f.; 1991: 27f.

21. "Gemeinsame Richtlinien der Landesmedienanstalten für die Werbung, zur Durchführung der Trennung von Werbung und Programm und für das Sponsoring im Fernsehen sowie im Hörfunk." 26 January 1993, reproduced in Evangelischer Pressedinst/Kirche und Rundfunk (*epd/KiFu*), No. 8, 3 February 1993.

22. See "ARD-Richtlinien für die Werbung, zur Durchführung der Trennung von Werbung und Programm und für das Sponsoring vom 24.06.1992," reproduced in *Media Perspektiven, Dokumentation,* 1992(3): 192ff.

23. As only one example, mention is made here of the former chairman of the Conference of Directors who stressed the necessity that the state media authorities "ensure liberty and maneuvering room for private broadcasters in structuring their programming" (Ring, 1990).

24. Keidel, 1993: 102ff. Political influence was also exerted at the time when commercial broadcasting was being introduced in the former GDR; see Kopetz, 1993: 87 ff, 95.

25. In addition, this practice has changed in part over time; see the example of the Hamburg HAM by Hoffmann-Riem & Ziethen, 1989: 236, 244f., 269f.

26. Compare Gebel, 1990: 53ff., who expressly refers to the testing of new advertising forms and shows how the rules of the state media authorities have been or are to be adjusted to meet the needs of the advertising industry and broadcasters.

27. To this extent, the case law is not yet conclusive. For interesting approaches, see VG München, NVwZ 1987: 435, 437, which has held the very refusal to undertake a formal administrative act to be an interference.

28. Particularly worthy of mention here are several information services, such as epd/KiFu, as well as the media sections of several newspapers, such as the *Süddeutsche Zeitung* and the *Frankfurter Rundschau.*

29. Some laws, such as in Hamburg and North Rhine–Westphalia, provide for media research; this is, however, only conducted to a limited extent. The most active is the North Rhine–Westphalian LfR. The 1991 treaty also provides for independent reports on the development of diversity of opinion and on concentration in private broadcasting.

CHAPTER 4

France

For a long time France has not played a preeminent role in the international media-policy discussion. For decades, it had simply proceeded too bluntly with its diverse governmental interventions in the broadcasting sector, which manifestly conflicted with the concept of editorial independence. In the phase of privatization as well, France initially did not play a trailblazing role. First, France did not have any noteworthy head start as opposed to other important countries; second, the opening to private enterprise apparently was also accompanied by governmental interventions. Furthermore, French developments lacked a clear concept comprehensible to outsiders. Changing political majorities, moreover, have led to alterations in the broadcasting order, which has precluded the system from being able to mature or even to serve as a model. However, in 1986–1987 the privatization of the most successful TV station, TF 1, created a stir and has become the point of reference for similar calls in other countries, such as Germany. In addition, European countries have paid close attention to the deliberate subordination of development to the technological and economic goals of developing an independent French media industry and ensuring an independent French program industry. This has also forced the European Community (EC) to take action, and other European countries had to weigh what they could do to increase the competitiveness of their own industries (for a discussion, see Lange & Renaud, 1989).

In light of this background, the stages of French broadcasting development as well as the present situation are outlined in the section that follows. With respect to the empiricism of broadcasting supervision, France has had various experiences with different supervisory authorities, all of which illustrate that the successful control of private broadcasting is very difficult in a country with a long tradition of governmental regulation of public broadcasting.

1. DEVELOPMENT AND STRUCTURE OF THE FRENCH BROADCASTING SYSTEM

1.1. State Broadcasting and Its Gradual Dismantling

The transmission monopoly, which was granted to the French government from the very outset, extended for many years into the programming area as well. Only during a brief interim phase from 1924 to 1925 were small, private radio stations in existence (Debbasch, 1991: Nos. 95ff.). The area was, however, otherwise consistently governed by public broadcasting committed to the concept of public service.[1] In the late 1950s, broadcasting was removed from the public sector and thereafter operated by a "public corporation with industrial–commercial character," closely supervised by the Ministry of Information (Institute National de l'Audiovisuel, 1988: 20ff.). In this manner, the information minister's authority to intervene was limited to control powers, although political influence did not disappear altogether (Turpin, 1988: 112). In the mid-1970s, the broadcasting company ORTF (Office de Radiodiffusion Télévision Française) was split up (Institute National de l'Audiovisuel, 1988: 37f.; Debbasch, 1991: Nos. 221ff.; Kuhn, 1985a; 1985b; 1988: 178). Thereafter, there was only one transmission company equipped with a transmission monopoly, the Établissement Public de Télédiffusion Française (TDF). This was joined by four national program companies—one for radio, La Société Nationale de Radiodiffusion "Radio France" (SRF), and three for television: La Société Nationale de Télévision "Télévision Française 1" (TF 1), La Société Nationale de Télévision en Couleur "Antenne 2" (A 2), and La Société Nationale de Télévision "France-Régions 3" (FR 3). The company shares belonged in their entirety to the French government. There was also a special production company (La Société Française de Production et de Création Audiovisuelles, or SFP), which produced programs for the television companies.

Financing was accomplished with fees—since 1978, solely as television fees (TF 1 began to carry commercial advertising in 1968; see Palmer & Sorbets, 1992: 68). The income from these fees flowed not only directly to the broadcasters but also to the government, which—with the approval of Parliament—reallocated them. This also ensured state influence (Teidelt, 1986: 532).

The government was responsible for supervision and had to see to the invio-lability of the state monopoly and broadcasting's fulfillment of its public service mandate. Interventions in programming were possible and, in fact, occurred.

Only in 1982, shortly after the election of President François Mitterrand, was the law amended (for the history, see Chevallier, 1982: 557ff.; Ott, 1990: 64ff.; Kuhn, 1985a), in particular, in reliance on the so-called Moinot Report (Commission de Réflexion et d'Orientation Presidée, Pierre Moinot, 1981), which had called for, among other things, more autonomy and stronger decentralization. At the same time, however, important forms of government influence remained intact. Public broadcasting was initially restructured, and alongside national institutions, regional television and radio companies were to be created. The transmission company TDF, the radio company SRF and the television companies TF 1, A 2, and FR 3 were retained, although in part with new structures. However, although the creation of "autonomous" regional entities did not occur in the television field,[2] such establishment did occur in the radio sector.

The broadcasting monopoly was abandoned. Since 1981, private local radio broadcasters with limited reach have been able to receive licenses. They were initially prohibited from using advertising as a source of financing, but this was then permitted in 1984.

The organization of television companies in privately owned forms began in 1984 with the founding of Canal Plus, a subscription television channel transmitted over the terrestrial network that mainly broadcasts movies (for Canal Plus, cf. Rozenblum, 1984: 123ff.; Ridoux, 1984: 46; Cousin, Delcros, & Jouandet, 1990: 127ff.). It holds a public service operating license. Adver-tising revenues also were permitted in 1986. Even with Canal Plus, the government initially retained influential power indirectly by way of corporate holdings.

The Bredin Report in 1985 spurred the next major change in media policy. It recommended the creation of two private, nationwide television networks with local window programming (Bredin Rapport, 1985; for the Bredin Report and its consequences, cf. Drouot, 1986: 47f.; Ott, 1988: 64ff.). During this period, conviction grew that France should make an effort to break the American dominance in the media sector and create operable European alternatives (see the remarks by Communications Minister Filliaud, in *Le Monde,* 29 November 1985).

Further developments, as well as public discussion, were marked by the surprising intervention of President Mitterrand, who created a sensation with his decision to permit the Italian multimedia entrepreneur Berlusconi together with two French industrialists to purchase a significant number of shares in a proposed fifth television network. The broadcaster commenced with program-ming in early 1986, at which time another national network, the music channel

TV 6, began operations. (For the controversies generated by the licensing of this broadcaster, cf. Turpin, 1988: 107; for the development of TV 6, cf. Cousin et al., 1990: 139ff.)

However, these activities also formed only a brief interlude. Once again, a change of government—the election victory of the center-rightist coalition headed by Prime Minister Jacques Chirac—created the incentive for charting a new course. Supported by a decision of the Conseil d'État (Council of State), means of transmission began to be reallocated under a new broadcasting law of August 1986. The first television channel TF 1—which also had the greatest reach—was privatized (for the consequences of privatization on programming, see Oehler, 1988b; Conseil Supérieur de l'Audiovisuel [CSA], 1991a). The fifth and sixth television channels were reallocated, with Berlusconi's partici-pation being reduced. The quite prosperous subscription broadcaster Canal Plus remained unchanged, although the privatization of the principal share-holder, the Agence Havas, resulted in the widescale elimination of state influence.

TDF, which was responsible for transmission operations, was trans-formed into a corporation, with provision being made for acquisition of shares by private parties (up to 49%). The production company SFP was likewise placed in part under private ownership. Moreover, it was charged with produc-ing programs for both public and private broadcasters. Public television—now reduced to A 2, FR 3, and SRF—remained the property of the government. Over the course of time, however, the influence of these broadcasters has waned considerably: Their viewer share has sunk by one-third. They are also caught up in a long-term financial crisis. For political reasons, broadcasting fees were not raised enough to allow them to keep step with increased expenses. This problem has been exacerbated by dwindling advertising reve-nues. As a result, in recent years public broadcasters have been dependent on ever larger governmental subsidies, which increases yet further the political dependence they already experience (for an overview of the situation, see Meise, 1992: 249ff.). By way of the act of August 2, 1989, A 2 and FR 3 were placed under the direction of a common president in order to make them more competitive with private commercial broadcasters (*Lettre du CSA,* No. 27, 1991). In 1992, they changed their names: A 2 became France (Télévision) 2, and FR 3 became France (Télévision) 3. By now, France 2 seems to be about to recover from the crisis it underwent (CSA, 1994b).

In addition to terrestrial broadcasting, cable networks also began to be expanded in France, although relatively slowly (for details, see CSA, 1990: 93ff.; *Lettre du CSA,* No. 33, 1992; Opitz, 1992: D 96; Kuhn, 1988: 181). This expansion, in part with fiber-optic technology, was stepped up somewhat in the early 1980s for reasons of industrial policy (Missika, 1986: 527; Simon, 1988: 169ff.; Forbes, 1989: 34; for the state of the law, see Turpin, 1988: 120ff.; Delcros, 1985: 243ff.), but it proceeded only sluggishly. In 1993 there were approximately 1 million subscribers out of a total of 4.6 million households in the areas covered by cable networks (*Lettre du CSA,* No. 43, 1993: 4; Lange, 1994: 74). This represents a connection density of around 22%. In order to increase this percentage, a reciprocal contract was agreed, primarily at the request of the CSA, between France Telecom and the cable companies whereby the tax burden of the latter was reduced so that the subscribers' fees could be lowered (*Lettre du CSA,* No. 43, 1993: 6). As a whole, although for the first time in the history of the cable industry, there are cautiously optimistic indications for the further development of the cable sector (*Lettre du CSA,* No. 45, 1993: 1), cable companies are still showing clear deficits (for further details and statistics, see Lange, 1994: 73f.). Nevertheless, this was one of the many French efforts to become a leading industrial power in the audiovisual sector. Cable disseminates not only French and some foreign programming but also programming from specialized cable broadcasters.

The field of satellite broadcasting has also been developed (for discussion and developments, see Kuhn, 1988: 176, 181ff.; Palmer & Sorbets, 1992: 71f.). France went to great lengths—above all, in cooperation with Germany—to play a leading role in the development of broadcasting satellites in Europe. A company (La Sept) was established for the creation of satellite programming, whose shares were held, in part, by the government and by public broadcasting companies (FR 3 and Radio France). La Sept was charged with broadcasting high-quality, cultural programming similar to that produced by the British Channel 4. In February 1989, La Sept began to broadcast its programming via the satellite *TDF 1*. This satellite used the D2 MAC norm, with which France and Germany prevailed in the late 1980s in their efforts to establish a high-definition television standard for Europe. This presented a burden for La Sept and other programming broadcast via *TDF 1*, since only very few private households were equipped with the required reception equipment. La Sept has since proved a failure and was integrated into the German–French cultural channel ARTE, which was founded to create a counterweight to mass-appeal programming of, in particular, American origin. The terrestrial broadcasting of this programming was made possible after the bankruptcy of La Cinq in 1992.

1.2. Supervision and Control

Although the recent history of broadcasting has been marked by the gradual retreat of government, the latter nevertheless still has important sources of influence. Only to a limited, although growing extent, has the government transferred powers to newly created supervisory and licensing bodies.

Between 1982 and 1986, the Haute Autorité was established as supervisory authority for public and infant private broadcasting. It granted broadcasting licenses to private parties interested in local radio and local cable services. Day-to-day supervision of public broadcasting programming and of private broadcasters also rested with the Haute Autorité (for details, see Section 2).

With the Broadcasting Act of 1986, the Haute Autorité was replaced by a new authority, the Commission Nationale de la Communication et des Libertés (CNCL). The CNCL was also in charge of public and private broadcasting; in comparison to the Haute Autorité, however, its duties were expanded (for details, see Section 3). In January 1989, the CNCL was, in turn, subsequently replaced by the Conseil Supérieur de l'Audiovisuel (CSA), which is also responsible, in part, for the entire broadcasting system (see Section 4).

1.3. Broadcasting as Public Service

As in other European countries, in France public broadcasting is conceived of as a public service. (For the structure of the public sector in the French broadcasting system, see, in particular, Drouot, 1987: 399; Debbasch, 1991: Nos. 305 ff; CSA, 1991a; Cousin & Delcros, 1990: 137ff.; a historical treatment with respect to the current situation is provided by Holleaux, 1987: 19ff.) However, the element of editorial independence anchored in, for example, the British and German traditions has been rather weak in France. Although provided for by law, the concept has been weakened by governmental intervention. For a long time, the respective administrations understood the public service orientation as governmental proximity, often coming into conflict with journalists, whose professional self-image included relative independence (see Bombardier, 1975).

Although private broadcasting is not public service, it is subjected to limited commitments regarding the common good (see Bullinger, 1987: 258f.; Debbasch, 1989: 308ff.). Accordingly, the Haute Autorité and the CNCL were, and the CSA is, empowered to conduct requisite supervision not merely in the public sector. For instance, the supervisory authority must ensure plurality in the programming structures of private broadcasters. In a landmark ruling of September 18, 1986, the Conseil Constitutionnel (Constitutional Council) specifically addressed the lofty value it attributes to the requirement that there be sufficient pluralism in giving coverage to the diverse streams of sociocultural opinion. Pluralism of opinion is held to be a constitutional value representing a fundamental condition for democracy.[3] The Conseil made reference to article 11 of the Declaration of the Rights of Man and of the Citizen of 1789, which can be deemed to be effectively ensured with respect both to public broadcasting and to the private sector only when it is guaranteed that opinions of different trends and different characters are expressed. Trust in the market alone is expressly rejected as being inadequate for this purpose. The conclusions that have been drawn from this ruling differ in part with respect to application in the public and private broadcasting sector, respectively. For the private branch, the significance of measures to combat concentration is particularly emphasized.[4] The constitutional requirement to combat concentration tended in the past and still tends to collide with the political interest in establishing and supporting a viable industry in the audiovisual sector.

1.4. The Financing of Broadcasting

Public service broadcasting is financed partly by audience fees. The television channels France 2 and France 3 are financed to a considerable extent through advertising. Private broadcasters are primarily dependent on advertising revenues. In contrast, Canal Plus is financed largely through subscriber fees. Most radio broadcasters are also financed through advertising. Other sources of funding—for instance, in 1987 TF 1 introduced teleshopping, which was later also picked up by various other broadcasters—play only a minor role.

Requirements imposed on advertising and program content are to aid in ensuring that advertising financing has only a limited effect on programming behavior. Furthermore, efforts are made to conscribe the dynamic of the economic market with the use of rules to restrict concentration.

2. LIMITS OF BROADCASTING SUPERVISION: THE EXAMPLE OF THE HAUTE AUTORITÉ

2.1. Duties and Powers

The Haute Autorité, which was set up between 1982 and 1986 as supervisory authority in the area of audiovisual communication, was entrusted with responsibilities in both public and private broadcasting. For public broadcasting, it was equipped with wide-ranging powers as guardian of independence and as supervisory authority for the purpose of maintaining programming commitments (ensuring pluralism, balance, human dignity, protection of juveniles, advertising restrictions, etc.; see, e.g., Chevallier, 1982: 567; Truchet, 1987: Nos. 18ff.). It was also entrusted with the selection of the president of the various public broadcasters.

Its powers over private broadcasting were, on the other hand, limited from the outset. For instance, it did not decide on the allocation of frequencies or the approval to construct transmission facilities. The granting of licenses for nationwide, terrestrial private television and for satellites rested solely with the government. However, it did license local private radio broadcasters, as well as cable services (after 1984, only in transmission territories up to 60 km; see Chevallier, 1982: 562). The government had the right to suspend the decisions of the Haute Autorité: It could demand a new vote on the decision. The Haute Autorité could enact binding guidelines for private broadcasting and monitor their observance as well as that of statutory restrictions.

The Haute Autorité was composed of nine members, of which three each were appointed by the President, the president of the National Assembly, and the president of the Senate. Appointments went to personalities with experience in the media field (for details on the personnel makeup, see Opitz, 1983: 97f.; Holleaux, 1987: 16), although attention was also paid to the respective political affiliation. For instance, the socialist government could count on a clear majority. A certain degree of dependence on the government was inherent due to the mode of delegation. Elements of pluralistic representation played no role; these were instead to be found, though in weaker form, in the Conseil National de la Communication Audiovisuel (National Broadcasting Council) (cf. Chevallier, 1982: 563), an advisory body that was unable to garner much significance and, since 1986, no longer exists.

The Haute Autorité quickly realized that effective application of its supervisory powers to private broadcasting was virtually impossible. Rules that ran counter to the economically motivated disseminational interests of broadcasters were usually unable to be successfully enforced in part because of the Haute Autorité's limited capabilities: For instance, it was not financially autonomous (its budget was a part of the Premier's budget). It lacked its own apparatus for monitoring programming and for technical control, making it reliant on the support of third parties, such as the transmission operations company TDF. It was also a prisoner of its own set of sanctions, which failed to provide sufficiently differentiated penalties for violations and whose use was dependent on participation by outside authorities, such as the government or the courts. This led to substantial violations by broadcasters that could not be stopped even under threat of sanctions. The weaknesses displayed by the supervisory authority were to all appearances a stimulus for private broadcasters to risk violating the law (see Cojean & Eskenazi, 1986: 205ff.).

The relative failure by the supervisory authority can be illustrated with several important examples. These remarks are based primarily on the annual activity reports by the Haute Autorité, which reveal (although usually with quite cautious formulations) the difficulties associated with supervisory work as well as on research by the author. Systematic analyses of the work of the Haute Autorité are to all appearances lacking in the literature.

In order to explain sufficiently the work of the Haute Autorité, it is helpful first to review the introduction of private broadcasting in France. Private broadcasting was born in postwar France in a state of illegality. A number of small radio broadcasters began to crop up in 1977, which were later given the catch phrase *"radio libres"* (Cazenave, 1984; Cojean & Eskenazi, 1986; Turpin, 1988: 103). The majority took a political profile, and many sympathized with a change in power. During the presidential elections in 1981, legalization was proposed in the event of a Mitterrand victory, a promise later made good on. Legalization was tied by the legislature to rather strict requirements in order thereby to keep development under political control and prevent commercialization.

As a result, radio concessions were reserved for noncommercial associations (Truchet, 1987: No. 31). Concentration was supposed to be prevented by prohibiting one individual from acquiring several licenses simultaneously. Municipal corporations and public institutions were only permitted to carry a maximum of 25% of the expenditures incurred in the establishment and on-going operations of broadcasters. Stations were prohibited from broadcasting advertising or from obtaining revenues from commercial advertising (Truchet, 1987: No. 31; Luppatsch, 1986: 790). Radio stations were financed by a support fund, into which flowed a portion of the advertising revenues generated by the public broadcasting companies. Broadcasters were to display a local flavor, and for this reason they were confined to transmitters with a small reach (500 W, 30-km maximum transmission radius). Programming and technical requirements were set forth in so-called duty booklets (*cahiers des charges*). In the event that these requirements were not observed or for reasons of public interest, the licenses could be revoked.

However, broadcasting reality soon rendered the concept outdated. The restriction to noncommercial associations and the prohibition on integration, the exclusion of commercial broadcasting, the limitation of transmission reach, and the set of sanctions all proved to be relatively ineffective.

2.2. Fields of Regulation and Supervision

Although some broadcasters did in fact observe the prohibition on commercially oriented broadcasting companies, the law of associations nevertheless made it quite easy for commercial undertakings to disguise themselves as associations. Broadcasters resorted to this strategy quite often (Haute Autorité, 1983: 39). The original concepts that broadcasting be used as a forum for associations and that social life be provided with chances for development in broadcasting were quickly outstripped by economic reality. Since this medium was well suited to commercial use, it gradually became usurped by the proponents of such interests. Private entrepreneurs displayed great ingenuity in conducting a covert trade in licenses. Licensees, particularly those experiencing difficulties, granted other natural or juridical persons the right to use the license without this being admitted publicly and without submitting the exchange to the Haute Autorité for approval (Haute Autorité, 1985: 34). It was difficult for the Haute Autorité to prove such trades, since the broadcasters were not subject to any duties of disclosure. The Haute Autorité was only able to elect to take a closer look at such circumstances prior to periodic reallocation of the licenses and, if necessary, to deny these. But after being replaced by the CNCL, the Haute Autorité no longer had recourse to these measures. All the same, it did succeed in 1984 in getting the legislature to require broadcasters to inform the Haute Autorité annually of the ten most important shareholders and to provide information on the composition of the management.

Particularly difficult to review was the rule prohibiting natural or juridical persons from directly or indirectly exercising directorial, managerial, or advisory functions or participating financially in more than one radio station. Intended as a measure to ensure economic and disseminational separation of the various broadcasters, this regulation, like many throughout the world is frequently violated; it may be observed that broadcasters very often refuse to accept such requirements. Financial and journalistic considerations argue, instead, in favor of diverse systems of contact, provisions for program exchange, and the creation of program supply systems (so-called networks). For example, radio stations that offer skeleton programming to other broadcasters began to emerge in France as well. A prominent example is the Parisian broadcaster NRJ (Luppatsch, 1986: 795), which distributed its programs to other radio stations with the aid of telecommunications satellites. The Haute Autorité could only intervene to the extent that such a system was based on cross- or multiple ownership relationships or cooperation in the managerial sector. However, it apparently was unable to uncover sufficient proof of such circumstances (Haute Autorité, 1985: 41), and it described the situation in its annual report merely as the "development of close relations between several radio stations" (Haute Autorité, 1984: 41).

The dilemma posed by cross- and multiple ownership was resolved to some degree by the legislature: In 1985 it permitted natural and juridical persons to obtain up to three licenses for the same sector (local radio, television, or cable) (Drouot, 1986: 49). Financial participation or other forms of influence were likewise permitted in up to three other stations. In this manner, the trend toward concentration was legalized in part—without, however, curtailing that trend. A 1986 analysis noted in this regard: "These rules were, however, unable to prevent the private radio market, in violation of relevant provisions, from becoming dominated today by large music chains and networks, which are mostly in the hands of press publishers, advertising agencies and the cinema industry" (Teidelt, 1986: 538).

The above-mentioned advertising prohibition was designed as an impediment to the commercialization of programming. Nevertheless, it was cleverly sidestepped, for instance, through repeated announcements, publication of free advertisement leaflets, or the "patronage" of broadcasts by wealthy shareholders. The Haute Autorité was helpless in the face of these actions (Haute

Autorité, 1983: 39). Proving and effectively enjoining such covert advertising practices took great and difficult effort. The Haute Autorité therefore recommended the path—later followed in the area of concentration as well—chosen on occasion in other countries, such as that followed by the U.S. FCC: The conflict between the norm and its observance was resolved by adjusting the norm to match reality (see Luppatsch, 1986: 791f.). The Haute Autorité recommended that the advertising prohibition be lifted altogether (cf. Haute Autorité, 1985: 36). The legislature followed this recommendation in 1984, making it possible for broadcasters to switch to advertising financing.

Private radio stations were subsequently allowed to finance their operations with advertising. But in order to reduce dependence on the advertising industry, it was stipulated that no more than 10% of revenues could stem from one and the same advertiser. In addition, a portion of advertising revenue was to be deposited in a fund designed to support noncommercial radio broadcasters.

However, permission to allow advertising created a new problem. With advertising financing, a broadcaster's transmission reach became the decisive criterion for success. The objective of maximizing returns collided with the requirement of transmitters confined to 500 W or a transmission radius of 30 km. Commercial broadcasters thus began to exceed permissible transmission strength (Haute Autorité, 1984: 37). The battle surrounding transmission strength was unleashed in the larger cities, particularly in Paris. After warnings had failed, the Haute Autorité decided in September 1984 to undertake a remarkable action: It imposed suspensions of ten to thirty days on several Parisian radio stations, depending on the severity of the violation.

In this case, however, it became apparent that successful broadcasting supervision is not simply a question of legal instruments of supervision: The affected radio stations were able to mobilize their listeners and to organize a public protest. On December, 11 1984, roughly 100,000 listeners, mostly teenagers, demonstrated against the Haute Autorité and for "their" radio stations. Since the radio broadcasters had their own transmitters, they could simply disobey the suspension without the Haute Autorité being able to take action and enforce its own powers. On the contrary, it was dependent on the transmission operations company TDF—and furthermore on the courts—for support for its sanctions, which, however, these bodies refused to offer in the face of the public protest. The stations continued to broadcast with prohibited transmission strength. Details have not been made public of the internal debates within the Haute Autorité and the substance of the interaction with the TDF (but see the listing of the procedure, in Haute Autorité, 1985: 40f.). Moreover, in its annual report, the Haute Autorité addressed its defeat only in guarded terms. The 100,000-strong public demonstration went unmentioned altogether. In summing up the conflict, the Haute Autorité cautiously wrote, "There is nothing to support the statement that the VHF band in Paris is today better protected against violations or that equal treatment among authorised radio stations is ensured" (Haute Autoritéibid, 1985: 41).

The Haute Autorité was thus just as powerless to thwart the unlawful extension of transmission strength as it was to eliminate illegal radio stations, some of which still existed after 1982. As a result, the Haute Autorité's structure for the radio sector was breached in part, and the reception of legal transmitters (e.g., the radio traffic of the fire department, etc.) was to some extent impaired. Although the operation of illegal radio stations qualified as a criminal offense (punishable with a prison term of up to three months), the Haute Autorité could not take action against this under its own power. Rather, it was the task of the TDF to control transmission operations and, when necessary, to resort to the courts. However, the demands by the Haute Autorité that action be taken against illegal radio stations (Haute Autorité, 1985: 35, Annex 25, p. 296) went unheard.[5] Its resignation in the face of requiring support from the TDF and the courts was cloaked by the Haute Autorité in cautious terms: "The legal proceedings instituted on the basis of charges filed by the TDF against these radio stations have not always been conducted as promptly and efficiently as one would desire."[6]

Events such as these reveal that the Haute Autorité set of sanctions was relatively ineffective. Its sole instrument initially was revocation of licenses in the case of violations of laws or requirements. This restriction to the most severe of conceivable sanctions crippled the Haute Autorité: Since it often seemed unreasonable to make use of this tool, it was not imposed. The Autorité therefore pressed for a differentiated, flexible set of instruments, with which it was in fact provided by the 1984 legislation. Thereafter, the Autorité was empowered to suspend licenses for six months (cf. Haute Autorité, 1985: 38). It was not, however, authorized to impose fines or criminal penalties. Although the maximum fine was raised to Fr 500,000, a court decision was necessary to levy the fine, as well as to impose penalties. In the event of violations of technical restrictions, the TDF was, as indicated above, called upon to step in.

3. A NEW START WITH OLD DRAWBACKS: THE EXAMPLE OF THE CNCL

It became increasingly apparent in the mid-1980s that the French broadcasting market had not been spared the industry drive toward concentration and commercialization. In the television sector, the government reacted to this by forming the private television networks France 5 and TV 6 (see Derkenne, 1986: 411; Ridoux, 1986: 42ff.; Drouot, 1986: 51ff; for the specifics of these decisions, see articles by Delvolvé, 1987; Moderne, 1987; Fornacciari, 1987) and shortly thereafter by privatizing TF 1 (Morange, 1987: 373). Changes were also imminent in the radio sector. It was once again a change of government—here, the election of President Chirac—that presented the formal occasion for the reform of the media system (see, in particular, Debbasch, 1987: 305ff.). The new rightist-conservative majority set its sights on stepping up the liberalization of the broadcasting market already under way, seeking at the same time to undercut "bastions of Socialist influence" (see *Le Monde,* 1986: 27). Both the abolition of the Haute Autorité and the privatization of TF 1 were thus to some extent also motivated by party politics. The allocation of the licenses for Channels 5 and 6 likewise aimed at altering the political orientation of mass media. But this political goal was also accompanied by deregulatory objectives, in particular, the desire to give the forces of the economic market more chances to develop (cf. Missika, 1986: 528f.).

These developments also had effects for the radio sector. The tailoring of transmission territories, financing, and the decision on the creation of networks were all no longer to be regulated rigidly but instead subject at most to flexible rules, with the pertinent wishes of broadcasters being taken into account by the yet-to-be-created supervisory authority. This was also intended to bring the state of the law into line with the reality of the economic marketplace. The result was the "Loi du 30 septembre 1986 relative à la liberté de communication" (for a commentary on this law, see Delcros & Vodan, 1987). It was amended on November 27, 1986, the Conseil Constitutionnel having declared important parts of the original version of the law to be unconstitutional.[7]

3.1. Duties and Powers

As mentioned, in the course of the reform, the Haute Autorité was replaced by the CNCL. Despite its relative lack of effectiveness, the Autorité was often lauded even by the opposition—particularly with regard to the supervision of public broadcasting—for its open criticism of the government in power at the time (Cotta, 1986: 258). All the same, when the opposition thereafter won the national elections, it announced plans to abolish this supervisory authority. A new authority was created with changed and in part broadened powers and was moreover composed in a different manner. The CNCL was comprised of thirteen members. Most of them were appointed by various state authorities, including the highest courts; others were co-opted (Morange, 1987: 374; Gebhardt, 1987: 124f.). The appointment method was able to ensure that eleven of the thirteen members were affiliated with the new rightist-conservative government.

Powers continued to extend to both public and private broadcasting (a summary of the CNCL's powers can be found in Morange, 1987: 378). In the area of private broadcasting, the CNCL was broadly charged with granting licenses for radio and television broadcasting via terrestrial, satellite, and cable transmission facilities.[8] The powers previously held by the TDF for allocating transmission facilities were transferred to the CNCL. The CNCL was authorized to enact rules on programming and to restrict advertising times; it was also charged with monitoring accuracy and plurality in informational programming and ensuring that dominating positions did not come about as a result of licensing decisions. The Conseil d'État supplemented this with rules on advertising, sponsoring, and the broadcasting of movies on television (Delivet & Rony, 1987: 15; Azibert, 1987: 31). The CNCL was endowed both with day-to-day programming control (e.g., with respect to advertising or the protection of juveniles) and with supervising the proper use of frequencies. The set of sanctions was refined, ranging from warnings, public admonishments, and suspensions of license up to revocation of license. Thus, a graduated use of sanctions was made possible (for doubts on their effectiveness, see Delivet & Rony, 1987: 137ff.). For administrative fines, action was required by the Conseil d'État; criminal sanctions could be imposed by the courts. These did not, however, relate to the disregarding of programming regulations but rather only to technical rules (e.g., those on permissible transmission strength, use of frequencies, etc.). The CNCL's budget was roughly ten times higher than that of the Haute Autorité (Turpin, 1988: 114) and was now set forth in the general federal budget instead of that of the Prime Minister.

3.2. Fields of Regulation and Supervision

In view of the CNCL's rather brief history, it is difficult to assess the effectiveness of its supervisory and sanctioning tools. The CNCL itself was cautious in making evaluations in its first annual report. For instance, it emphasized that in the event of violations of the rules by broadcasters, it preferred to begin "softly," that is, to give those broadcasters called upon to remedy violations the opportunity to do so (CNCL, 1987: 213). It asserted that it had sufficient tools for sanctioning private broadcasters, to the extent that coordination with the courts functioned well.[9] At the same time, it stressed that it was virtually impossible to prohibit private broadcasters from televising: Punishment was thereby imposed solely on viewers (CNCL, 1987: 215f.). It argued that broadcasting prohibitions, or the temporary or permanent revocation of licenses, were only called for in the event of particularly grievous violations and, in addition, only when the pertinent broadcaster was relatively unimportant, that is, for a local station but not for a nationwide broadcaster.

In its second annual report (CNCL, 1988), the CNCL went to great lengths to furnish detailed evidence of its successful activities. This aggressive depiction was likely motivated by the decision that had since been made to abolish the CNCL and replace it with the CSA. The CNCL called attention to its intensive, effective use of the set of sanctions at its disposal (CNCL, 1988:

181ff., especially 189ff., 193ff., Working Booklet b also 108ff.). Pursuant to article 42 of the act of 30 September 1986, these included such administrative sanctions as temporary suspension or revocation of license, which in each case was normally to be preceded by a warning. Further measures (e.g., the threat and imposition of administrative fines) were possible in conjunction with action by the Conseil d'État. Also available were criminal sanctions (criminal fines, in some cases even imprisonment), which were principally employed to combat illegal radio and television stations (CNCL, 1988: 187ff., 189). A large number of administrative sanctions (the CNCL listed fifty-five cases with respect to radio broadcasters and six with respect to television broadcasters) were imposed in order to punish violations of license conditions, such as using an unassigned frequency, exceeding the permissible transmission strength, or failing to commence broadcasting operations (CNCL, 1988: 192ff., 195ff.). On this point, the CNCL noted that it had been relatively easy for it to monitor radio broadcasters in metropolitan Paris but more difficult with those operating in the provinces (CNCL, 1988: 194). Sanctions were likewise addressed to television broadcasters, particularly in order to enforce advertising-time restrictions, quota regulations, and precautionary measures for the protection of juveniles (CNCL, 1988: 195ff.). However, the Conseil d'État did not always issue the desired statute in every case.

The CNCL's duties also included programming supervision for the economic consequences for private television broadcasters (see Bullinger, 1988: 55ff.). This function was carried out by the Service d'Observation des Programs (SOP), which for this purpose employed broadcasting protocols and monitored programs. The SOP was also responsible for monitoring balance in accordance with a rule in force since 1969, which provides that majority and opposition parties be entitled to equal broadcasting time.

One of the most strongly advocated rules in France is that regarding quotas for films and other productions of French or European origin. The corresponding requirements (an overview may be found in Blaise & Formont, 1992: 197ff., 201ff.) have thus far not been fulfilled by private television broadcasters. For instance, TF 1 has not imposed the self-restrictions stipulated by Bouygues in the public hearing prior to awarding its license, such as the annual limit of 104 movies that may be broadcast during the prime viewing time of 8:30 P.M. It has also been determined that TF 1 as well as other broadcasters have continuously exceeded the prescribed maximum advertising time (twelve minutes per hour).[10] Due to the failure of its own efforts in this case, the CNCL even turned to the Conseil d'État, which felt itself compelled to impose a fine of Fr 480,000 on TF 1 (CNCL, 1988: 112f., 195 & Annex No. 104). Violations were also ascertained with regard to the restrictions on commercial interruptions for television (e.g., repeated interruption of televised films). Finally, quota regulations (see, e.g., CNCL, 1988: 93f., 96, 196) and precautionary measures for juvenile protection (see CNCL, 1988: 76, 77) were also violated.

Plans announced by TF 1 to introduce teleshopping (télé-achat) were met with reservations by the CNCL (on the development of teleshopping in France, see Woldt, 1988: 427ff.). Nevertheless, TF 1 launched teleshopping in the fall of 1987 despite an express prohibition by the CNCL. However, the CNCL did not follow up on its threat to interrupt TF 1's programming because of forbidden teleshopping broadcasts. Following heated public discussions, the CNCL was obligated by statute in early 1988 to enact special rules on teleshopping. It complied with this duty by way of a decision on *télé-achat* of February 4, 1988, which was later modified (these decisions are reproduced in CNCL, 1988: 382f., 383; see also CNCL, *Lettre d'Information,* No. 10, 1988). Teleshopping programs were thereby made subject to minimum and maximum times: They may not be shorter than thirteen minutes or longer than ninety minutes per week. They may only be broadcast between 8:30 and 11:30 A.M. or after conclusion of the normal broadcast day. They are also subject to substantive limits: Children may not be included in the broadcasts; goods and services must be depicted, both quantitatively and qualitatively, as comprehensively as possible; and prices, warranties, and other terms of sale must be indicated in a clear, comprehensible manner.

The CNCL decided to review the allocation of radio frequencies and, where necessary, to reassign licenses. It started with this task in Paris. In so doing, it made an effort to leave intact not only the large networks but also smaller stations, including club radio stations and "opinion radio stations" (*radios d'opinion*), which advocate a specific stance. Although the CNCL's frequency plan met with difficulties, it was successfully implemented. Legal action had to be taken against fourteen unauthorized stations. Transgressions of permitted transmission strength were rectified following requests by the CNCL. To this extent, the CNCL was more successful than the Haute Autorité, although it also had to refrain from conflicts with popular stations. However, as mentioned above, the CNCL had difficulties with respect to monitoring in the provinces. In seeking to put a stop to illegal radio stations, it was not always able to count on the necessary support of the Public Prosecutor's Office, although it was supported by the courts (CNCL, 1988: 187f.). The amnesty granted by the act of 20 July 1988 did not prompt all illegal radio stations to terminate their operations (CNCL, 1988: 188).

Special difficulties became evident in the effort to enforce statutory restrictions on concentration (cf. Boudet, 1987: 83ff.; Schulz, 1990: 128ff.). Although the 1986 act was in some respects more tolerant with regard to concentration than the 1982 act, it also established limitations on cross- and multiple ownership (for an overview, see Turpin, 1988: 119f.; Boudet, 1987: 66). In combating concentration, the CNCL was, in particular, faced with the considerable problem of proof.[11] For instance, it was particularly difficult for it to stem the illegal trade in radio licenses, which was possible because the operating company could be purchased without the ownership relations among the purchasers always being made entirely clear. It has been asserted that figureheads and complexly structured corporations were set up for this purpose without any recognizable punishment by the CNCL (see "Die heimlichen Wellenreiter," in *Die Zeit,* 20 November 1987: 47).

In the television sector, the CNCL proved to be more the patron of concentration. Seemingly motivated by the effort to stem the tide of American media companies with the aid of efficient French firms, it supported the emergence of national multimedia groups (for the development of concentration in the 1980s, see Schulz, 1990: 175). Yet it was called upon to implement statutory limits on concentration (CNCL, 1988: 127, 128ff.). It also made an effort to preclude monopoly positions among program producers (CNCL, 1988: 134ff.).

A particularly important function of the CNCL consisted of the reallocation of licenses for the national television networks 5 and 6 and the privatization of TF 1. Its decisions cast the CNCL into the arena of political conflict. In some cases, it boosted multimedia concentration (as with the decision on the licenses for Channels 5 and 6); but in others, it surprisingly gave preference to companies new to the field (e.g., to the construction entrepreneur Bouygues pursuant to the privatization of TF 1). The CNCL made political headlines as a result of suspicions of nepotism: One of the members was alleged to have intervened for the benefit of friends in the course of allocating a radio frequency (*Le Monde,* 29 October 1987, December 11, 1987). But even this was overshadowed by the headlines surrounding President Mitterrand's public chiding of the CNCL in September 1987, when he attacked not merely the result of licensing decisions but also the political one-sidedness of the CNCL in general (*Le Point,* 27 September 1987: 26). His criticism continued unabated and became decisive for the plan to replace the CNCL with the CSA, which was implemented following Mitterrand's reelection in 1988.

3.3. Assessment of Supervision by the CNCL

The foregoing serves to show how difficult it was for an authority that not only lacked a reputation bolstered by political consensus and a tradition of independence but also, because of its composition and method of election, could not shield itself from political attacks. Such weak points made the already difficult task of controlling private broadcasters more difficult, a task further complicated and problematic for the simple reason that governmental regulation normally runs counter to the economically motivated disseminational interests of broadcasters. However, the legislature did not provide for any sort of active regulatory role for the supervisory authority; on the contrary, it conceived of the CNCL mainly as a law-applying authority (Guillou & Padioleau, 1988: 133).

Although some of the Haute Autorité's shortcomings were eliminated—for instance, the excessively weak powers were enhanced—important powers were lacking, such as adequate possibilities for the control of concentration. However, one major shortcoming was not remedied, and this has marked the development of French broadcasting: the lack of confidence in the independence of broadcasting by the government. Thanks to its early demise, the CNCL was spared the review of how successful a supervisory authority can be in this situation.

The available literature does not clearly reveal the intensity of the ties that existed between the various supervisory authorities and the supervised broadcasting industry. Starting points for such an evaluation, while lacking for the Haute Autorité, can be found for the CNCL's work. The CNCL reported that an ongoing dialogue with representatives of the broadcasting industry had come about, which was said to have made it easier to implement programming control (CNCL, 1987: 173). Such a dialogue was beneficial for both sides. The CNCL was dependent on the willingness of those supervised to cooperate, just as, in the reverse direction, the broadcasting industry had an interest in taking part in the CNCL's decisions. The decision-making practice by the CNCL showed that the supervisory authority had learned to treat the interest structures of private broadcasting adeptly (cf. Guillou & Padioleau, 1988: 129; Belot, 1988: 5, doubts that the CNCL was ever taken seriously by those responsible for programming). To this extent, one also finds in France that steps are increasingly being taken to establish a cooperative network between the supervisory authority and private broadcasting.

4.   A RENEWED EFFORT: THE CSA

A new beginning was made in January 1989: The CNCL was replaced by the Conseil Supérieur de l'Audiovisuel (for the background to the demise of the CNCL, cf. Chevallier, 1989: 63f.). As early as Mitterrand's campaign for reelection, notice was given that this restructuring of the French media system would be carried out in the event of a victory. In his Open Letter to all French Citizens, Mitterrand complained of the CNCL's shortcomings and called for the creation of a new, independent supervisory authority to be set down in the French constitution (cf. Chaumont, 1989: 8; Morange, 1989: 236).

4.1. Duties and Powers

As with the Haute Autorité and the CNCL before it, the CSA was given the legal form of an independent administrative authority (*autorité adminis-*

trative indépendant) (Huet, 1989: 181; Debbasch, 1991: Nos. 274). It is charged with safeguarding the liberty of "audiovisual communication" within the framework of the requirements set forth in the 1989 Media Act (for the constitutional issues pertaining to freedom of opinion, see Debbasch, 1989: 305). Specifically, the CSA is obligated to ensure compliance with the doctrine of equal access, to guarantee the independence and impartiality of public broadcasting, to promote free competition, and to monitor quality and diversity in programming, the development of national productions, and the dissemination of French language and culture.

The CSA's size and the manner in which its members are appointed is modeled on the Haute Autorité: It is composed of nine full-time members, of whom three each are appointed by the national President and by the presidents of the Senate and the National Assembly (cf. Chaumont, 1989: 10). In contrast to the CNCL, the president of the CSA is designated by the French President. One-third of the CSA members are replaced every two years. A glance at the specific appointments reveals that the effort to give expertise priority over political affinity has apparently met with success.[12] Participation in the CSA is incompatible with the assumption of an electoral mandate (Chaumont, 1989: 11). In order to safeguard its work methods, the CSA has set its own rules of procedure (*Lettre du CSA,* No. 3, 1989, 2f., 5).

The act of January 17, 1989[13] did not fundamentally alter the existing media constitution (as pointed out by Huet, 1989: 182; Nevoltry & Delcros, 1989: 5; Truchet, 1989: 209). Rather, the state of the law in force since 1986 was supplemented or amended in only a few areas (for a comparison of the functions of the CSA and the CNCL, see Coppey et al., 1989: 15f.). In relation to the CNCL, responsibilities were primarily changed in three areas: First, as of March 31, 1990, the CSA lost its power in the telecommunications sector (see Delivet, 1989: 5; CSA, 1991b: 43ff.). Second, it was given responsibility for supervision of the broadcaster Canal Plus (see Morange, 1989: 249). Third, the CSA was forced to relegate the power to monitor conditions of fair competition in the media sector to the Conseil de la Concurrence (Council of Competition). The CSA was henceforth charged only with applying the anticoncentration rules of the Media Act (for details, see Nevoltry & Delcros, 1989: 150 ff; Cousin & Delcros, 1990: 347ff.; Spitz, 1991: 33ff.). Pursuant to the original draft law, the CSA was to be endowed with the power to draw up general rules (*règles déontologiques*) for setting program and production quotas and for the areas of advertising and sponsorship (*parrainage*) (for the original version of article 11 of the Media Act, cf. Nevoltry & Delcros, 1989: 223). However, the Conseil Constitutionnel struck down this transfer of law-making functions (*pouvoir réglementaire*) to the CSA, arguing that this would constitute an impermissible interference with the government's constitutional rights.[14]

Significant modifications have been undertaken in the licensing system.

Private radio and television broadcasters must now negotiate a contract (convention) with the CSA setting the terms for use of the license, which is granted for either five (radio) or ten years (television).[15] Negotiations take up the length and substantive direction of programming, in-house production quotas, quotas for French and EC productions, coverage of regional concerns, advertising restrictions, and other issues (Delivet, 1989: 3f.; Mestmäcker, Engel, Gabriel-Bräutigam, & Hoffmann, 1990: 113). In order to ensure that these obligations are in fact implemented, conventional penalties may also be agreed upon. The conclusion of a convention makes it possible to set differentiated stipulations, which are likely to be superior to rigid statutory, and thus abstract, general duties. It is also intended to enable the CSA to call upon broadcasters to operationalize the promises they made upon receiving the license (for this legislative intention, cf. Huet, 1989: 184; Nevoltry & Delcros, 1989: 122). Moreover, this strengthens the cooperative ties between industry and supervisory authority: In this way, a communicative relationship aiming at negotiation and compromise can be built up from the very outset, which will undoubtedly have an effect on the further administration of broadcasting supervision.

The previous licensing procedure has been supplemented with a number of detailed regulations. For instance, in awarding private radio licenses, the CSA is charged with setting the geographical transmission territories and the category under which broadcasters fall (Huet, 1989: 185; for classifications undertaken by the CSA, see CSA, 1990: 148ff.). In order to be able to take into account the special requirements associated with the supervision of local radio stations, the 1989 Media Act provides for the establishment of decentralized monitoring committees (Comités Techniques Radiophoniques, or CTRs) (Delivet, 1989: 6). The CTRs are an integral component of the CSA. The main duty of these advisory committees is evaluating the applications for licenses and preparing recommendations for the central office in Paris. In addition, they review whether the obligations set down in the convention are being complied with and report possible violations (CSA, 1990: 153f.; CSA, 1991b: 131ff.; Cousin & Delcros, 1990: 95). The CTRs themselves have no power to impose sanctions.

Among the central improvements in the 1989 Media Act is the further refinement and extension of the set of administrative sanctions previously available to the CNCL (for a summary of the new regulations, see Morange, 1989: 243ff.; Coppey et al., 1989: 19ff.). The CSA is now able to direct not only revocation or suspension of the license or individual programs but may also curtail the length of the license term and impose criminal fines.[16] Compliance with the terms set forth in the license agreement may furthermore be enforced with the aid of the above-mentioned conventional penalties.[17] In the event a sanction is imposed, the offender may call upon the Conseil d'État to review its legal permissibility (Huet, 1989: 186; Nevoltry & Delcros, 1989: 175ff.).

A violation of the provisions of the 1989 Media Act or of the license agreement does not in practice necessarily lead to the imposition of coercive measures. The CSA makes an effort to fulfill its monitoring duties first with informal approaches and persuasion (CSA, 1990: 23, 303; CSA, 1991b: 20; CSA, 1993: 27; CSA, 1994a: 25). In so doing, the threat of imposition of coercive measures becomes a bartering chip for enforcing statutory requirements. One example for this approach is offered by the negotiations between the CSA and TF 1, which took place when this broadcaster was unable to fulfill certain conditions that had been placed upon it, namely, to broadcast programs originating in France or the EC and to produce its own programs. In order to compensate for these deficits, TF 1 committed itself (in addition to existing requirements) by year-end 1990 to purchase French productions for Fr 50 million, to broadcast sixty-nine hours of programming for juveniles, and to invest Fr 8 million for support of the program industry (CSA, 1990: 284f.). In exchange for these compensatory steps, the CSA refrained (for the moment) from the use of coercive measures.

4.2. Fields of Regulation and Supervision

The choice of cooperative approaches does not, however, mean that the CSA refrains altogether from using the set of sanctions provided to it by statute. Already, the CSA's 1989 report and statement of account lists for the area of local radio alone sixty-four warnings, eleven suspensions, and one revocation of license (CSA, 1990: 312, Annex 74). Most of the sanctions were imposed in cases involving transgression of permissible transmission strength.[18] However, administrative action was also taken—in continuance of proceedings initiated by the CNCL—against the television stations TF 1, La Cinq, and M 6 for failure to fulfill the fixed transmission and production quotas. (For the obligations currently imposed on various broadcasters, see CSA, 1994a: 248ff.; 257ff.). In so doing, it often imposed substantial fines, which were calculated on the basis of the number of hours lacking to fill the quota. The fine of Fr 20,000 to 60,000 that would have been the legally imposed sanction for TF 1's violation of quota regulations in 1990 was, as mentioned above, not imposed by the CSA only because of TF 1's far-reaching promise regarding its future behavior (CSA, 1990: 313ff.). Since TF 1 did not comply with its obligations in the following year, CSA imposed a fine of Fr 30,000. M 6 and La Cinq have also repeatedly violated quota rules.

The rules dealing with quotas for European and for French productions have been modified over time, in part due to pressure from the EC Commission. The broadcasters have attempted to avoid the nevertheless high quotas by meeting their obligations mainly at night between 1:00 A.M. and 6:00 A.M., that is, during periods with few viewers. In reaction to this, the CSA decided in 1990 that quotas had to be fulfilled between 6:00 P.M. and 11:00 P.M. (on Wednesdays between 2:00 P.M. and 11:00 P.M.). The distribution of the required programs, however, has remained a problem (CSA 1993: 265f.).[19] Several national television companies have in addition been subjected to financial penalties or admonishments in a number of cases for violation of provisions on juvenile protection and quantitative advertising regulations (for details, cf. CSA, 1990: 317ff.; CSA, 1993: 464ff.; CSA, 1994a: 460ff.).

It is noteworthy that public broadcasting has done a considerably better job of fulfilling programming and other obligations imposed on it. For instance, the CSA was able to state in its report for 1989 that Radio France had fulfilled all its programming requirements (CSA, 1990: 275). Antenne 2 has essentially discharged its obligations as well. Shortcomings were ascertained merely with regard to origin quotas in the area of juvenile programming and the type of sponsoring.[20]

In carrying out its functions, the CSA is well aware that, despite the existing set of sanctions, the requirements of media law can only be enforced when private broadcasters enjoy sound economic health. In view of the fact that the broadcasters La Cinq and M 6 were booking losses,[21] the CSA made an effort to improve prerequisites to the sale of advertising time. For instance, the reach of these broadcasters was increased, and permissible maximum advertising time was extended (CSA, 1990: 76). The CSA justified these measures by pointing to the unabated price increases in the area of program acquisition. As a result, said the CSA, private broadcasters were forced to make increasing use of cheap American productions in order to cut their costs; it asserted that this practice necessarily brought with it the danger of slackening program quality and an undercutting of cultural objectives (CSA, 1990: 78f.).

It should be noted that both the CSA and the government have rejected further relaxation of the rules, allowing La Cinq to go bankrupt in the process. This was at least in part motivated by the feeling that there were more broadcasters in France than could be supported by the advertising market. But another factor was probably the desire to be able to put the terrestrial frequencies awarded to La Cinq to other use (specifically, for ARTE).[22] The concern about the economic health of the broadcasters and the assumed negative effects of very restrictive advertising rules on programming are also reflected in the CSA's propositions concerning the future legislative framework. The CSA suggests allowing a second break for advertising in films (*Oeuvre audiovisuelles*). The rationale is to prevent films being substituted for reality TV and other programs that can be interrupted more frequently or broadcasters, because of financial restraints, using cheap American programs (CSA 1994a: 98).

A central facet of CSA work is monitoring compliance with the programming obligation placed on both private and public broadcasters. An important function in this regard relates to ensuring diversity of opinion. The CSA evaluates diversity of opinion mainly by performing a quantitative analysis of the speaking time of the members of the various political camps (government, parliamentary majority, parliamentary minority, political parties not represented in parliament) in national television and by analyzing the complaints received (CSA, 1993: 226; CSA, 1994a: 221; for figures see, e.g., CSA, 1993: Annex 64; CSA, 1994a: Annex 71). It publishes the results of the quantitative analysis monthly and, if necessary, demands remedy of substantial imbalances (CSA, 1994a: 222f.).

In the view of the CSA, the "3/3 rule" (*règle des trois tiers*), under which the government, the opposition, and the majority parties are each to be provided with one-third of broadcasting time, has proved to be a failure. The CSA alleged that under this rule the government and the majority parties supporting it would receive exceedingly good opportunities for self-depiction (CSA, 1994a: 200). Moreover, CSA determined, the rule is not sufficiently flexible, since self-depiction opportunities virtually always depend on the given political issue (see CSA, 1994a: Annex 58). Despite these shortcomings, the 3/3 rule continues to represent a central criterion for determining balance (cf., e.g., CSA, 1994a: 222), since the CSA has been unable to develop any more reliable indicator of comparable simplicity (CSA, 1991b: 181). However, the CSA seems to have introduced a certain degree of flexibility in the way it handles the 3/3 rule when evaluating the results of its quantitative analysis (CSA 1994a: 223).

In addition, the CSA also makes an effort to protect the truth or reliability of information. For instance, the CSA once objected that pictures of a demonstration targeting German poison gas exports were, in the same broadcast, claimed first to have been taken in Jerusalem and later were said to be from Bonn (*Lettre du CSA,* No. 28, 1992: 5). As sanction, the CSA made public the rule violation in press notices. The CSA also took action to enforce military censorship during the Gulf War.[23]

In 1989 a dramatic rise in programs unsuitable for juveniles that were broadcast before 10:30 P.M. was noted in private television.[24] At the initiative of the president of Antenne 2, experts got together with representatives of private broadcasters to work out a juvenile protection directive, which was published in February 1989. TF 1 was the only broadcaster that refused to accede to this code of conduct. It instead opted to introduce its own protective measures, such as labeling programs unsuitable for juveniles with a blue triangle (CSA, 1990: 222). In spite of these measures, the CSA in May, 1989 published its own directive on juvenile protection (CSA, 1990: Annex 60). This obligates broadcasters to refrain from broadcasting depictions of violence or programs of an erotic nature between 6:00 A.M. and 10:30 P.M. For violations of this directive, the broadcasters M 6 and La Cinq have been subjected to sanctions and violations by other broadcasters have been punished as well (CSA, 1990: 224 ff.; CSA, 1991b: 228). In applying sanctions, the CSA also makes use of its power to oblige a broadcaster to point out its own violation in its programming (for more on this power, see Blaise & Formont, 1992: 244). In 1991, as a condition of license, La Cinq even set up an ethics commission to ensure that during prime time neither sex nor inappropriate violence was broadcast. Such measures taken by the CSA appear to have had a temporary effect; with regard to 1992, CSA noted that by and large, the broadcasters performed considerably better than they had before (CSA, 1993: 242f.). As early as the following year, however, the CSA noticed a "certain regain" of "violence and voyeurism" with respect to overall television programming, even though not a single program triggered a sanction (CSA, 1994a: 235).

By 1989, enforcement of advertising requirements already made up a considerable amount of the CSA's activities (CSA, 1990: 18). In so doing, the CSA was able to resort to the monitoring system developed by the CNCL (for details on control practice, see *Lettre du CSA,* No. 6, 1990: 1ff.). To this end, the CSA is allowed to conduct advance control, that is, a sort of precensorship measure. The special committee set up for this purpose reviewed a total of 8,408 programs in 1989 (6,178 in 1988). Objections were lodged against 22% of the commercials, thereby indicating the need for modifications. However, only thirty-four commercials (i.e., 0.4% of all such broadcasts) were rejected in their entirety (CSA, 1990: 230). Thereafter, 4,429 commercials failed to pass subsequent conformity control, in which compliance with the endorsement previously issued is reviewed. At this stage, objections were raised in only 270 cases (CSA, 1990: 233). In view of this enormous control effort, it comes as little surprise that the CSA simplified its procedure as of September 1990: Advance control was abolished and replaced with the obligation to announce the broadcasting of the commercial and submit a copy of it to the committee.[25] From February 15, 1993, even advance declaration was no longer required (CSA, 1994a: 339).

The CSA has complained of substantial deficits in the implementation of not only the above-mentioned quota rules but also the provisions on sponsorship and combating concentration (for details on the legal possibilities, see Blaise & Formont, 1992: 248ff.). It has accused private television broadcasters of general (global) disregard for existing regulations (CSA, 1990: 242). For instance, M 6 and La Cinq have had informational programs sponsored although this is expressly prohibited. Furthermore, the escalating use of advertising elements in sponsored programs has led to an increased mingling of sponsorship and advertising, which threatens to undermine advertising regulations. Not in small part as a result of the CSA's energetic supervisory activities, broadcasters have tailored sponsoring more closely to the legal guidelines (CSA, 1991b: 257f.; e.g., in 1993, CSA took measures with respect to sponsorship in private television in three cases only; CSA, 1994a: 356). All the same, it is still possible to identify the sponsor in sponsored programs, as long as this is "periodic and discrete."

In fighting concentration in the media, the CSA is faced with problems similar to those faced by the Haute Autorité and the CNCL. It also has to deal with the seemingly endless conflict between the fight against concentration and the support for media companies that are competitive, and if need be, competitive at an international level. Also difficult to solve are the problems of interpreting norms to fit the needs of a given case, as well as those of collecting the statistical data needed for applying them. The concentration rules in the 1989 Media Act apply, for example, not simply to potential licensees but also to those companies that "control" them (*"contrôle"* or *"sous son autorité ou sa dépendance"*).

Although over the years the CSA has not been freed from the task of fighting concentration, 1994 did see a certain loosening of the concentration rules. Above all, the former limitation of shareholders to 25% of the shares of a TV company was seen as an impediment to the development of communication groups (cf. *Lettre du CSA,* No. 54, 1994).

The CSA has dealt actively with both television and radio. In the radio sector, the CSA began in 1989 to develop special activities. Special emphasis was placed on ensuring the nontransferability of frequencies and bringing about diversity of programming in radio. Priority was also given to ensuring the survival of noncommercial and independent radio stations (CSA, 1990: 149ff.; CSA, 1994a: 145ff.).

In the authority's view, cable was not widespread enough to be of use in disseminating satellite television.[26] This prompted the CSA to make an effort to accelerate the approval of new cable networks. In addition, the use of the satellite *TDF 1* was regulated (for the allocation procedure, cf. CSA, 1990: 125ff.), the introduction of new technologies pushed forward and the monitoring criteria further differentiated (CSA, 1990: 104ff.). With the 1990 law dealing with the regulation of telecommunications, the basic legal conditions as well as the CSA's supervisory powers in the cable sector have now been specified in more detail (CSA, 1991b: 43ff., 95ff.). However, even the CSA cannot change the fact that the development of cable in France has been disappointing.[27] As a result, no attractive offerings had been made via cable until very recently, and the future of this sector still seems uncertain (cf. CSA, 1994a: 121ff.).

4.3. Assessment of Supervision by the CSA

It is difficult to attempt an overall appraisal of the CSA's activities. In comparison to those of the Haute Autorité and the CNCL, significant improvements have been made in the CSA's set of sanctions. Nevertheless, in attempting to exercise control over private broadcasting, even the CSA has found that the possibilities for controlling private broadcasting are limited. Because of the opposition of private broadcasters, substantial deficits can be noted in the implementation of important rules of the 1989 act, such as quota regulations and the provisions on commercials and sponsorship. Proceeding from the assumption that these problems are primarily engendered by the high costs associated with complying with the rules, the CSA has sought, particularly with M 6 and La Cinq, to create more latitude for advertising deals and thus for revenues as well. The question of whether the desired improvement of the economic situation will lead to stronger compliance with the rules remains unresolved, since these rules commonly conflict with the broadcasters' economic interests.

The bankruptcy of La Cinq (see Meise, 1992: 242ff.) has made it clear that a broadcaster's economic survival will not become the primary purpose of media policy. However, it should be noted that La Cinq was unable to achieve success with the public or to garner political support. Many of France's media politicians had, at any rate, reached the conviction that there was one television broadcaster too many. To all appearances, the government did not act to help avoid the bankruptcy. In fact, public complaints have been voiced that it pressured banks to refuse to extend further credit to La Cinq. The CSA declined to approve the release from programming commitments demanded by La Cinq. But this would have been equivalent to a concession of all programming supervision. In sum, the bankruptcy of La Cinq represents, in an international comparative setting, a remarkable example of how a supervisory authority placed a broadcaster's survival interests behind that of upholding other obligations it had entered into. At the same time, the demise of La Cinq has also meant improved earnings prospects for competing broadcasters, such that the bankruptcy essentially served the viability of the overall television order in France.

It has become apparent in several areas that the monitoring of broadcasting is impeded by limits on the applicability of the instruments of control or that these are largely unsuited to achieving certain set objectives. For example, a number of existing anticoncentration rules have come to be thwarted by the difficulty of obtaining sufficiently reliable data. Other rules have also proved to be of limited suitability for achieving the desired objectives. For instance, with regard to political programs, the 3/3 rule became increasingly inappropriate as a means of ensuring a reasonable portrayal in broadcasting of the spectrum of political opinion.

At present, the CSA can be seen as employing two main strategies in order to deal better with its supervisory dilemma: First, it constantly attempts to refine and modify the set of instruments at its disposal. Second, as with the CNCL before it, the CSA has further developed forms of cooperative and negotiable supervisory conduct, although as an aid to achieving this, it has made use of the power to threaten that emanates from formal sanctioning instruments. All the same, the number of violations the CSA formally admonished—the extent of informal action is of course not fully documented—is surprisingly high (Blaise & Formont, 1992: 244). Also unmistakable is the discrepancy between private and public broadcasting: The latter is apparently more easily able to operate in conformity with the norms than are private broadcasters, who are locked into a stiff battle for revenue and audiences.

The CSA's formal and informal interventions relate at most only indirectly to the quality, in particular the substantive diversity, of programming. A favored object of broadcasting supervision is the observance of easily measured "target values" and in particular the warding off of "dangers," including possible violations of privacy. Supervision of broadcasting is unable to commit itself to the positive assurance of diversity in programming generally but must instead limit itself to specific fields of regulation—such as program quotas, combating concentration, and so forth—as well as rectifying violations of the law (relating to advertising restrictions, juvenile protection norms, etc.). It goes without saying, however, that the rectification of such violations by no means ensures that programming will thereby meet substantive target values, such as those of quality, diversity of communication, or the portrayal of specific, French cultural values.

Since to all appearances the possibility for effective control is strongly contingent on the basic economic and legal structure of the media system, it is doubtful whether it makes sense to entrust one and the same authority with monitoring broadcasting as both a public service and a commercial undertaking. But it is conceivable that the factors and considerations decisive for commercial broadcasting will in this manner have a stronger effect on the public sector. In France, where public service broadcasting was substantially reduced in favor of commercial broadcasting, this risk is particularly large. It seems that in supervising private broadcasting, the CSA mainly accepts the latter's conditions and to a great extent rejects the idea of continuing the older public service tradition.

The appraisal of the effectiveness of broadcasting supervision in France is made more difficult by the fact that the CSA's work was not accompanied by any broad-based, critical media research, studying, for example, the implementation of supervisory norms and their suitability for enforcing media-policy objectives. Among the isolated critical remarks, one finds the assertion that the CSA's activities principally belong in the category of symbolic policy-making (Wolton, 1990).

4.4. Discussion of the Future

In spite of sluggish development of the cable networks, discussion on the consequences of a technically possible future multiplication of programs has already begun in France. At the moment, however, there are more questions than answers, and the CSA has not yet taken up a particular stance. A working group, composed in part of representativess of the affected branches, has formulated a paper in which at least some of the pertinent questions are addressed; the CSA has submitted this working paper to the government (CSA 1994c: 1ff.). This report points out that the present regulatory framework needs to be adapted to the exigencies of a multichannel environment. The regulations concerning licensing, for example, suppose that one frequency can carry only one service and is therefore based on an assumption that will be outdated as soon as data compression is introduced. With respect to other regulatory fields, such as anticoncentration regulation or must-carry rules, the report has only pointed at the fact that it remains to be decided to what degree a multichannel environment is to be regulated. The main task, according to the working group, will be to define the role of the various distribution systems and to design the framework in which they will compete. In the face of internationalization of competition in the sector, French regulators appear to aim for a framework that is favorable to the French audiovisual industry. Once more, the discussion of broadcasting policy tends to be heavily influenced by objectives of economic policy, which are also linked to notions of cultural policy. The discussion on information superhighways (*autoroutes de l'information*) that has begun in France, as well as in other industrialized countries, will in all likelihood also be characterized by these considerations.

5. SUPERVISION OF BROADCASTING IN THE POLITICAL ARENA

The history of supervision of broadcasting in France makes clear that it is not easy in that country to break from the traditional mode of direct governmental supervision and establish an authority that is, on the one hand, independent but, on the other, in a position to exercise effective control over broadcasting. Since, in keeping with its centralist tradition, France has opted to organize supervision centrally, it was from the outset possible for the government to exert influence on the supervisory authority. Over the course of time, however, a variety of political camps intensified efforts to strengthen the independence of broadcasting supervision,[28] which also met with relative success with the establishment of the CSA. Centralization has as such remained intact,[29] and in the face of concentration in the media sector and the admission of national broadcasting networks, it has found additional legitimization. Nevertheless, it has remained controversial.

With the exception of Italy, France restructured its media order in the 1980s in a more radical manner than its neighbors (this is also the opinion of Kuhn, 1988: 181). The forced privatization in France was coupled with the effort to avoid the rampant development so characteristic in Italy and to set framework requirements at the state level. However, France was forced to learn that the market mechanism often prompted broadcasters to act in a way that conflicted with the requirements set by the state. Normative requirements were disregarded on various occasions. Norms that conflict with the economic and disseminational interests of the broadcasters had in the past and have now, at best, limited chances of compliance.

Thus far unique in Europe, the French experiment with privatizing the most popular public television broadcaster has been a success for private television: TF 1 is clearly the dominant player in the market, and as a result of its prominent position, it enjoys an advantage over all private and public competitors. However, this privatization move delivered to the public broadcasting system a blow from which it could not recover. France 2 particularly sought to compete with TF 1, though without much luck, by trying to copy the latter's programming concept. However, duplication could not ensure the same success with public broadcasting. Since its programming is regionalized, France 3 is in an easier position because it can work to achieve complementary programming. But both public broadcasters are suffering from financial constraints. The CSA has been unable to do anything to stop the decline of public broadcasting.

If one leaves aside compliance with individual rules and instead addresses in a broad, overarching manner the attainment of the objectives set for media policy, the result is most sobering. France has achieved to only a limited degree the objectives it has pursued with such vigor, namely: ensuring a specific cultural identity and expanding national and European program production, as well as, pursuant thereto, warding off foreign, in particular, American, programs. However, quota policies have had a massive impact on media-policy discussion in France. By no means have they been inconsequential; rather, they have made a contribution to strengthening the French audiovisual industry.

The efforts to permit wider access to broadcasting licenses and thus to make television more plural have been successful only to a limited extent. Moreover, the wave of concentration has not stopped in France, and France like other countries has been unsuccessful in preventing broadcasters from locking into international, multimedia conglomerates. It has also discovered that competition among four, large private television broadcasters—three of these financed with advertising—and two public broadcasters results in skyrocketing program costs (cf. CSA, 1994a: 325ff.) but does not sufficiently ensure substantive differentiation of programming, leading instead to somewhat homogeneous programming.[30] The French advertising market is also apparently too small at the moment to ensure that such a number of competing broadcasters will operate profitably. The two private broadcasters La Cinq and M 6 were operating with substantial deficits, which meant that economics alone reduced their ability to broadcast high-quality programming. In the end, La Cinq went bankrupt.

In contrast, Canal Plus has been an unqualified success. After start-up difficulties and with strong governmental support (Jezequel, 1991: 3ff.), this pay-TV channel is firmly established in the market and is operating profitably (Palmer & Sorbets, 1992: 68; Meise, 1992: 252ff.). Canal Plus usually scrambles its broadcasts, primarily for movies, sports events, and entertainment programs; during midday and in the evening (i.e., 6:30 P.M. to 8:30 P.M.), however, Canal Plus broadcasts unscrambled programs with commercials. Canal Plus has received a substantial number of privileges, such as preferred allocation of terrestrial frequencies. Special contracts with the French film industry provide it with a time advantage for the broadcasting of French films. In return, the French film industry has a relatively secure purchaser of its own film products. Canal Plus is one of the most profitable media companies in Europe, with holdings in many other French and foreign media companies; it has become an international, multimedia enterprise. In this way, France has been able to establish a strong French company in the European media market. Thus, a specifically French variant of media policy has succeeded, one that aims at strengthening the French media industry and the cultural influence of France.

The founding of La Sept, begun as early as 1986, was another effort to ensure at least a culturally innovative supplement to the palette of programming. This attempt at the provision of additional, culturally oriented programming was initially burdened with using the satellite *TDF 1*: The fact that the transmission norm D2 MAC was forced upon it for reasons of industrial policy put a considerable damper on customer acceptance. Whether the merger of La Sept into the German–French cultural channel ARTE will improve the situation is still open.

State supervision of broadcasting is only one of many actors in media development and is by no means the most influential. The noteworthy activities of the CSA do call attention to the existence of legal commitments tied to the operation of broadcasting, and they indisputably have—as with the Haute Autorité and the CNCL before it—aided in preventing the emergence of Italian or American program conditions in France. But detaching the development of broadcasting from the trend toward adjustment to advertising needs that was triggered by commercialization and from homogeneity in the palette of programming, will require greater power than mere supervision.

NOTES

1. For a brief overview of the development of the French broadcasting system, see Institute National de l'Audiovisuel, 1988; Palmer & Sorbets, 1992: 52ff.; Noam, 1991: 95 ff; Jezequel & Pineau, 1992: 429ff.; Lange, 1994: 64ff.

2. Nevertheless, there have long been—as with FR 3—regional studios or regional information offices.

3. Décision No. 86-217 DC, 18 September 1986, J.O. 19. September 1986, 11294ff.

4. Décision No. 86-217 DC, 18 September 1986, J.O. 19 September 1986, 11294, 11297. For the significance of anticoncentration rules in the area of external pluralism, see also Drouot, 1987. The trend toward concentration has also taken on large dimensions in France. See Schulz, 1990: 175ff.; Spitz, 1991: 33ff.

5. M. Cotta, at that time president of the Haute Autorité, called for "an effective policing of frequencies."

6. Haute Autorité, 1985: 39. A similar finding was made in Italy. The Italian radio network Radio Radicale had for years violated the prohibition of live broadcasts. Only much later did the PTT take the matter to the courts. The judgment rendered in 1983 was, however, not enforced by the public administration. See Pace, 1983: 630ff. One reason for this was likely that the network was quite popular among listeners.

7. Décision No. 86-217 DC, 18. September 1986, J.O. 19.September 1986, 11294. For details on this judgment, see Boudet, 1987: 66ff.

8. However, the government has been endowed with several, important powers: See the cable ordinance of 29 September 1987. Cf. also Delivet & Rony, 1987: 135f.

9. CNCL, 1987: 215. It did not, however, take action against all violations: See, for example, Oehler, 1988a: 359 (on heeding the obligation to broadcast local programming).

10. For the relevant values, see the list in CNCL, 1988: 108. The values are not identical for the various broadcasters. For details on exceeding limits on advertising time (by public broadcasters as well), see the references in CNCL, 1988: 112. On other violations of advertising restrictions, such as, the requirement that advertising be separated from programming, see CNCL, 1988: 113.

11. Its mandate in this area has, however, been relaxed in that it not only tolerated but even promoted the creation of networks. For the development of networks in radio, see Oehler, 1988a: 358ff.

12. However, the appointment of the CSA president, Jacques Boutet, met with criticism; he has been accused of having provided for an expansion of the influence of leftist parties during his tenure at TF 1 (at that time a public television broadcaster).

13. The following remarks concentrate on the most important changes implemented in the media order by the new statutory regulation. For a broad depiction of current broadcasting law, see Gavalda & Boizard, 1989.

14. The Conseil Constitutionnel found here a violation of article 21 of the French constitution. Décision No. 88-248 DC, 17 January 1989, J.O. 18 January 1989, 755. For the discussion of this that followed in the literature, see in particular Genevois, 1989: 215.

15. The conclusion of a convention presupposes that a license has been awarded. The duality of license and specifying contract is a common means of control in French administrative law. See Nevoltry & Delcros, 1989: 122; Truchet, 1989: 213. According to the amendments to the Media Act of 1994, the license is renewed by CSA unless the frequency has been ascribed to another use, plurality of the overall programming is not ensured, or the broadcaster was subject to sanctions by CSA that, because of the gravity of the broadcaster's failure, make an automatic renewal unsuitable.

16. For the constitutional problems associated with the extension of the set of sanctions, see Conseil Constitutionnel, Décision No. 88-248 DC, 17. January 1989, J.O. 18. January 1989, 755, 756f.

17. However, revocation of the license may not be agreed upon as a conventional penalty. For the relationship between administrative and contractual sanctions, see Huet, 1989: 187.

18. The situation was similar the following year; CSA, 1993: 467ff. and Annex 96. In 1993, however, the reasons for imposing a sanction covered a broad range; CSA 1994a: Annex 104.

19. Despite the difficulties in implementing the quotas, legislators still use them as a regulatory instrument. According to the 1994 amendment to broadcasting legislation, radio operators are obliged to fulfill a quota of 40% of "*chansons d'expression française*" (songs of French expressions) with respect to their programs of mixed music before January 1, 1996.

20. CSA, 1990: 278. Similar results were also arrived for France Region 3; see CSA, 1990: 282.

21. M 6 has since become more popular, especially among young people, as well as profitable.

22. With regard to the attribution of the frequency to ARTE (7:00 P.M. to 1:00 A.M.) and the governmental influence concerning this decision see CSA, 1993: 98ff.

23. For more examples, see: CSA, 1993: 231ff.; CSA, 1994a: 225ff.

24. Under the CSA's analysis, whereas twenty-two such films were broadcast in 1985, the number climbed to 106 in 1988. Of these, ninety-two were broadcast before 8:30 P.M. This represents an increase of 380%. CSA, 1990: 222.

25. For details, see CSA, Décision No. 90-182, 19. June 1990, J.O. 24. June 1990, 7351ff. See also CSA, 1991b: 246ff.; Phillipe, 1991: 8.

26. The responsibility for the expansion of the cable network lies with the municipalities.

27. By early 1993, cable had only some 1.1 million subscribers and a penetration rate of 24%. See *Lettre du CSA,* No. 43, 1992, 6; CSA, 1994a: 16.

28. Delivet & Rony, 1987: 128, stressed that the principle of independence in supervision was no longer fundamentally controversial in France.

29. Apart from the establishment of the CTRs, which do not, however, have any great practical importance.

30. For the effects of competition on programming, see, in particular, Forbes, 1989: 32ff. According to a study commissioned by the CSA, the programming of leading private broadcasters has become more and more alike, even though it might not be said that it is homogeneous; see CSA, 1991a: 110f. The study determined that A 2 has adopted—though always several years behind—TF 1's programming orientation; CSA, 1991a: 109. It also ascertained that in competition among various broadcasters, differences or identities in the type of programming apparently depend on whether the broadcasters are fighting for market predominance (e.g., TF 1 and A 2) or adopt a complementary role (e.g., FR 3 and private broadcasters); CSA, 1991a: 68, 102ff., 110.

CHAPTER 5

Canada

Bernd Holznagel

anada is the classic spill-over market for U.S. broadcast programming. Since roughly 80% of the population lives in a 300-km-wide strip of land along the U.S. border, American broadcasters early on took steps to increase their viewer and listener numbers by broadcasting their programs into Canada. For their part, Canadian broadcasters are glad to make use of programs produced in the United States. In this way, they are able to save on the high costs for in-house productions and reduce the risk of low audience acceptance. Since its very beginnings, Canadian broadcasting policy has sought to counteract cultural alienation stemming from U.S. programming. Promotion of "Canadian identity" by broadcasting programs of Canadian origin was and is its primary concern. For this reason, instruments of broadcasting law, such as quota regulation or subsidization of domestic television productions, were tried out at a very early stage in order to increase the share of Canadian programs in overall programming. The various regulatory approaches each had to be coordinated with the different economic conditions facing private and public broadcasters. Moreover, the special interests of French-speaking Quebec had to be taken into account, as promoting both of Canada's official languages is another characteristic of broadcasting policy here.

1. DEVELOPMENT AND STRUCTURE OF THE CANADIAN BROADCASTING SYSTEM

1.1. The Canadian Broadcasting Corporation

The dual broadcasting order, which only began to be set up in Europe in recent years, has a long tradition in Canada. The Canadian Broadcasting Corporation (CBC) was founded in 1936 along the lines of the British BBC (on the origin of the CBC, see Gerlach, 1990: 22ff.; Peers, 1969: 192ff.). It was set up in the form of a federal Crown corporation, since with the exception of educational programming,[1] the federal government had jurisdiction over broadcasting.[2] Its most important function involved the construction of a national transmission network in sparsely populated Canada—a project that could not be financed by existing private broadcasters in view of the high investment costs required. The CBC originally exercised a double function: It broadcast national radio programming while it supervised private broadcasting activities (see Raboy, 1990: 60; Wiesner, 1991: 67ff.). These functions were first split in 1958, with the CBC and private broadcasters being subject to a common, neutral supervisory authority (see Raboy, 1990: 131ff.; Gerlach, 1990: 38ff.; Ellis, 1979: 45ff.; for the Broadcasting Act of 1958, see Peers, 1979: 152ff.).

The CBC's duties do not differ essentially from those of public broadcasters in Europe (for the functions of the CBC, see CBC, 1988: 8ff.; CBC, 1990: 8ff.; Task Force, 1986: 261ff.). In order to fulfill its programming mandate, which is oriented to the public service concept, the CBC operates eighteen regional television stations and forty-seven regional radio stations. CBC Northern Radio and Television Service transmits broadcast programming for the Yukon, the Northwest Territories, and the northern reaches of Quebec. These programs are often aired in the languages of the Inuits and other Native Americans living in these regions (Köbberling, 1984: 341). In recent years, the CBC has gone to great lengths to improve its information and news programming. Since July 1990, CBC has operated the 24-hour cable news channel Newsworld, a competitor of the U.S. cable broadcaster Cable News Network (CNN). The commencement of a CBC news channel is an expression of a change in programming strategy undertaken by the CBC in the early 1980s. After competing in the 1970s for audience acceptance with mass-appeal programming of mainly U.S. origin (see Hardin, 1985: 116ff.; Juneau, 1989: 27; Federal Cultural Policy Review Committee, 1989: 673ff.), the CBC began to make an increasing effort under President Juneau to broadcast genuine Canadian programming (see Raboy, 1990: 290ff.; Juneau, 1985: 12ff.; Juneau, 1984: 19f.). This strategy includes expanding the production of films, strengthening regional CBC centers, and increasing the share of Canadian programming broadcast during prime time.

In view of the difficult financial conditions currently being encountered by the CBC, however, it is doubtful whether the course thus taken can be maintained in the future. With the election of the conservative government of Brian Mulroney in 1984, parliamentary subsidies, which had been used to finance nearly 75% of the CBC budget,[3] were reduced by 12% as a consequence of battling the high federal deficit (for details, see Enchin & Winsor, 1990: A 10.5; McPhail & McPhail, 1990: 165). Additional curtailments have already been adopted (CBC, 1990: 5; Vardy & Siklos, 1990: 3.1). As a result, the CBC saw itself forced to reduce personnel considerably and to close several of its regional television stations. Furthermore, it announced additional cuts in its administrative department (CBC, 1990: 5). The financial problems may be expected to grow even larger if the recommendations of the Task Force on the Economic Status of Canadian Television are put into practice. In order to improve on the poor economic situation of many private television stations, the Task Force has proposed reducing the CBC's maximum per-hour advertising time from 12 to 8 minutes (Task Force, 1991: 104). The financial restrictions placed on the public broadcaster would thereby reach new levels of stringency.

1.2. Private Terrestrial Broadcasting

1.2.1. Historical Development

The first commercial radio station in Canada commenced operations in 1920, at the same time as radio service began in the United States (Bird, 1989: 30). Although private radio did not experience as dramatic growth in Canada as in the United States, there were already seventy-five licensed broadcasters with 300,000 listeners by 1927. However, the development of private radio was restrained by the establishment of the CBC. The predominant position occupied by the CBC in the Canadian broadcasting system began to experience a setback only with the introduction of television (see Widlok, 1984: 402; Collins, 1990: 61; Wiesner, 1988b: 708). Even though public television programming had yet to be broadcast throughout the country, commercial broadcasters were able to apply for regional broadcasting licenses beginning in 1954. In contrast to the situation with radio, the CBC thus was not granted priority. By 1958 there were thirty-six private television stations as opposed to eight stations operated by the CBC (Peers, 1979: 46). In 1961 the national television network, Canadian Television (CTV), was licensed (see Rutherford, 1990: 103ff.). As a result of new technological transmission opportunities, the 1970s and especially the 1980s saw a wave of new licenses that considerably raised the number of commercial television and radio broadcasters.

1.2.2. Television

The largest Canadian broadcaster of English-language television programming is CTV, although its audience share is surpassed by that of the U.S. networks (Task Force, 1991: 55). The largest rates of growth have been obtained in recent years by the so-called independent stations. The broadcaster GLOBAL has operated particularly successfully in recent years, achieving a viewer share of 7% (Task Force, 1991: 55). GLOBAL broadcasts its programming in densely populated southern Ontario and was licensed as a counterweight to U.S. stations, which had targeted this region for especially intensive broadcasting (see Hardin, 1985: 123ff.). With respect to French-language television, Télé Diffuseurs Associées (TVA), first licensed in 1963, is ahead of the CBC. Télévision Quatre Saisons is the second largest private broadcaster (Task Force, 1991: 55).

The private television sector is currently caught up in a serious economic crisis. The profits of commercial stations have continued to drop over the past several years (Task Force, 1991: 21f.). The development can be attributed to the increasing battle for viewers, which has substantially intensified with the licensing of new broadcasters (see Task Force, 1991: 51ff.). Since 1979 the number of television stations has increased from sixty-seven to more than 100. In particular, the introduction in the early 1980s of pay-TV, with its diversified programming, as well as the growing licensing of independent stations, together have led to a scattering of the viewing audience and to a relative reduction in advertising revenues. The biggest losers are the established terrestrial stations.

Another consequence of increased competition is the disproportional increase that can be observed in day-to-day operational and programming costs (for details, see Task Force, 1991: 36ff.). In acquiring particularly attractive programming, a mutual overbidding may be observed among broadcasters. Prices for the broadcasting rights to popular U.S. programs or sporting events, such as reports on the World Cup in ice hockey, have thus risen some fourfold since 1985. Furthermore, the expenses for selling advertising time have also increased dramatically. This increase can be traced back to costs for the now common use of brokers who specialize in the acquisition and resale of programming rights. Finally, it must be noted that financial outlays are incurred in producing and broadcasting Canadian programs that broadcasters feel compelled to present because of the existing quota regulations and increased demand among viewers.

1.2.3. Radio

In radio as well, private broadcasters predominate over the public sector. They operate more than 600 radio stations (Canadian Radio–Television and

Communications Commission [CRTC], 1990: 82), with an audience share of nearly 90% (CBC, 1988: 12). The Canadian radio landscape is characterized by the absence of significant networks (Task Force, 1986: 120). Another important difference between radio and television is the low acceptance of U.S. programming for radio (Task Force, 1986: 122). Although U.S. broadcasters can be received in most urban areas, Canadians tune into the programs they broadcast only to a minor extent.

The most significant change in the last fifteen years in the radio sector has been the decreasing importance and number of AM stations (see Task Force, 1986: 122, 391ff.). Although today there are two times more commercial AM stations than FM stations, the latter have an audience share of 40% nationwide; in urban centers, this share is usually even higher. The profit margins of these stations are much greater than for broadcasters operating in the AM field. Without governmental protectionist measures, the collapse of the AM market would probably have been even worse. Wide attention has been given to the expansion of noncommercial radio stations in recent years (for details on this radio sector, see Wiesner, 1988a: 229; Task Force, 1986: 515ff.). This broadcasting sector is primarily financed by donations and state subsidies.

1.3. New Transmission Technologies

1.3.1. Cable Television

Canada follows Belgium as the country with the second highest rate of cable penetration in the world. At present, 80% of Canadian households are connected to cable networks (for data, see Task Force, 1986: 71, 551ff.; CRTC, 1989: 27; CRTC, 1990: 81). Of these households, 76% use the increased television offering. This corresponds to a quite remarkable 61% of the overall population. The rapid development of cable television had one important reason: For Canadians, cable television is the medium by which they can receive popular U.S. programs without interference (see Wyman, 1983: 172; CRTC, 1975: 2; Collins, 1990: 46). The very first cable systems that were set up in the larger cities of Montreal, London (Ontario), and Vancouver had the task of pulling in the signals of U.S. stations with large antennas and then retransmitting these to cabled households. In the early 1970s, the transmission of programming by the three U.S. networks (NBC, ABC, CBS) and the Public Broadcasting System (PBS) was allowed. Thereafter, there was a veritable boom in the cable industry (Shaw, 1975: 656; for data, see Task Force, 1986: 552). Increases of 20% in new subscribers were not uncommon. Between 1970 and 1974, the number of cabled households rose from 42% to 61%. Only after the most pressing demand was satisfied in the mid-1970s did the rate of growth begin to sink.

Of late, cable television has begun to experience a renewed boom in Canada, which can be attributed to the introduction of pay-TV (for the difficulties associated with the introduction of pay-TV, see Meier, 1984: 528ff.; Hardin, 1985: 293ff.; McPhail & McPhail, 1990: 195ff.). Pay-TV offerings normally cover movies and similar productions. Since being licensed in 1984, specialty services, which pay-TV viewers can acquire as a packet for an additional charge, have been particularly successful. These currently include sports, music, weather, news, health, and children's programming services (see CRTC, 1989: 27ff.; Collins, 1990: 85). The licensing of specialty services was designed to provide effective Canadian competition to American programmers active in the market (see Raboy, 1990: 316ff.; Task Force, 1986: 482 f.). This strategy is quite evident, for example, with the licensing of The Sports Network (TSN) and MuchMusic, which were intended to prevent a loss of viewers to the U.S. sports channel, ESPN, and the music channel, MTV. Currently, specialty services have an audience share of less than 5% (for the most recent upswing in speciality services, see Task Force, 1991: 64 f.). The losses these have brought to the terrestrial programmers have resulted in structural changes whose specific consequences cannot yet be determined.

1.3.2. Satellite Broadcasting

In light of Canada's size, it comes as no surprise that this country took steps at a very early stage to promote satellite technology and is now one of the world's leaders in this field. In order to ensure comprehensive provision of broadcast programming in outlying regions, Canadian Satellite Communications (CANCOM), a commercial enterprise, was licensed in 1981. CANCOM was intended to serve as a counterweight to the diverse U.S. satellite broadcasters (Vipond, 1989: 135f.; McPhail & McPhail, 1990: 194f.), which at that point could only be illegally received, primarily in the northern parts of the country (Melody, 1983: 6). CANCOM originally broadcast eight radio and four television channels. American programmers were not included. Since CANCOM's programming—in contrast to the aforementioned freely receivable American competition—could be obtained only in exchange for a monthly fee and since potential subscribers found it somewhat too high, the first business year produced an unsatisfactory rate of subscription. Therefore, in 1983 CANCOM was permitted to broadcast the programming of the three major U.S. networks and PBS (Task Force, 1986: 606f.). The service thereby offered the same programming as the cable systems in the north of Canada. This measure resulted in considerably greater subscriber figures. Nevertheless, because of the high costs associated with broadcasting to remote areas, CANCOM long accepted considerable losses. The position of the satellite service today is considered so solid that it was granted a five-year license extension in 1989 (CRTC, 1990: 60).

2. BODIES AND INSTRUMENTS
OF BROADCASTING SUPERVISION

2.1. The Canadian Radio–Television
and Telecommunications Commission

In order to realize optimally the goals of broadcasting policy set forth in the Broadcasting Act of 1991, the legislature charged a single independent public authority with regulating and supervising the Canadian broadcasting system (for details on this comprehensive mandate, see Broadcasting Act, sec. 5(1), 3(2); 1991 Statutes of Canada, chap. 11; and Trudel & Abran, 1991: 206ff.). This task is exercised by the federal authority, the Canadian Radio–Television and Communications Commission (CRTC), which is also responsible for the telecommunications sector.

The composition of the CRTC, determined by the governor in council, has been amended by the new Broadcasting Act. Previously, the supervisory authority was made up of nine full-time and ten part-time members; by way of the 1991 Act, the number of full-time members has risen to thirteen, while honorary officers have been reduced to six (Scott, 1990: 27). The term of office is now a uniform five years. However, the governor in council is entitled to replace CRTC members in justified cases prior to expiration of their term. Requirements for nomination are Canadian citizenship, residency in Canada, and no conflicts of interest with the telecommunications industry.

The CRTC is headed by a chairperson and two vice-chairs. The chairperson is authorized to form panels with three commission members, who may make decisions in the name of the CRTC. An additional amendment enables the governor in council to set up regional offices; these offices are then headed by a full-time member of the CRTC. This new regulation has been criticized in the literature as possibly impeding uniform regulatory action in the future (Scott, 1990: 28ff.). The commission employs more than 400 persons (CRTC, 1990: 27), and its annual budget amounted to Can$31.1 million for the year 1989–1990 (CRTC, 1990: 92).

The CRTC has had a somewhat uneasy relationship with the Department of Communications (DOC), which reports to the Minister of Culture and Communications and has been responsible for drafting legislation. The DOC and CRTC have not always pursued the same policy direction.

2.2. Instruments of Control

In order for the CRTC to carry out its mandate effectively, it has a number of significant instruments of control at its disposal. From time to time, the

CRTC publishes statements on broadcasting policy and on the interpretation of the Broadcasting Act. Although these pronouncements have no legal effect, they have a great impact in practice on the conduct of those to whom they are addressed.[4] In its binding regulations, the CRTC sets general rules for broadcasting activities (see Babe, 1979: 32ff.; Trudel, 1984: 303ff.; Trudel, 1990: 11ff.). In fact, the CRTC is virtually unique among Canadian administrative agencies in having the power to promulgate subordinate regulation. In contrast to broadcasting laws commonly enacted in European countries, it is here—and not in the Broadcasting Act—that, for example, the provisions dealing with advertising and concentration restrictions are found. For individual cases, the CRTC can make use of the broadcasting license, which can (subsequently) be made subject to conditions of license (see Babe, 1979: 34ff.; Trudel, 1990: 13f.). The authority to impose conditions of license is worded very broadly. The CRTC can take such measures as it deems "appropriate" in order to implement the policy objectives of the Broadcasting Act.[5] In practice two types of conditions can be distinguished: measures aiming at the particular situation of a given licensee and measures by which subordinate goals of the CRTC are made binding on all (or at least most) licensees.

As may be found in every country with a common law tradition, control through rules of procedure also plays a significant role in Canadian broadcasting law (see Trudel, 1984: 299ff.). In general, the CRTC has discretion to decide whether a hearing is necessary for reasons of the common good. Particularly since the courts attribute such great importance to these hearings (see Scott, 1990: 29ff.), wide use is made of this procedure, for example, when regulations are to be amended or policy statements drafted. Even though the hearings are held under a highly formalized procedure—with rules of evidence and confidentiality—the rules of procedure provide for sufficient latitude for informal action by the CRTC (see Babe, 1979: 37f.).

2.3. Enforcement

Adherence to regulations and conditions of license is monitored by program recordings covering all broadcasts (see Babe, 1979: 35). A broadcaster is required to submit recordings to the CRTC once a month and to maintain these for one year. In addition, the CRTC reviews a limited number of programs on its own. When a broadcaster violates any duties placed on it, the CRTC can impose sanctions. Under the previous legal regime, a violation of regulations represented an offense punishable by fine. Transgression of a condition of license might result in suspension or revocation of that license. However, only limited use was made of the latter measures because of their drastic consequences (Scott, 1990: 46; Babe, 1979: 185ff.; Trudel, 1984: 372; Task Force, 1986: 178; Vipond, 1989: 176). One example is the three-day suspension of the license of the FM radio station CKFM in Toronto, which had repeatedly violated programming duties (CRTC, 1989: 21). The Broadcasting Act of 1991 has increased the spectrum of possible sanctions. In the event of nonobservance of conditions or regulations, it now provides equally for fines as well as suspension or revocation of license (see Scott, 1990: 45ff.). Furthermore, the commission is given power to impose mandatory orders forbidding any act that is contrary to the legislation, regulation, license, or CRTC decision. Such orders may be enforced by the courts.

3. FIELDS OF REGULATION AND SUPERVISION

3.1. Licensing

Allocation of scarce broadcasting frequencies in Canada has not been left to the market but is the responsibility of a state supervisory authority. In exchange for being granted a license, the licensee must—in addition to paying a small fee—promise to observe all public service commitments. Broadcasting licenses are awarded by the CRTC in a comparative selection procedure for a period of seven years (prior to the Broadcasting Act of 1991, the term had been five years). In order to ascertain which candidate best meets the licensing requirements, the CRTC holds public hearings, at least with initial licensing (for administrative practice, see Babe, 1979: 37ff., 194f.). This enables local initiatives and individual citizens to comment on an application. In reaching a decision on the license, the CRTC's discretion is limited by the directives issued by the governor in council. For instance, the CRTC is prohibited from granting licenses to public juridical persons or to companies in which a foreigner holds more than 20% of the shares (see Trudel, 1984: 342, 365f.; CRTC, 1979: 77ff.; Miller, 1973: 194f.). It must also observe important selection criteria. For example, the CRTC is supposed to favor the applicants' various experience in broadcasting, present projects that are economically viable, and plan programming that reflects local and Canadian content. In order to assure fulfillment of these criteria, broadcasters must make corresponding "promises of performance" in the license application; these are often subsequently incorporated as conditions in the final license (Babe, 1979: 194f.; Trudel, 1984: 368). The private broadcaster GLOBAL, for example, promised prior to obtaining its license to limit advertising to eight minutes per hour and to commission new programs from independent producers (Babe, 1979: 187; for a further example, see Trudel, 1984: 368ff.). In practice, the CRTC constantly reviews whether a new licensee might threaten the economic viability of broadcasting stations already licensed for the transmission territory (Babe, 1979: 156ff.; Trudel, 1984: 367; Hoskins & McFayden, 1982: 355). For this reason, licenses are made subject to the condition that commercials may only relate to certain regional markets.

The most important instrument of pressure to ensure that applicants observe their promises of performance is the decision on renewal of the license. In regulatory practice, however, existing licensees have been granted de facto protected status, particularly in the television sector (see Babe, 1979: 173, 182, 185; Babe, 1976: 569; Task Force, 1986: 178). Only in exceptional cases have licenses not been renewed (for an example of recent administrative practice, see CRTC, 1988: 34). Usually there is not even a hearing held for the applicants competing for frequencies. When the CRTC wants to make clear that it is unhappy with the fulfillment of conditions and promises of performance, it generally grants a renewal for only a limited period instead of the maximum term.[6] In 1987 this measure affected more than thirty radio stations that had violated quota rules as well as the obligation to record their programs. Instead of the customary five years, the licenses were renewed for a period of eighteen months to three years, depending on the severity of the violation (CRTC, 1988: 33).

3.2. General Programming Principles

General programming principles can be found in section 3(1) of the Broadcasting Act. This is the central provision of the act, containing the legislature's main objectives of broadcasting policy (for the origin and inter-pretation of this provision, see Frémont, 1986). It obligates broadcasters to disseminate programming of the highest quality (see Trudel, 1989: 203ff.). It requires programs to be balanced in informing, instructing, and entertaining viewers; the interests of men, women, and children of all ages and tastes are to be taken into consideration. Furthermore, with regard to topics of general interest, there must be adequate opportunity for audience members to obtain information on opinions prevalent in society. Finally, there is an obligation to broadcast local and educational programming.

Additional principles on programming can be found in regulations. The television regulations, for example, prohibit programs that violate statutory provisions, that defame individual persons or groups on the basis of their race, ethnicity, nationality, gender, physical deformities, and so forth, or that use obscene language, or spread untrue reports.

3.3. Limits on Concentration

In order to prevent the emergence of private monopolies over opinion, the CRTC has made some efforts to monitor processes of economic concentration in commercial broadcasting. (An overview of ties between companies in the broadcasting sector may be found in Task Force, 1986: 619ff.; McFayden, Hoskins, & Gillen, 1980: 19ff.) In contrast to the United States, however, Canada lacks a detailed body of rules setting limits on the various types of concentration in broadcasting, such as multiple ownership in broadcasting stations, cross-ownership between press and broadcasting, and so on.

Although Canadian broadcasting law does not set forth a prohibition on multiple license ownership, in awarding licenses the CRTC does attempt to limit the number of licenses in the hands of one person. For instance, it ensures that, in a given market, a broadcaster does not own more than one AM and one FM license (Task Force, 1986: 621). Moreover, operators of cable networks are unable to receive a television license for their transmission territory (Task Force, 1986: 644). The awarding of a broadcasting license to a press company was regulated in 1982 by a government directive that essentially prevents a daily newspaper from acquiring a license when its territory of distribution basically overlaps with that of the television or radio station (Matte & Jakhu, 1987: 79).

Furthermore, the CRTC limits the transfer of shares and controlling rights in broadcasting companies. In the 1960s and 1970s, it awarded licenses only under the condition that they not be transferred without its permission (see CRTC Public Notice 1990-81: 2; Trudel, 1984: 342f.; Matte & Jakhu, 1987: 79). It was not until the 1980s that this transfer prohibition was adopted in a modified version in regulations. The new version initially provides for a reporting duty when more than 20% of the voting shares are to be acquired or when an acquisition passes 40% of total equity. In addition, the CRTC must give its advance approval to any action or agreement that results in a change in the ownership structure or in the company's effective control possibilities. The same policy applies when control is obtained over 30% of the licensee's voting shares or over a juridical person who exercises "effective control" over it.

The essential criterion for granting permission is that transfer of the license is in the public interest (see, e.g., CRTC Decision 1986-367: 8f.; Decision 1986-586: 5ff., 11ff.; Decision 1986-789: 4ff.; Decision 1987-62: 6ff.; for literature on the subject, see Trudel, 1984: 345). In conducting this required balancing of interests, the CRTC operates by the principle that concentration in private broadcasting is not per se detrimental. In its opinion, a certain company size may indeed be desirable in order to remain competitive against foreign broadcasters and to expand the economic basis for the production of Canadian programs (see Matte & Jakhu, 1979: 79). However, an acceptable limit on concentration is said to be reached when diversity of opinion in the affected region is unreasonably restricted (see especially CRTC Decision 1986-586: 11ff.). In practice, this has been the case when multiple ownership would have arisen. Empirical studies show that the CRTC has approved the vast majority of applications for transfer.[7]

In some cases, the CRTC has used the requirement of permission as an effective bartering tool in getting its programming objectives enforced (for discussion of this new strategy, see Janisch, 1990: 11). The spectacular takeover of Télé-Métropole, which owned CFTM-TV, the most important French-language television station, by Vidéotron, a national cable broadcaster, was tolerated only because the buyer committed itself to a number of "counterperformances" valued at Can\$5 million (CRTC Decision 1987-62). These had to do with the promotion of French-language films, music productions, and informational programs. The CRTC hoped that the merger would facilitate the production of competitive French programs for the world market. In order to counter the detrimental effects on diversity of opinion in Quebec, Vidéotron agreed to a number of guarantees regarding internal plurality. For instance, a control body was created with three independent experts, who were to monitor every complaint dealing with access to cable networks and business conduct and to report to the CRTC.

Critics point out that the flexible approach (for a justification of this administrative practice, see CRTC Public Notice 1990-81: 2) taken by the CRTC in battling concentration cannot be a replacement for a well-thought-out concept (Task Force, 1986: 644). The current wave of takeovers in the television sector—which is apparently motivated by dwindling profit prospects and which has affected more than twenty-five important stations since 1986—has seemingly led to a reversal of opinion at the supervisory authority (for a list of affected companies, see Task Force, 1991: 24f.). In August 1990, it was announced that the regulatory practice would be reviewed and that interested parties should submit comments (see CRTC Public Notice 1990-81; CRTC Public Notice 1989-109: 6). However, further measures have yet to be taken.

3.4. Limits on Advertising

In enacting advertising rules, the CRTC practices great restraint. This policy follows from the fact that general restrictions on advertising are contained in a number of federal and provincial laws (an overview of these is provided by Trudel, 1984: 425ff.). An example of such restriction is Quebec's Consumer Protection Act, which prohibits advertising aimed at children under age thirteen (Loi sur la Protection du Consommateur, sections 248, 249, R.S.Q., ch. P-40.1). The Supreme Court of Canada decided that the Act also applies to broadcast commercials (see *Irwin Toy Ltd. v. Quebec A.G.* (1989) 58 D.L.R. 4th, 577). It held that the Act represents neither a violation of federal competence to regulate broadcasting nor of the basic right of freedom of expression set forth in the Constitution. In addition, the CRTC relies broadly on self-regulation by commercial broadcasters with regard to advertising (for the phenomenon of self-regulation in Canadian broadcasting law, see Trudel, 1990; Trudel, 1984: 507ff.). The CRTC has therefore limited its activities to four central fields of regulation: setting permitted advertising times, establishing minimum standards for election advertisements, limiting advertising aimed at children, and restricting drug and alcohol advertising.

Maximum advertising time allowed for private television stations amounts to twelve minutes per hour. For FM radio stations, commercials are limited to 15% of the weekly broadcasting time. The restrictions for AM stations, which were allowed to broadcast 250 minutes of commercials per day, were abolished in 1986 (see CRTC Public Notice 1986-248: 5ff.). This was intended to compensate for the financial losses resulting from the rising popularity of FM broadcasters. The Canadian Association of Broadcasters (CAB) has since established advertising guidelines for AM stations.

An interesting mix of self-regulation and government control is practiced in the area of children's advertising and alcohol and drug advertising. In order to protect children under twelve against the influence of excessive advertising, as early as 1971, the advertising industry and broadcasters developed as early as 1971 the Broadcast Code for Advertising to Children. In 1974 this code was made binding on all broadcasters by the CRTC as a condition of license. Compliance with these provisions, which were amended in 1987 (Canadian Association of Broadcasters, 1987), has from the outset been monitored by the Canadian Advertising Foundation through a nine-member committee composed of representatives from consumer associations, the CRTC, private broadcasters, the advertising industry, and advertising agencies. To ensure enforcement of the guidelines, the committee conducts a "preclearance procedure," that is, review prior to a commercial's broadcast. This committee is not endowed with the power to sanction. Rather, broadcasters are obligated not to run "objectionable" commercials. In addition, the CRTC and the company commissioning the advertisement are informed of the results of the review. In 1988, the CRTC determined that a number of broadcasters had been violating the established code and thus a condition of license. Pursuant to a special monitoring action, the CRTC ascertained that only one of fifteen monitored broadcasters had been in compliance with all provisions (see CRTC Public Notice 1991-11: 1f.). The requirement that the same product not be advertised twice within thirty minutes was particularly widely ignored. The Canadian Advertising Foundation has proposed new guidelines, which the CRTC has yet to approve. (The draft of these new guidelines appears in CRTC Public Notice 1991-11, Appendix.)

Particularly onerous restrictions apply to advertising for alcoholic drinks. In essence, a prohibition applies against commercials aimed at the general promotion of alcoholic beverage consumption or commercials that sell "hard" liquor. Commercials for wine and beer may be broadcast only when it has been determined in advance that they are compatible with the requirements found in the Code for Broadcast Advertising of Alcoholic Beverages (CRTC Public Notice 1986-247, Appendix). The obligatory preclearance review is conducted by the CRTC. An attempt by the CRTC to ease these restrictions and to provide for self-regulation failed in the mid-1980s. No agreement could be reached on a mechanism for supervision (CRTC Public Notice 1986-247: 3).

Election advertisements are also subject to special supervision. Broadcasters must treat all political parties and candidates equally and fairly in their informational programming and in election broadcasts.

3.5. Measures to Counter Gender Stereotypes

Equality of men and women has been a topic addressed by Canadian broadcasting policy since the 1982 report of the Task Force on Sex Role Stereotyping (for the task force's recommendations, see CRTC Public Notice 1986-351, Appendix B). The task force determined that the depiction of women in broadcasting programs was incompatible with the objective of equality in a number of respects. To counteract this situation, it recommended that rules of conduct to be drafted and monitored by the CAB, be established. However, this approach has not proved effective in practice (CRTC Public Notice 1986-351: 10ff.). The CRTC thus called for an amendment to the CAB guidelines and decided to make them binding as a condition of license (see CRTC Public Notice 1990-99: 2ff.).

The current guidelines, which entered into effect in 1990 with the approval of the CRTC, are directed toward the objective of equal and fair depiction of men and women in television and radio broadcasts. Programs are to portray the whole range of behavior of both sexes. Especially in news and information programs, women as well as men are to be shown in the function of speaker, expert, and decision maker. One-sided sexual stereotypes, such as the portrayal of housework as a typically female occupation, are to be avoided. The objective of equality is also to be expressed in the type of language used. Moreover, the diversity of Canadian family life is to be documented. The various lifestyles of ethnic and other minorities must be portrayed. The disadvantaged position currently experienced by women vis-à-vis men is to be corrected through greater inclusion in broadcasting activities.

Furthermore, the principles laid down in these guidelines must be observed both in the production of new programs and in the purchase of foreign ones. Commercials also come within the province of the guidelines. Their enforcement is the responsibility of the Canadian Broadcast Standards Council created especially for this purpose (see CRTC, 1989: 33).

4. MEASURES TO ENSURE NATIONAL AND CULTURAL IDENTITY

4.1. Normative Bases

The Canadian broadcasting order is designed as a public service that has the task of maintaining and promoting national identity and cultural sovereignty. The legislature sees this idea best realized through a "single system," that is one composed of public, private, and local elements (for more on the "single system" concept, see Ellis, 1979: 75ff.). All the same, the CBC is subject to stricter requirements than its private competitors. Whereas public broadcasting programs must be "predominantly and distinctively Canadian," private broadcasters are only required, within their financial means, to contribute "significantly" to broadcasting Canadian programs.

It is thus clear that a core concern of broadcasting supervision and regulation is the effort to ensure the production and dissemination of Canadian radio and television programming. The reason underlying this concern is the traditionally high amount of American programming aired, particularly in private broadcasting. The set of instruments to ensure these guidelines as spelled out in the Canadian Content Regulations ranges from the setting of quotas to the provision of subsidies. These instruments have, however, been of varying success.

4.2. Quotas for Canadian Programs

4.2.1. Set of Control Instruments

Traditionally, the most important tool for program control has been the establishment of quotas for television and radio broadcasts of Canadian origin. Quota regulations are set in two ways. First, they are found in the regulations in force for television and radio. These regulations establish what percentage of Canadian programs must be shown for overall broadcasting time and during prime time, and they define the term "Canadian program" and determine the period during which quotas must be fulfilled. Quotas are also imposed by way of condition of license. For this approach by the CRTC, see especially CRTC Public Notice 1987-8: 3ff.; see also Task Force, 1986: 467ff.; Janisch, 1990: 10f. For the associated legal issues, see Supreme Court of Canada re *CTV Television Network Ltd. v. Canadian Radio–Television and Telecommunications Commission et al.* (1982) 134 D.L.R. 3d 193.

4.2.2. Radio

Quotas have been in place for radio since 1970. The first regulations referred exclusively to the AM stations then predominant and only placed quotas on music programs. Although half of the programming of private AM stations was music, only 4 to 7% of this programming constituted Canadian productions (CRTC Public Announcement, 1979: 74). The share of Canadian music broadcasts was therefore set at 30% for the time between 6 A.M. and midnight.

Two important structural changes in the Canadian radio market led to an amendment of these provisions in 1991. First, FM stations, which were primarily used in the 1960s for the hi-fi broadcast of AM music programs, experienced an enormous commercial upswing in the past two decades. Furthermore, the Canadian music industry went through a phase of expansion. In the past, the CRTC often had to take into account deficiencies in some categories of Canadian music when setting quotas (Task Force, 1986: 408f.). For the FM stations affected, the market situation often determined the type of quotas, which frequently were lower than for AM broadcasters (CRTC, 1979: 75; CRTC, 1980: 9). This phase of scarcity has been overcome today (CRTC Public Notice 1990-111: 3f.).

In the new regulations, therefore, program quotas for AM and FM stations are identical. Under the rules, the quota for the categories "pop, rock, and dance" and "country music" is 30% (CRTC Public Notice 1990-111: Appendix III). The quota for "traditional and special interest music" is set at 10% and that for "ethnic music," at 7%. All quotas refer to a broadcasting week, although programs must be included in overall programming in a "reasonable manner." The daily calculation method previously in place for AM stations was abandoned as result of efforts at harmonization. A program is deemed to be "Canadian" when it fulfills two of the following four criteria: the musician, the lyricist, or the composer is Canadian, or the music was recorded in Canada or is broadcast live from a concert within Canada. When instrumental music is involved, it is sufficient when the composer is Canadian.

For FM broadcasters, the quota rules are not limited only to music programs. The new regulations require of broadcasters that 15% of weekly broadcasting time belongs to the category "spoken word."[8] By this is meant broadcasts that are neither music nor commercials. Moreover, the CRTC imposes by way of condition of license that at least three hours per week of local news be broadcast (CRTC Public Notice 1990-111: 34). This measure is attributable to an awareness that in the course of the FCC's deregulatory efforts, American stations have continually given a smaller share of broadcasting time to local news (see Task Force, 1986: 126f.).

The success of Canadian quota rules in radio is uncontested as regards the fulfillment of objectives of cultural policy (Task Force, 1986: 408ff.;

CRTC, 1979: 74; CRTC, 1980: 8f.). These rules are considered the principal reason why—in contrast to the situation in the United States—stations broadcast balanced, diverse programming in accordance with broadcasting's trustee role (this was the express opinion of the Task Force, 1986: 400ff., 408ff.). The implementation difficulties experienced mainly in the 1970s have today essentially been overcome. During that period, FM stations opposed the complex formulation of the quota rules as well as the costs associated with them (CRTC, 1980: 12). In contrast to the case of television, programming control for radio is thus not a focal point of media policy discussions.

One of the primary reasons for the positive regulatory experience in radio is the special economic condition accompanying radio operations. For example, the costs for a Canadian music program are today no longer so high that broadcasters are forced to make extensive use of American offerings (see Task Force, 1986: 408). This state of affairs has continued to improve over recent years as the Canadian music industry has expanded (for the relationship between quota regulation programs, see Section 7, below). Moreover, none of the American-style programming schemes has prevailed in Canadian radio: Radio in Canada is primarily locally broadcast. There is no nationwide broadcasting of American radio programs.

4.2.3. Television

4.2.3.1. Regulatory Approaches of the CRTC

The CRTC has adhered to the instrument of quota regulation, despite the lack of success of its predecessor, the Board of Broadcasting Governors (BBG), which began experimenting with quotas in 1959 (for the experiences of the BBG, see Fowler, 1965: 49; Grant, 1968: 322ff.; Beke, 1972: 104ff.). The 1970s were characterized by the Trudeau government's policy of national independence and the reduction of American influence on Canadian society. It was particularly during this period that quotas came to be considered the most important regulatory means of counteracting the "Americanization" of Canadian television (see Hardin, 1985: 15ff.). As a result, the 1970 television regulations increased the quota of Canadian programming to 60%. The share of programs that could originate from any one given foreign country amounted to 30%. These rules also applied to "prime time," that is, the period between 6:30 P.M. and 11:30 P.M. Calculation was made every three months. In addition, although it had been possible to fulfill quota obligations with programs originating from Commonwealth or French-speaking countries, this policy was done away with. However, opposition by broadcasters prevented enforcement of these strict regulations, and they were gradually relaxed over the following years (see Babe, 1979: 141ff.). To begin with, "prime time" was extended from 6:00 P.M. to midnight, and the period for calculation was stretched to one year. Furthermore, the quota for prime-time programming was reduced to 50%. And the 30% rule, which was in actuality directed against the dissemination of U.S. programming, was rescinded by the CRTC. Finally, the CRTC also allowed Canadian commercials to be included in the programming quota.

In the early 1980s, the CRTC once again sought to promote the broadcasting of Canadian programs by means of quotas. Once the revenues of commercial broadcasters began to improve significantly in the 1970s as a result of more licenses and the expansion of cable television, this prompted the CRTC to undertake greater efforts in the programming area. However, earlier experiences had shown that quotas based solely on quantity were incapable of fulfilling the supervisory authority's objectives of broadcasting policy (see CRTC, 1979: 99ff.; CRTC, 1980: 25ff.). Particularly in the categories of English-language drama and children's programs, the previously applied regulatory regime had proved unable to ensure the dissemination of a sufficient amount of high-quality Canadian programming.

To redress these shortcomings, the CRTC introduced two important changes in its quota policy (see the central document: CRTC Public Notice 1983-18). First, a point system was adopted in the regulations in which the term "Canadian program" was more precisely defined and open to qualitative aspects (CRTC Public Notice 1984-94: Appendix). This system remains in force today. It means that a program is classified as being of Canadian origin when it obtains at least six points. A Canadian director is worth two points. A Canadian-born screenwriter, leading actor, artistic director, main cameraperson, composer, or editor is worth one point each. The system also determines when a coproduction or a television series classifies as being Canadian. But one of the most important functions of the system is that it forms the basis for creating incentives to broadcast previously ignored categories of programming. Every film that receives ten points and is broadcast between 7:00 P.M. and 10:00 P.M. is credited with a time bonus, that is, it counts for 150% in determining the share of Canadian programming broadcast. A children's program can likewise receive a bonus when it is broadcast during a time at which children commonly watch television.

In addition, the CRTC is making increased use of conditions of license in order to set program quotas.[9] In comparison to regulations, this instrument of control, first used in relation to CTV in 1979 (for the associated legal problems, see also McPhail, 1986: 41ff.), makes it possible to promote individual program categories more intensively and to adjust the burdens that arise to the broadcaster's economic viability. For example, CTV was required to broadcast at least 2½ hours of self-produced films each week (CRTC Decision 1987-200: 16ff.). This share then rose thirty minutes each year, that is, to 4½ hours by expiration of the license. GLOBAL was required to broadcast at least 200 hours of new first-run Canadian drama, variety, children's, and documentary programs (CRTC Decision 1986-1086). The former had to be broadcast between 8:00 P.M. and 11:00 P.M.. In the pay-TV sector, the CRTC has imposed quotas solely by way of condition of license. Quotas amount to 20% overall and 30% for prime time.

The adoption of the television regulations in 1987 has not resulted in a change in this regulatory practice. It sets the share of Canadian programming at 60%. However, private broadcasters need only fulfill a quota of 50% during prime time between 6:00 P.M. and midnight. The period for calculation is one year.

4.2.3.2. Practical Experience

In spite of this diversified set of instruments, the CRTC's quota policy has not led to satisfactory results (for more on this estimation, see Janisch, 1990: 9ff.; Hoskins & McFayden, 1989: 174ff.; Collins, 1990: 75ff.; Vipond, 1989: 170ff; but see, more optimistically, Romanow, 1976: 36f.). For example, in 1991 the share of Canadian programs between 7:00 P.M. and 11:00 P.M.—the time with the largest viewing audience—amounted to only 31.6% in English-language television. At 76.4%, the share was considerably higher in French-language television (for this data, see Task Force, 1991: 53f., 119). But it must be noted that the majority of Canadian programming was offered by CBC (see Task Force, 1986: 91ff.). The dissemination of Canadian dramas and children's programs continued to be particularly disappointing. In the category of "drama," the choice of the majority of viewers each evening, only 5% of all programs were of Canadian production on English-language television from 7:00 P.M. to 11:00 P.M.. The analogous share on French-language television was 16%.

The central cause for deficiencies in certain programming categories remains the high cost of Canadian productions (see Task Force, 1991: 126f.; Task Force, 1986: 433; Hoskins & McFayden, 1989: 178; Collins, 1990: 71f.; Department of Communications, 1988: 20). For domestic broadcasters, it is, for instance, roughly ten times less expensive to purchase broadcasting rights for U.S. programs than to undertake an in-house production. According to information supplied by CTV, replacing a one-hour U.S. entertainment program with a Canadian offering can spell a loss in profits of up to Can$5 million. The amount of these opportunity costs is due not simply to the low purchase price for U.S. productions but also to reduced advertising revenues linked to dwindling viewer acceptance of Canadian programs (Task Force, 1991: 39). Thus, there are no economic incentives for broadcasters to produce and broadcast Canadian programs. They instead attempt to fulfill their legal obligations in such a way that costs remain as low as possible, and as a result, they have developed a number of techniques to exploit the leeway in the regulations.

A majority of these practices involve artful placement of Canadian productions in overall programming (see Shedd, Wilman, & Bunch, 1990: 65ff.; Hoskins & McFayden, 1989: 176f.; Babe, 1979: 75ff., 144f.; Vipond, 1989: 270f.; Johnson, 1982: 2). For instance, Canadian productions are often relegated to the start or end of the 6:00 P.M. to midnight prime time. During the periods of maximum viewing, more profitable U.S. programs are broadcast. Since showing reruns counts in fulfilling quota obligations, many broadcasters have opted to rebroadcast their Canadian programs almost immediately. In addition, these productions tend to be shown during the summer months when ratings are traditionally lower than during the winter. In this way, broadcasters can at once compensate for low ratings and fulfill their quota obligations.

When broadcasters do produce their own programs or commission them, they attempt to keep financial expenditures as low as possible. As a result, quota obligations tend to be fulfilled by broadcasting news, information, and sports programs (see Task Force, 1986: 96f., 106f.; Babe, 1979: 79f.). Moreover, game shows and talk shows can frequently be found in the programming structures of private broadcasters. Special problems are raised in this context by coproductions, which, under certain conditions, can be classified as Canadian programming (see Task Force, 1986: 114ff.; Freiman, 1983: 26ff.; Hardin, 1985: 31; Babe, 1979: 144). In order to reduce production costs and make the best use of foreign distribution channels, Canadian broadcasters often collaborate with their U.S. counterparts to produce together new television programs. Because the coproducers seek to raise the chances for success in the U.S. market, the ties to Canadian culture in these productions are often barely discernible or are sometimes intentionally avoided altogether. There is thus hardly any difference between these programs and their American models. A well-known example is the coproduction between a pay-TV broadcaster and the U.S. Playboy Channel in which Canadian "playmates" participated in order to fulfill quota requirements (McPhail & McPhail, 1990: 196).

A further reason for the deficits in quota regulations can be attributed to the conduct of the CRTC. Criticism of the CRTC focuses on three levels: To begin with, critics say, the CRTC is not willing to enact and enforce the measures needed to ensure national and cultural identity as called for in many government reports and other position papers. These critics have accused the CRTC both of failing to enact stricter quota rules and of ignoring strategies that undermine existing rules (for the relevant proposals, see, e.g., Johnson, 1982: 2). Further, critics assert that at the moment there are no quotas for individual programming categories and that neither the calculation period nor the definition of prime time has been tailored to meet these categories. They also allege that there have been virtually no cases in which a (co)production was rejected as "Canadian programming" solely on account of qualitative aspects. Moreover, critics contend that the CRTC bows much too quickly to broadcasters' opposition to regulations they claim are "too strict." These critics cite the example of the battles surrounding the calculation period, which has repeatedly been amended to the benefit of broadcasters (Babe, 1979: 142ff.; Johansen, 1973: 193f.). Finally, the critics say that the CRTC has failed to impose sufficient sanctions for violations of quota regulations (see Howse, Prichard, & Trebicock, 1990: 525; Janisch, 1990: 13; Hardin, 1985: 40). The existing legal sanctions allegedly have been applied only rarely and, often, half-heartedly. Failure to fulfill quotas, these critics further maintain, has normally not led to denial of an application for license renewal (Howse et al., 1990: 525; Hoskins & McFayden, 1982: 350). Rather, the CRTC is said to have relied on informal approaches, such as the effect of moral persuasion and threats, to put through its interests (Law Reform Commission of Canada, 1986: 24; Babe, 1979: 147f., 227; Hardin, 1985: 40).

In view of these experiences, those responsible for media policy now take the view that traditional quota regulation alone is not enough to ensure a sufficient amount of Canadian television programming. The use of additional means of control has, in recent years, led to a shift in the function of quota regulation. This is discussed in detail in the concluding section. First, we turn our consideration to descriptions of the various instruments.

4.3. Improving Profitability

The CRTC has traditionally used measures to improve the profitability of domestic broadcasters as a means of programming control. The commission expects that profits earned be utilized to fulfill quota obligations (Janisch, 1991: 11f.; Hoskins & McFayden, 1989: 177).

Protection against American competitors is afforded to Canadian broadcasters above all by Canadian tax law. According to article 19.1 of the Income Tax Act (1974–1976 Canadian Statutes, ch. 106), expenditures for advertising that targets Canadians are deductible only if the commercials are broadcast on domestic stations. This rule came into force in 1976 and thereafter caused a considerable dispute with the United States (see Swinton, 1977: 580ff.; Stoler, 1979: 46ff.). The rule lessens the incentive for the advertising industry to have American stations advertise its products (for the meaning and purpose of the rule, see Matte & Jakhu, 1987: 81f.; Stoler, 1979: 43). Empirical studies for the year 1984 have shown that a similar tax rule applied earlier to the press led to an increase in the advertising revenues of domestic broadcasters of Can\$35.8 to Can\$41.8 million (Task Force, 1986: 461; see also Donner & Kliman, 1984: 1095). In practice, the stations lying on the border, whose existence is most threatened, have especially benefited from this rule.

Improving profitability is also the objective of simultaneous program substitution rules applying to cable television (for the legal problems of the substitution rules, see Federal Court of Appeal re *Capital Cities Communica-*

tions Inc. et al. v. Canadian Radio Television Commission (1975) 52 D.L.R. 3d, 415; Supreme Court of Canada (1975) 81 D.L.R. 3d, 609). The basis for this regulatory approach is a list setting forth the rank of programs to be fed into cable networks. When two or more television broadcasters show the same programs at the same time, a local or regional station can apply to have a cable network broadcast its program (including commercials) in the place of that of a lower-ranked—usually American—station on all affected channels (for the way in which the substitution rules work, see Task Force, 1986: 460f., 572; CRTC, 1971: 27f.). The substitution rules create an incentive for the advertising industry to broadcast its commercials on domestic stations (for the meaning and purpose of the rules, see Swinton, 1977: 567; Shaw, 1975: 671f.). It has been conservatively estimated that thanks to this rule, in 1990 Canadian television stations had increased revenues of some Can$99.2 million. This amount corresponded to roughly 8% of the entire advertising revenues of private television in Canada (Task Force, 1991: 81).

These steps have not had the result of considerably increasing investment in Canadian productions (Hoskins & McFayden, 1989: 177; Hoskins & McFayden, 1986: 36f.). However, laws of economic necessity, which are responsible for the purchase and broadcast of American productions, cannot be suspended in this way. Ironically, the substitution rules have also created an incentive to make selective purchases of popular U.S. programs and to broadcast these at the same time as American stations. The result is adaptation of the Canadian broadcasting schematic to customs common in the United States (Task Force, 1986: 98, 460f.; CRTC, 1979: 102). In spite of these negative effects on cultural policy, no serious consideration is being given in the broadcasting policy debate to abolishing these rules. In fact, they can strengthen the economic foundation of private broadcasters, which in turn—or so goes the predominant philosophy—is the prerequisite for realizing the cultural policy objectives of broadcasting regulation (see, e.g., Task Force, 1986: 461). For the same reason, the Task Force on the Economic Status of Canadian Television has recently proposed even broadening economic protection through control and substitution rules (Task Force, 1991: 81ff.).

4.4. Subsidizing Program Production

Since the early 1980s, subsidies have been increasingly used to promote the production of radio and television programs. In spite of quota regulations, the programming industry in both broadcasting sectors had not developed to an extent able to satisfy demand for domestic productions at that time, as well as for projected future demand (for the causes of this, see CRTC, 1980: 9; Task Force, 1986: 401; Hinkson & Krasikovsky, 1982: 211). In several music categories, such as country music and French chanson, there was a lack of high-quality recordings. For this reason, the CRTC saw itself forced to lower even existing quotas (Task Force, 1986: 408f.). In the television sector, it was predicted that the need for programming would explode within a few years as a result of new technologies. The Department of Communication thus feared that commercial broadcasters would soon only be able to fulfill the quotas imposed on them with cheap productions. This would, of course, mean that viewer acceptance of Canadian productions would decline even further (Department of Communications, 1983: 6ff.). Therefore, in March 1983 the Canadian Broadcast Development Fund (CBDF) was founded to counteract such possible developments (see Department of Communication, 1983: 9f; for the subsidization measures undertaken with respect to radio, see Task Force, 1986: 409ff.; Wiesner, 1991: 154f.).

The CBDF is charged with supporting the production of Canadian television programs belonging to program categories with little available material, such as dramas, variety shows, documentaries, and children's television (for the current goals of the CBDF, see Task Force, 1991: 109). Under the CBDF, which is administered by Téléfilm Canada, independent production firms are to be given primary consideration in awarding subsidies. It has sole power to decide whether a project should be supported with a loan, a loan guarantee, or a stipend. The conditions for subsidization have been amended a number of times in recent years (for earlier development, see Task Force, 1986: 361ff.). To receive a subsidy, it is first necessary for an institution that is responsible for educational television to declare its willingness to accept the subsidized production of its programming. In addition, the financial feasibility of the project must be described. During the initial years, Téléfilm Canada assumed roughly one-third of production costs. At the moment, the authority may contribute up to 49% of the investment amount for English-language productions (Hoskins & McFayden, 1989: 179; for the rate of contributions in recent years, see Hoskins & McFayden, 1989: 183; Téléfilm Canada, 1990: 19). However, the project must then fulfill all ten criteria listed in the CRTC's point system for recognition as a Canadian program (for some form of subsidization, it is usually enough to have six points).

The work of Téléfilm Canada has generally been deemed successful (see especially Hoskins & McFayden, 1989: 182ff.). The Broadcast Development Fund was even spared curtailments in the latest round of budget cuts. Téléfilm Canada initially had difficulty in finding enough applicants, but this situation was able to improve gradually through increased participation in production costs (see Task Force, 1986: 365f.; Hoskins & McFayden, 1989: 184f.). During the fiscal year 1989–1990, Can$75.7 million was invested in the development and production of 203 new television films (Téléfilm Canada, 1990: 17). In the private sector, CTV was the primary beneficiary of the subsidies awarded (see Janisch, 1990: 16). Whereas in 1984–1985 CTV was involved in less than 1% of the production of all English-language television programs, this share had reached 27% by 1987–1988. At the same time, praise has been heaped on the high quality of some programs, which would otherwise not have been funded without the assistance of the CBDF.

Critical voices, however, point out contrary examples. In some cases, the subsidized productions were said to be nothing more than "American clones" (Hoskins & McFayden, 1989: 183; Task Force, 1986: 369ff.). Téléfilm Canada was therefore called upon to place greater emphasis on cultural concerns in awarding future subsidies. Furthermore, the authority has made headlines because of a number of internal problems, which led to the well-publicized resignation of its management (Hoskins & McFayden, 1989: 184). Accusations have been raised that Téléfilm Canada gives preference to a few, large producers and that some of its employees have close ties to the subsidized industry. The fund's future will probably depend to a large extent on whether sufficient incentives are offered for producing programs. Television broadcasters will attempt to receive the largest possible subsidies in order to keep their own production costs low. But when the average share of costs assumed crosses a certain threshold, there is a danger that subsidy programs will become inefficient (see in particular Janisch, 1990: 17f.; for possible counterstrategies, see Hoskins & McFayden, 1989: 186f.).

4.5. Production Obligations

In recent years the CRTC has made increased use of production requirements in its conditions of license. Such conditions address a variety of issues: First, it sets the number of hours of Canadian programming that a broadcaster must produce. For instance, CTV was required to produce an additional twenty-six hours of Canadian programs annually (see CRTC Decision 1987-200: 16). Second, the CRTC fixes a monetary amount that must be invested in in-house productions or for the procurement of Canadian programs. The extent of expenditures may be increased over the years or be linked to the broadcaster's economic success. CTV, for example, was required to invest Can$68.4 million in 1987–1988, Can$80.2 million in 1989–1990, and Can$93.3 million in 1991–1992 for Canadian programs (CRTC, 1988: 46). Pay-TV broadcasters were required to spend 20% of their annual gross revenue on domestic productions (see Raboy, 1990: 312f.). Third, the condition of license can also serve to promote certain categories of programming. For example, GLOBAL was required to invest an additional Can$5 million in drama productions (CRTC Decision 1986-1086; for other examples, see Task Force, 1991: 43ff.).

According to estimates by the CRTC, this approach will create an investment volume of some Can$2 billion between 1989 and 1994. For the television sector, however, it must be noted that production obligations (see

Acheson & Maule, 1990: 287 n. 10) were imposed in only seventeen cases via condition of license. This affected commercial broadcasters that had earned more than Can\$10 million in advertising revenues by the end of the fiscal year 1987–1988.

As of yet, there are no empirical studies on the success of production conditions. The use of this regulatory instrument was particularly advocated by the Caplan-Sauvageau Commission, which had hoped for especially effective implementation of the Broadcasting Act's objectives for cultural policy (Task Force, 1986: 471). Nevertheless, the same objections that apply to quota obligation can be raised regarding production obligations. To begin with, the production conditions make no provision for the placement of these broadcasts in overall programming. As a result, programs produced in accordance with these regulations can still be shown outside the hours with the largest viewing audience. The conditions are therefore most likely to fill the function of supporting the domestic film and television industry. This theory is also underscored by the fact that the CRTC first began to make use of this tool only after expanding subsidized programs. It is doubtful, however, whether production obligations are truly enforceable during times of economic hardship. The Task Force on the Economic Status of Canadian Television recommended that the current rules be suspended for at least a certain period of time citing that the conditions had been imposed during a phase of economic prosperity that is no longer existent today (Task Force, 1991: 45). For this reason, in 1992 the CRTC improved the flexibility of the regulatory scheme. It now permits a licensee to underspend with respect to Canadian programming by an amount representing no more than 5% of the minimum amount otherwise established for that year by the conditions of license. If a licensee takes advantage of this possibility, the amount by which the licensee has underspent must be added to the amount required to be spent for the following year (CRTC Public Notice 1992-28).

4.6. License Fee as Incentive for Program Production

The Broadcasting Act of 1991 introduced a new instrument of control exhibiting elements of traditional regulatory policy as well as creating economic incentives for broadcasting Canadian programming. The CRTC will now be able to impose a license fee on every private broadcaster at the beginning of each year (see Janisch, 1990: 18f.; see also the theoretical considerations by Shedd et al., 1990: 67ff.). As long as the broadcaster fulfills its programming obligations, the fee will be refunded; should this not be the case, it will be retained. However, the CRTC has shown no great enthusiasm for implementing the new instrument.

5. BROADCASTING SUPERVISION IN THE POLITICAL ARENA

The CRTC is not subject to the imperatives of a minister. As an authority that is required to reach some of its decisions in the manner of a quasi judicial tribunal, the CRTC enjoys a certain degree of autonomy (see Janisch, 1979: 83ff.; Scott, 1990: 52ff.; Raboy, 1994: 17ff.). All the same, the government has a number of possibilities for exercising control and direction, which were expanded under the Broadcasting Act of 1991 (for criticism of this new power, see Scott, 1990: 51ff.). First, the governor in council can enact binding directives. The Act initially provides for directives serving to specify general broadcasting policy and the CRTC's regulatory style. In addition, directives are enacted with regard to specific fields set down by law. For example, the number of frequencies available in a given region or incompatibility rules for those applying for licenses are stipulated by directive.

Furthermore, the governor in council has the right to set aside or refer back all CRTC decisions regarding the license that are incompatible with the guidelines set forth in section 3(1) of the Broadcasting Act (see Conklin, 1992: 297). However, this possibility has only rarely been resorted to (between 1968 and 1988, there were only four cases: Morrison, 1988; see also Kaufman, 1991: 115ff., with an analysis of the cases). The government and the CRTC tend to coordinate their actions informally (Kaufman, 1991: 112, 116). This reluctance can be explained by the Canadian tradition of keeping broadcasting free of governmental influence (see Bureau, 1988: Nos. 21f.; Morrison, 1988).

6. BROADCASTING SUPERVISION AMONG INCOMPATIBLE OBJECTIVES

Over the past two decades, the number of broadcasters has increased considerably as a result of the introduction of new transmission technologies. In the early stages of this development, the CRTC played more of a passive role. Both cable and satellite television were licensed in Canada because Canadians would have accessed the American palette one way or another (i.e., illegally if necessary). Under this situation, a continued wait-and-see approach would have made it more difficult for domestic broadcasters to enter the market. Moreover, the licensing decision was influenced by two other important factors. First, the CRTC hoped to be able to enforce its objectives for cultural policy with new distribution technologies as well (see in particular Raboy, 1990: 318ff.; this mechanism has been applied above all with the introduction of pay-TV; see Wiesner, 1991: 308ff.; Task Force, 1986: 477ff.). Second, interested industrial circles exercised political pressure in order to accelerate their introduction. Only during the past few years have the CRTC and the Department of Communication agreed to pursue an active licensing strategy. The most important American program offerings were to have Canadian counterparts (see Collins, 1990: 81). This approach becomes most evident with the licensing practice for speciality services: Newsworld was to counter CNN; MuchMusic, MTV; and so forth. From an industrial policy standpoint, this strategy even appears to have worked: After initial difficulties, Canadian broadcasters are making more and more progress in the market.

Overall, however, the development of new broadcasting technologies has led to an undermining of the economic conditions for traditional programming control (Collins, 1990: 85f.; Melody, 1983: 6f.; McPhail & McPhail, 1990: 201f.). The propagation of commercial broadcasters has led, particularly in television, to a strong demand for mass-appeal programming. In view of the relatively small market, it was from the outset to be anticipated that demand could not be satisfied with products from Canada. The CRTC could at best expect that because of the high cost, television production would intensify when broadcasters enjoyed economic prosperity. However, it turned out that this strategy was weakened by the fragmentation of viewers and the associated loss of advertising revenues. Moreover, increased demand resulted in considerable price increases for programs. Under these conditions, a type of program control based to a large extent on quotas and that required those regulated to act contrary to their own economic interests had little prospect of success.

As an alternative strategy, the set of control instruments was significantly expanded in the 1980s. No longer was the broadcasting of Canadian programming subject merely to quotas but, instead, it was also subject to other measures designed to promote production. Examples include the initiation of program subsidies and the increased imposition of production requirements. Broadcasting regulation meshed more and more with traditional film and television support. The development of the Canadian culture industry thereby became one of the principal goals of broadcasting policy (see Starowicz, 1990: 9f.; Department of Communications, 1983; Fox, 1983). The success of these efforts can be seen in the upswing of this industrial sector, which for several years has recorded above-average growth rates and offers employment for many creative people (Kleinsteuber & Wiesner, 1988: 344).

Associated with this changed strategy is a departure from the traditional tools of regulatory policy. The lengthy experience with quota regulations shows that a set of instruments based exclusively on prohibitions and requirements cannot prompt commercial broadcasters to produce and broadcast a sufficient number of Canadian films and children's programs (see in particular LaPointe & Le Goff, 1988: 248; Grafstein, 1988: 283ff.). On the contrary, these broadcasters resort to avoidance strategies and apply political pressure to get the rules relaxed. In addition, the CRTC's discretion to act must be considered limited under this regulatory approach. Quota regulations are based on the premise that, by making use in part of American programming, profits will be reinvested in the broadcasting of less profitable Canadian programs. This concept of cross-subsidization (see Wyman, 1983: 174f.; Denton, 1984: 394f.) requires that the supervisory authority also help ensure the economic stability of broadcasters (Hardin, 1985: 26f., 40; Collins, 1990: 78; Vipond, 1989: 171; Babe, 1979: 231f.). These interrelationships find expression, for example, in the CRTC's licensing practice, which is characterized by giving protection to established broadcasters.

Since compliance with quotas is normally associated with considerable costs, the CRTC's decision to relax its rigid definition and to forgo blind enforcement is yet another example of this protective policy. It therefore comes as no surprise that in recent years the CRTC has decided not to test the limits of what commercial broadcasters can endure. As a result, the commission has been accused of failing to use the latitude it has in applying quotas (for the accusation that the CRTC is a "captive" of the private broadcasting industry it is supposed to regulate, see Hardin, 1985: 40f., 102ff., passim; Vipond, 1989: 171f., 176; Task Force, 1986: 177f.) or of conducting only symbolic policy (this was voiced by former Director of the CRTC Meisel, 1989: 2, 11).

With the expansion of subsidy programs, however, quota regulation has experienced a change in function and significance. It has become an important prerequisite for the success of production support (for radio, see Task Force, 1986: 408; Department of Communications, 1988: 18f.; for television, see Task Force, 1986: 408). Even with only minimal participation by Téléfilm Canada in production costs, it can be expected in many cases that the acquisition and broadcasting of (above all) American films and series are more profitable. The obligation to broadcast Canadian programs in 50% of prime time thus represents an incentive to accept subsidization offers altogether. Moreover, this increases the willingness to purchase Canadian productions that are priced similarly to U.S. programs and can expect comparable viewer acceptance (see Task Force, 1991: 126). As a result of the successful work of Téléfilm Canada, such programs have of late been turning up on the market with greater frequency.

Thus, the diversity of supervisory instruments has become the most important characteristic of broadcasting supervision and regulation in Canada in the early 1990s (Janisch, 1990: 9). The introduction of a license fee, repayable in the event that programming obligations are fulfilled, nevertheless shows that policy makers must continue to experiment with unconventional tools in order to increase the share of Canadian programs in private broadcasting. An approach that involves a variety of diverse tools is thus furthermore confronted with the complexity of control obligations (for these difficulties, see Task Force, 1991: 163ff.; Janisch, 1990: 9ff.). This involvement of course impedes the development of a consistent approach; this is all the more so the case when there are strong interactions between broadcaster categories (for recent data, see Task Force, 1991: 51ff.). For example, a change in the regulation for pay-TV and speciality services can have substantial economic effects on terrestrial television broadcasters. This was the reason, for example, for the disputes that occurred between terrestrial and cable broadcasters on the one side and pay-TV broadcasters on the other when MuchMusic and TSN were switched to basic cable service. The CRTC thus becomes more and more of an allocator of market and profit opportunities. With a change in prospects for profits, however, it can be expected that broadcasters who are detrimentally affected will offer considerable opposition. These aspects will likely make effective program control even more difficult in the future.

It is difficult to predict which track the coming development of the Canadian broadcasting system will take. Certainly, the relationship between satellite and cable television will play an especially important role. The mass distribution of private satellite dishes will likely interfere with the profit potentials for cable television (see Denton, 1984: 395f.; Task Force, 1991: 156). It remains to be seen whether the CRTC will soon afford cable television protection against satellite broadcasting as it previously did for the benefit of established terrestrial television broadcasters (see Wyman, 1983: 173f.; Collins, 1990: 73f.; Wiesner, 1991: 270ff.; CRTC, 1979: 2ff.). A particular threat for the Canadian broadcasting system will also soon follow from U.S. satellite programming, the "death stars" (Task Force 1991: 147ff.; CRTC Public Notice 1993-74: 1ff.). The commencement of operations of directly receivable satellites, planned for the mid-1990s, would make large numbers of U.S. stations accessible to wide parts of the Canadian viewing audience without this access being covered by Canadian broadcasting regulation. Even the slightest loss of viewers to U.S. broadcasters could threaten the economic viability of domestic stations due to loss of advertising revenues and could substitute in some measure for cable services.

Recently, Canada has entered the debate concerning convergence between broadcasting and telecommunications. In its framework decision on telecommunications, the CRTC established rules governing the future entry of cable carriers and others into the telecommunications market. Furthermore, the Minister of Industries has directed the CRTC to consider the possibility of permitting telecommunications carriers and others to enter the cable industry. The trend is very clearly toward greater competition and convergence of technologies (CRTC Telecom Decision 1994-19, September 16, 1994, Regulatory Framework; see also Intven, 1994: 131).

Should the current experiments with new instruments of control fail to be successful, it cannot be ruled out that the basic debate concerning the limits to and possibilities for the supervision and regulation of private broadcasting will recommence. In the course of the recent disillusion over the possibilities for successful quota regulation, a number of proposals have been developed aiming at a fundamental modification of the current system. Concepts calling for broad deregulation of private broadcasting have garnered special attention (see, e.g., Hoskins & McFayden, 1982: 355; Janisch, 1987: 43; LaPointe & Le Goff, 1988: 250ff.). Instead of demanding the fulfillment of programming obligations in exchange for the license, the government might, for example, auction it to the highest bidder. The proceeds could be used to finance subsidies for the production of radio and television programs (see Hoskins & McFayden, 1982: 355) or the activities of the CBC (Janisch, 1987: 43). In view of the failure of public regulation, this is posited to be the sole, realistic possibility for ensuring the development of an independent broadcasting culture. Public broadcasting, which many in Europe have termed outdated in times of large numbers of broadcasters and programs, would in the latter case be revitalized significantly in Canada.[10]

NOTES

1. Educational programming plays an especially important role in Quebec in that it helps ensure the cultural identity of this French-speaking province. See Trudel, 1984: 356ff.

2. See the landmark decision by the Privy Council re *Regulation and Control of Radio Communication, A.G. Que. v. A.G. Can. et al.* (1932) 2 D.L.R., 81ff. See also Ontario Court of Appeal re *C.F.R.B. v. Attorney General of Canada et al.* (1973) 38 D.L.R. 3d 335; and Supreme Court of Canada re *Public Service Board et al., Dionne et al. v. Attorney General of Canada et al.* (1977) 83 D.L.R. 3d, 178 for cable television.

3. The remainder of the budget comes from advertising revenues. For data, see Task Force, 1991: 102. Financing through broadcasting fees was abolished in 1953. See Frémont, 1986: 18.

4. For examples, see Babe, 1979: 36. Although the study by Babe is more than fifteen years old, it continues to represent the most comprehensive analysis of the regulation and supervision of private broadcasting.

5. The validity of this broad power to act has been upheld by the Supreme Court. See Supreme Court of Canada re *CTV Television Network Ltd. v. Canadian Radio–Television and Telecommunications Commission* (1982) 134 D.L.R. 3d, 193. See also the decision by the Federal Court of Appeal (1980) 116 D.L.R. 3d, 741. Russel takes a critical view (1993: 185ff).

6. According to empirical studies by Babe, 1979: 183f., of license renewals for the period 1969 to 1975, 95% (1,087) were renewed for the period envisioned by statute, 4.3% (49) were extended for two years or less, and only 0.7% (8) were denied altogether. Renewal was primarily refused when the broadcaster transferred shares in the company or rights of control without first informing the CRTC and obtaining the necessary permission. For more on this administrative practice, see also Babe, 1976: 567ff.; Trudel, 1984: 372; Beke, 1972: 121, for the equivalent practice by the Board of Broadcasting Governors.

7. In the period 1968 to 1975, the CRTC processed 515 applications for permis-

sion to transfer a license. In 423 cases, the application was approved. It was denied in 92 cases (18%). See Babe, 1976: 569f.; See Babe, 1979: 180f.

8. CRTC Public Notice 1990-111, 33ff., Appendix III. This replaces the former obligation to broadcast "foreground programs," broadcasts that contained high-quality programming for a certain length of time and without interruption.

9. This regulatory approach was not without controversy within the CRTC. See the comments of the CRTC members Gagnon and Grace, in CRTC Public Notice 1983-18.

10. For the proposal to allocate a second national television channel to the CBC, see Hoskins & McFayden, 1982: 356f.; Task Force, 1986: 351. For criticism of this proposal, see Grafstein, 1988: 292ff.

CHAPTER 6

Australia

A ustralia's broadcasting order has thus far been the subject of little attention in non-Australian literature. But a closer look reveals that it is worthwhile to focus on Australia, for a comparative analysis of its system with those of other nations opens additional vantage points.

Among "Western"-oriented, industrialized nations, Australia takes on a special status. Its geographically isolated location in the Southern Hemisphere, relatively low population in proportion to its territory (about 17 million), and the continent's wealth of natural resources and agricultural products have all aided Australia in developing a remarkable standard of living at peace with its neighbors. Despite the isolated location, the country has oriented itself strongly according to other nations—first the former colonial power Great Britain, then the United States—and only in recent times has it undertaken great efforts to build up a separate national identity and to give greater consideration to its geographic proximity to Asia.

1. DEVELOPMENT AND STRUCTURE OF THE AUSTRALIAN BROADCASTING SYSTEM

Such basic conditions have also strongly influenced the Australian broadcasting system. In Australia as in other leading Western countries, radio was introduced in 1923. Shortly thereafter—following experiments with various possible structures—this led to Australia's establishment of a dual broadcasting system. Alongside the publicly financed Australian Broadcasting Commission (ABC), created in 1932 on the model of the BBC and later reformed as the Australian Broadcasting Corporation, there also existed commercial radio broadcasters that financed their operations from advertising revenues and were based on the model of American broadcasting.[1] In this manner, dualism of structures became a fundamental element of the broadcasting order in Australia earlier than in Great Britain. In 1956 television was introduced in Australia and structured according to the model of the dual system.

The authority to regulate the broadcasting system within Australia's federation lies exclusively with the commonwealth (i.e., federal) Parliament rather than with the six states (see Armstrong, 1991b: 115). Federal regulatory power is part of Parliament's authority to control postal, telegraph, telephone, and similar services (Commonwealth Constitution, s. 51(v)). The Australian Constitution does not generally guarantee the protection of human rights. Rather, the Constitution envisages that human rights will be safeguarded by the democratic process. Consequently, the Constitution does not contain an explicit basic right of freedom of communication or broadcasting, nor does it even give special mention to broadcasting. The main written right of liberty in the Constitution that is applicable to broadcasting relates to the liberty of "trade, commerce and intercourse" between and among the states (Common-wealth Constitution, s. 92; for significance of this clause, cf. Lane, 1990: 204ff.), a rule comparable to the U.S. Interstate Commerce Clause. In 1992, the High Court[2] recognized for the first time a constitutionally guaranteed right to communicate, limited to communications that are of a political nature. It conceded that since representative democracy is, as an institution, entrenched in the Constitution, then all those individual rights and freedoms which constitute an essential component in the political process must also be pro-tected under the Constitution. However, this individual right exists only for the benefit of the national community.

In 1942, the basic legal conditions were augmented by statutory bases for the broadcasting system in the form of a unified act, the Australian Broadcast-ing Act of 1942, which has since been amended on numerous occasions. Its name as well has likewise undergone many changes, and it was finally termed the "Broadcasting Act." Amendments were often reactions to current devel-opments, but they were apparently not aimed at developing a transparent, internally consistent system of broadcasting law. Measured against the Euro-pean standard of legislation, the Australian Broadcasting Act was a rather inaccessible instrument. The Australian Broadcasting Corporation Act of 1983, regulates the work of the ABC.

In 1992, extensive changes in broadcasting law took place. Beginning in the mid- to late 1980s, the Australian government made a pragmatic shift toward deregulation. Market forces and self-regulation became the slogans. The shift in general policy paved the way for new developments in the broadcasting field as well. The government sought a broadcasting industry that was able to respond to the opportunities created by technological developments and to the complexities of the marketplace (for the object of the new legislation, cf. Broadcasting Services Act 1992 of ss. 3–4).

These concerns culminated in the Broadcasting Services Act No. 110 of 1992, which took effect on October 5, 1992. However, the Senate failed to approve part 7, dealing with subscription television. This led to the December 1992 enactment of the Broadcasting Services (Subscription Television Broadcasting) Amendment Act of 1992 and a series of major amendments in May 1993 (Broadcasting Services Amendment Act of 1993; Broadcasting Services Amendment Act (No. 2) of 1993), which regulate the introduction of subscription television. The management and allocation of radio spectrum were also reregulated with the Radiocommunications Act of 1992. It placed great reliance on "market-based" allocation of the radio spectrum, creating a new Spectrum Management Agency.

The Broadcasting Services Act of 1992 fundamentally altered the way in which broadcasting is regulated. Previously, four different sectors were distinguished: two national public broadcasters (ABC and SBS), commercial broadcasters, and "public stations." The new system comprises six categories, each subject to a different mode of regulation. These are:

1. National (ABC and SBS)
2. Commercial
3. Community (covering the earlier "public stations" designation)
4. Subscription broadcasting
5. Subscription "narrowcasting"
6. Open "narrowcasting"

The first three categories continue to be the most important.

1.1. The Australian Broadcasting Corporation

The ABC was founded as a public broadcaster whose programming mandate was oriented toward the public service concept.[3] It broadcasts a national television program, supplemented with regional window programming, that is receivable throughout Australia. The radio sector consists of two national networks and a number of regional broadcasting networks, which also include among others, one for young persons broadcast in metropolitan areas (for the ABC's offerings, see Bureau of Transport and Communications Economics, 1990: 58f.; ABC, 1990: 2f., 12ff.). Thus, most of the population receives four ABC radio services and one ABC television service.

The ABC'S management, the board of directors, is composed of seven to nine directors. With the exception of the managing director and a director to be elected by the ABC personnel, all directors are officially appointed by the General Governor, who actually implements cabinet decisions (Armstrong, 1987–1991: No. 3055). The managing director is appointed by the board. The National Advisory Council, which is appointed by the board, serves as an advisory body in programming issues.[4] In contrast to the board, it is composed of socially representative groups.

In contrast to the earlier tradition of selecting directors closely associated with the government, greater discretion began to be exercised in the early 1980s (Armstrong, 1987–1991: No. 3055) with the issuance of the Dix Report (Committee of Review of the Australian Broadcasting Commission, 1981). Although the ABC is assured independence in programming matters by statute,[5] this does not rule out perceptible efforts to exert influence, either directly or indirectly. These attempts by the government to intervene have been of varying intensity. Usually, they began with informal warnings (see Armstrong, 1987–1991: No. 3030) or movements to control financing (see the examples in Armstrong, 1987–1991: No. 3045; Kleinsteuber, 1989: 212). The ABC's financing ensues via allocations from the general federal budget. Until their abolition in 1974, broadcasting fees had also been levied, the revenues from which were included in the federal budget rather than used to fund the ABC directly. Moreover, the ABC does not raise revenues from the sale of advertising.

1.2. The Special Broadcasting Service

In Australia, a multicultural and (to some extent) multiracial society with important ethnic minorities, a growing need developed to take account of the user interests of autonomous cultural groups and ethnic minorities. In light of the impression that the ABC was not dedicating sufficient attention to this task, a new institution, the Special Broadcasting Service (SBS) was created in 1977 (for details on the SBS, see Jakubowicz, 1987: 18ff.)

The SBS was charged with the duty of broadcasting multilingual program services as well as multicultural programs intended for all Australians, thereby supplementing the offerings of commercial broadcasters and the ABC. The SBS operates a television network covering nearly all metropolitan areas (there are plans to extend it) and four radio stations, including one each in Melbourne and Sydney. In some regions in which SBS radio cannot be received, the broadcaster makes a portion of its programming available to community stations (Bureau of Transport and Communications Economics, 1990: 62). As with the ABC, the SBS is financed from the federal budget.[6] However, in 1992, financial realities led to the sale of broadcasting time for commercials (Special Broadcasting Services Act of 1991, s. 45; Minehan, 1993: 213).

1.3. Commercial Broadcasting

From the outset, the commercial sector of the broadcasting landscape in Australia was vigorous and popular.[7] Commercial television was originally supposed to ensure regionally structured programming, for which reason the licenses were distributed among independent local groups. Over the course of time, however, networks have come about that now dominate nearly the entire television sector and considerable amounts of the radio field (Bonney, 1987: 43ff.; Armstrong, 1987–1991: No. 115). In the area of television, comprised of roughly fifty television stations, three networks, which are named according to their transmission channels (i.e., Seven, Nine, and Ten), predominate.

Whereas the ABC was (and still is) obligated to guarantee national programming provision for all segments of the population and thus also to service the sparsely populated rural areas, commercial broadcasters were licensed and permitted to concentrate their programming in lucrative areas—specifically the metropolitan areas—and to restrict themselves to satisfying user interests there that are attractive from a commercial perspective.

In the large population centers, commercial broadcasting consists of television channels offered by all three networks and a number of radio stations, which normally broadcast group-specific programming. In the more sparsely populated regions, program offerings are considerably restricted, although at least one commercial broadcaster is receivable everywhere (see, in particular, Bureau of Transport and Communications Economics, 1990: 62ff., especially 75ff., 97ff.). In order to enlarge the amount of receivable programming in these regions, the policy of so-called equalization has been introduced, which aims at providing three commercial TV outlets in sparsely populated regions as well. Under this policy, the regional markets have been enlarged in the hope that the amount of advertising revenues that can be obtained will be able to suffice for financing several broadcasters. Its objective is to increase programming here as well to three television channels (for details, see Brown, 1987: 175ff.; Armstrong, 1987–1991: No. 115). Commercial broadcasters finance their operations with advertising, whereby, apart from some recent difficulties, considerable profits have nearly always been obtained (see figures in Bureau of Transport and Communications Economics, 1990: 79ff.).

1.4. Community Radio

A new broadcasting model entered the field in 1974 with the community radio sector, originally known as public broadcasting. This third broadcasting pillar, which has thus far been limited to radio, is principally designed to focus on locally rooted interests, including those of minorities. Following rapid growth in recent years, it is now comprised of 100 nonprofit broadcasters (Armstrong, 1991a: E 45). In program production, a wide number of voluntary, unpaid assistants create a significant amount of the programs, particularly in the creative area (for figures and data, see Bureau of Transport and Communications Economics, 1990: 109f.). Community licenses are granted to serve the needs of particular communities, and licensees must continue to represent those community interests (Broadcasting Services Act of 1992, s. 84, schedule 2 cl. 2(b)).

Financing is accomplished mainly through donations and membership fees, but also through governmental subsidies. Although advertising is prohibited as a source of financing, limited sponsoring is tolerated.

1.5. New Transmission Technologies and New Services

As a result of its isolated location, Australia does not experience any significant overspill of television or radio programming from other countries.[8] For the use of terrestrial frequencies as well, there is no need for complicated coordination processes with other nations. Consequently, most technically usable transmission frequencies are basically available for radio and television.

The government did, however, make an effort to take into account in its frequency plans the capabilities and thus the survival interests of licensed broadcasters, particularly the networks (this objective was expressly recognized in the Broadcasting Act of 1942, s. 81(2)), and for this reason it long practiced restrictive frequency planning and licensed new services only sluggishly. The decisive consideration was that in view of its low population figures and the accompanying limited advertising revenues,[9] Australia would hardly be able to support additional broadcasters. Possibilities for creating new transmission avenues—such as with cabling—went undeveloped, even in densely populated metropolitan areas. The government also proceeded slowly with DBS (satellite) technology. Its use could have led to better provision for remote areas, but it would undoubtedly have resulted in pan-Australian transmission of programs of the leading networks while it ignored respective local concerns.[10] The satellites of the Australian domestic satellite system (AUSSAT), which commenced operations in the mid-1980s, were used for transmitting programming to the various broadcasting stations, but except for remote areas, they did not broadcast directly receivable programming.[11] The Australian Sky Channel, which is broadcast only by satellite, was thus not generally accessible but instead was licensed only for public places, such as clubs and hotels.[12] The government also proceeded slowly with the introduction of satellite pay-TV,[13] partly because existing broadcasters, including the networks, might lose a portion of their advertising revenues. The Broadcasting Services Act of 1992, however, paved the way for the introduction of pay-TV, regardless of delivery technology.

As a result of the Broadcasting Services Amendment Act of 1992, three licenses for satellite subscription television have been issued. Each license permits the provision of subscription broadcasting services via satellite. Until 1997, satellite services must use the AUSSAT satellite system, which was privatized in 1992 and is operated by the new telecommunications carrier, Optus Communications. In addition, licenses may be granted for the distribution of subscription television by cable or microwave, provided the license applicants satisfy a competition test administered by Australia's competition regulator, the Trade Practices Commission (Broadcasting Services Act of 1992, ss. 96, 97).

The two "narrowcasting" services mentioned above—subscription and open—allow broadcasting with limitations on reception. These target special interest groups, limited locations, or use during a limited period to cover a special event. In contrast to the subscription category, open narrowcasting does not require subscribers.

Australia has not been immune to international debates about what has become popularly known as the information superhighway. Australia's communications infrastructure has a number of unique features that have conditioned this debate. Although the national telecommunications network is quite sophisticated, with 50% of Australian homes within 700 meters of optical fiber, there is no established cable television industry. Very few Australian homes are connected to a network capable of supplying cable TV. By late 1994, however, the duopoly telecommunications carriers had both announced ambitious plans to construct broadband networks.[14]

Telecom planned to develop a broadband cable network, using a variety of technologies, to pass 1.1 million homes by the end of 1996. Optus, for its part, joined forces with the U.S. cable company, Continental Cablevision, and the Nine and Seven commercial broadcasting networks. The consortium planned to build a hybrid fiber-coaxial network to link 2 million homes within four years, thus establishing a national network to compete with Telecom's. At the same time, the government was reviewing much of the current regulatory framework. It appeared likely that many of the issue, including the important issues of access to broadband networks, would be taken up by a review of the telecommunications duopoly, announced in mid-1994. The government had also established an expert group to investigate the implications and opportunities of broadband services for Australia. The Broadband Services Expert Group (BSEG) released a draft report in July 1994,[15] and was scheduled to release a final report by 1995.

2. MARKED STRUCTURAL DIVERSIFICATION OF PROGRAM OFFERINGS

As a result of the diversity of models in the broadcasting order reflected in its different sectors, Australia displays a system of structural diversification unique in the world. The differences among the four traditional sectors result from the combination of various requirements, especially at the levels of program mandate, legal form, and financing. These factors have in practice proved to be relevant as elements of control, which have led to fundamentally different programming conduct in the various pillars. To this extent, Australia exhibits an especially broad range of programming with respect to substance.

The distribution of various program mandates to different sectors of the broadcasting order has led to reciprocal relief for each of the other sectors. For instance, by the complementary program offerings of the other pillars, commercial broadcasting has in particular been relieved of an otherwise likely duty to serve minorities or to broadcast programs with low mass appeal. For its part, the ABC has not been relieved by the SBS of its duty to reflect in its programming the cultural diversity in society. But it has been spared the obligation of having to broadcast specific minority programming in languages other than English, and it is allowed to consider the output of the SBS in planning its formats. In the radio sector, the existence of community broadcasters relieves the ABC, at least de facto, of having to give intensive consideration to local or political minority interests, as long as public radio attends to these. To this extent, structural diversification has been enriched through the compensation and supplementation concept.

The relative diversity in programming brought about by structural diversification has not, however, meant equal audience shares. The clear "winners" by this measure are the commercial broadcasters, particularly the three leading television networks[16] (Channels Seven, Nine, and Ten), whose programming is oriented exclusively toward mass appeal and thus is strongly homogenized. Despite efforts to strengthen Australian components, the recipe for success, especially in the selection of formats, has been derived from American television. Foreign observers have often had the impression that the type of program offered is nearly identical with that in the United States (cf., e.g., Kleinsteuber, 1989: 213). This programming strategy has been accepted by the public. ABC television, by contrast, oriented itself along the public service concept and usually has lower audience shares than commercial TV. Many Australians still view ABC programming each week (cf. data on audience reach, in ABC, 1990: 7), but ratings have generally favored commercial television. All the same, the ABC is still able to claim 16 to 20% ratings in radio, though only 10 to 12% in television. These numbers are still remarkably high if one takes into consideration that ABC (as well as SBS) were introduced subsequent to commercial television and never attempted to usurp commercial broadcasting's role in general popular entertainment. The SBS with its metro-politan-concentrated television stations obtains 1 to 2% ratings. The signifi-cance of community radio is different from city to city and region to region, but it remains within quite limited figures.[17]

In view of the limited earning power of the advertising market, the number of competing commercial television broadcasters is remarkable, par-ticularly as regards the emergence of the three large networks. The scarcity of resources has, however, historically encouraged the purchasing of a large amount of relatively inexpensive foreign, particularly American, programs. In view of its financial situation,[18] the ABC is also under heavy pressure to import its programming; in so doing, it makes great use of British productions. The import of foreign programs in all television sectors is thus considerable. In the early days of television, nearly all programs were imported, and it has been a great struggle to reach the point where a substantial number are Australian. Consequently, the proportion of imported programs on Australian television has been steadily decreasing since the 1950s.

3. BODIES AND PROCEDURES OF BROADCASTING SUPERVISION

3.1. The Australian Broadcasting Control Board, the Australian Broadcasting Tribunal, and the Australian Broadcasting Authority

The commercial pillar of the Australian broadcasting order is clearly based on the American model of commercial broadcasting. At a quite early point—in the mid-1930s—commercial broadcasters expressed the desire for the establishment of an independent licensing and supervisory authority, which primarily followed from the wish to stem governmental influence with the aid of this body (Armstrong, 1987–1991: No. 345). But only in 1949 was this authority created in the form of the Australian Broadcasting Control Board (ABCB), which nevertheless displayed institutional proximity to the govern-ment (see Armstrong, 1987–1991: No. 345; Armstrong, 1979). Endowed with the "power to control programs," it only had competence to make recommen-dations in licensing matters. In 1977, building on the recommendations in the Green Report (Postal and Telecommunications Department, 1976), it was replaced, though with the same staff (Harrison, 1980: 34), by the Australian Broadcasting Tribunal (ABT), an independent statutory corporation responsi-ble for radio and television. The ABT was charged with the power to grant, renew, suspend, and revoke licenses; to authorize transactions in relation to licenses; to determine the program standards to be observed by licensees; to assemble information relating to broadcasting, and so on. In particular, it was responsible for that part of the broadcasting order dependent on licenses, that is, for commercial broadcasters and public radio. With regard to the ABC and the SBS, it had only marginal powers. The ABT was composed of a chairperson and up to seven members, who were appointed by the government for a period of up to five years. It relied on a staff of some 120 assistants for support.

From the very outset, the ABT sought to remain independent of both the government and the broadcasting industry. As a result, it was often exposed to heavy criticism, and its competences were curtailed over time. Moreover, in accordance with the policy of de- and reregulation conducted with greater intensity in the late 1980s, the ABFT saw itself called upon to relax some of its rules and in particular to create increased room for self-regulation within the broadcasting industry (see ABT, 1977: of programmatic importance later; see also Postal and Telecommunications Department, 1976). In the early 1990s, controversy again surrounded the question of whether it should continue to exist in its current form.[19]

Finally, in October 1992, the ABT was replaced by the Australian Broadcasting Authority (ABA). The ABA has a broader charter than did the ABT, particularly in the planning and technical area (cf. Armstrong, 1987–1993: No. 5015), combining the functions of planning, licensing, and program regulation. Its governing board is composed of a full-time chairperson and deputy chair and up to five other (full- or part-time) members. Maximum membership is two terms of five years each. The staff was selected mainly from the previous ABT staff. However, the planning group was drawn from the station planning branch of the Department of Transport and Communication, which previously had been in charge of planning.

3.2. Procedures and Enforcement

3.2.1. The Australian Broadcasting Tribunal

Australian broadcasting law made a particularly great effort to couple the ABT's activities with procedural guarantees, such that the risks of abuse of supervisory power were probably relatively low. For instance, the ABT's procedures were more strongly formalized than in other countries under study here and have been likened in many areas to court-like procedures. Above all, the "substantive power inquiries" were carried out in a quasi-judicial manner as they had to be conducted prior to any measures directly affecting broadcasters.[20] These were accompanied by "general inquiries"[21] in which more general questions of media policy were discussed, such as "TV Violence in Australia."

Inquiries were conducted in the form of public hearings and/or less formal conferences.

In 1986 the procedural requirements began to grow more formalized and detailed, particularly for the substantive power inquiries of interest here.[22] Anyone commencing the inquiry process had a nominal right to participate.[23] A particular personal interest was not required. The quasi-judicial nature of the inquiries process led to a situation where lawyers dominated the proceedings, making use of methods common in Anglo-Saxon trials for dealing with individual disputes, particularly the procedure of cross-examination.[24] With this abundance of legalization and increased emphasis on procedure, it was hoped that the transparency of decision-making processes would be raised, and unfairness, of which the ABT had sometimes been accused, would be avoided.[25] As a result, the ABT was generally, but also especially in the course of inquiries, cautious in using informal communication networks. The informal cooperative relations and forms of conflict management observed and widely exploited in other broadcasting orders were resorted to most infrequently in Australia. This reluctance was also influenced by the fact that the ABT was viewed by the broadcasting industry as an opponent that aimed at acquiring information to be used against the broadcaster's interests. Accordingly, the readiness to provide information in informal contacts was quite low.

Adversarial trials common in Anglo-Saxon law also have the effect that conflicts are usually carried out as disputes over the facts. This lessens the chances that the parties exchange information on the essential values and perhaps approximate one another. One critic of the ABT referred to a "reductionist tendency to isolate contested issues, to enumerate the facts and to test the validity of any proposition by cross-examination. These reductionist tendencies have discredited the concept of rights to contribute to decision making" (Armstrong, 1991b: 117, 119).

However, this procedural methodology avoided the risks of informal cooperation that might particularly have led to so-called agency capture.[26] But also lost were the advantages used elsewhere of the informal, cooperative search for consensus and implementation. Moreover, the formality of the procedure increased the expenditure of time and resources and led to some delays, which then provided a reason to criticize the ABT. Since the ABT could apply only limited influence on its procedures, it remained exposed relatively helpless to such criticism.

The ABT had at its disposal a graduated set of measures to record and sanction violations of broadcasting norms or license conditions. First it should be noted that the ABT did not regularly tend to take action on its own. Above all, programming was not constantly controlled (for criticism of this, see Saunderson Report, 1988: 92ff.); particularly precluded were advance controls.[27] However, at the initiative of third parties or in the course of an application for license renewal, the ABT could take note of possible violations, which it then investigated. One example of a legal violation occurred when the licensee breached its duty to monitor programming in such a way that program-related duties were violated in negligent fashion.[28] Since in these cases it would normally be unreasonable to resort to such action as revocation of the license, broadcasting law provided for graduated measures, ranging from a reprimand or admonishment to a requirement to publish an admonishment. In reality, however, the ABT was most reserved when it came to implementing such measures (Armstrong, Blakeney, & Watterson, 1988: 170). Also possible was the more severe directive to "suspend from broadcasting the person who was responsible for a program which contravened the standards" (see, including for practice, Armstrong et al., 1988: 170 f.). This power was virtually never used. Only the ABT's predecessor, the ABCB, recommended a temporary suspension of a TV license (in 1975), a suspension that was implemented by the Minister of Transport and Communications (Minehan, 1993: 215).

Violation of a programming rule would not directly constitute an offense; rather, this was only possible once an inquiry had been conducted and the ABT had then issued a directive to ensure compliance with the rule (see Walker, 1989: 272, 283f.; for the reserved practice, see Armstrong, Blakeney, & Watterson, 1988: 170). For violations of rules other than the program standards, comparable, though modified, regulations applied. Particularly worthy of mention is the limited power to punish in the event of violations of provisions of the Broadcasting Act on combating concentration. Enforcement of the limitations on media ownership found in the Broadcasting Act focused on prosecution, which required written consent of the Attorney General or a delegate (Broadcasting Act of 1942, s. 92R(4)). The ABT's role was to report on contraventions, not to initiate prosecutions. Despite apparent breaches of the law, however, not one prosecution was ever initiated for contravention of the ownership and control provisions (Armstrong, 1987–1993: No. 7285).

The ABT's sanctioning power was by no means limited to immediate reaction to violations. Rather, an important element here was the procedure for extending or renewing the license. For instance, the ABT not only had the legal authority to refuse to renew the license or to suspend or revoke it, but it was also able to renew the license for less than the usual five-year period (see, in particular, ABT, 1990a: 76 f.). Shortening the license period was used as a sanction for violations of license conditions in cases in which refusal to renew was deemed unreasonably severe. This possibility was sometimes resorted to in practice (see the examples set forth in ABT, 1990a: No. 5), and it enabled the ABT to react flexibly to violations.

3.2.2. The Australian Broadcasting Authority

The Broadcasting Services Act of 1992 has substantially changed the procedural requirements and enforcement measures formerly administered by the ABT. The ABA is more flexible, less legalistic, and relies extensively on informal contacts with the broadcasting industry. The ABA is responsible for monitoring the broadcasting industry and investigating complaints. The government can direct the ABA to conduct an investigation into or hold a hearing about any matter connected with ABA functions.

The preceding Broadcasting Act of 1942 had been criticized for not allowing sufficient flexibility in enforcement, which was limited to suspension or revocation of license. Now, under the 1992 Act, "The Parliament intends that the ABA use its powers, or a combination of its powers, in a manner that, in the opinion of the ABA, is commensurate with the seriousness of the breach concerned" (Broadcasting Services Act of 1992, s. 5(2)).

Accordingly, the ABA has been equipped with a wide variety of new disciplinary powers, including:

- Discussions with the service provider with a view to effecting rectification of any problem;
- Imposition of a license condition;
- Initiation of direct federal court action through the Director of Public Prosecution, which may result in fine;
- Issuance of a notice to take specified action within a specified period of time;
- Initiation of federal court action to enforce compliance with a notice, which may result in a fine
- Imposition of an administrative penalty, such as barring the broadcasting of advertisements for a specified period of time;
- Procurement of a court order to prevent provision of an offending service;
- Suspension of license;
- Revocation of license.

There is a considerable variation as to the amount of fines, with a maximum amount of A$2 million for an individual and A$10 million for a company.[29]

The Broadcasting Services Act of 1992 also encourages the ABA to be less formal than the ABT, providing it with extensive investigative powers, especially with respect to its monitoring functions. As set forth in clause 5 of the explanatory memorandum:

> It promotes the ABA's role as an oversighting body . . . rather than as an interventionist agency hampered by rigid, detailed statutory procedures, and formalities and legalism as has been the experience with the ABT. It is intended that the ABA monitor the broadcasting industry's performance against clear, established rules, intervene early when it has real cause for concern, and has effective redressive powers to act to correct breaches.

Already the ABT had reduced the use of public hearings in favor of written processes. In conformity with its power to inform itself in the manner it considers to be the quickest and most economical under the circumstances, the ABA may extend these possibilities. It may choose to consult with persons or groups, call for public submission, hold hearings, or conduct investigations. Thus, the ABA has the discretion to decide whether formal or less formal means should be employed to achieve the desired result. In conducting hearings, the ABA also has broad discretion to set the procedures to be used, including the decision of whether a hearing will be held in private. The rationale behind private hearings is to enable broadcasters to protect their confidential commercial information against competitors. Conferences may also be held within the hearing framework in order to negotiate solutions to problems.

4. FREQUENCY PLANNING AND USE

Australia did not opt to entrust broadcasting regulation and supervision as a whole to a single independent institution. Apart from the broadcasting regulatory authority, a number of important powers were exercised directly by the responsible minister and the Department of Transport and Communications.

Prior to enactment of the Broadcasting Services Act of 1992, the government, acting through the Minister of Transport and Communications, handled the use of transmission possibilities in the following manner: The Minister decided on general frequency planning, individual broadcasting zones, and details on use of the spectrum and certain frequencies (see Armstrong et al., 1988: 157, 180). He or she thus also decided on the introduction of new communications services and, as a result, ruled on the extent to which competition was to be created for existing broadcasters. The Minister was also endowed with initiating whether available frequencies were to be used and then opened to bidding by the ABT.

The Broadcasting Services Act of 1992 transferred the planning role to the ABA, providing for a more public process. The ABA is required to prepare frequency allotment plans and license area plans for the broadcasting bands referred to it by the Minister. The ABA is to perform its functions in accordance with policies dictated by the government, and it is subject to directions given by the Minister of Transport and Communications. The government continues to retain the power to determine which frequencies should be reserved for national or community broadcasting services, and it has supervening authority with regard to decisions by the ABA on planning activities. It may also give directions to the ABA on the preparation and alteration of frequency allotment plans.

5. FIELDS OF REGULATION AND SUPERVISION BY THE ABT

Regulation of broadcasting, including license requirements, is justified in Australia as in other broadcasting orders with consideration of, on the one hand, the scarcity of transmission possibilities and, on the other, the power of broadcasting to influence public opinion.[30] The privilege bestowed with the license is taken as grounds for submitting the licensee to special public accountability, which is reflected in the statutory obligations and in the license conditions.

Since it is still too early to analyze the experiences under the new Broadcasting Services Act of 1992, the following addresses regulatory practice under the Broadcasting Act of 1942. A later section (see Section 6) deals with the most important changes under the 1992 Act.

5.1. Licensing

Commercial broadcasters and operators of community radio stations were (and still are) required to have a license, which was granted solely to locally or regionally operating stations. Suggestions, such as those by the ABT, to introduce a network license were not put into effect.[31] The ABT could only pressure the networks in the course of decisions relating to their individual stations.[32] The licensing procedure took the form of a "substantive power inquiry." Following the public tender of the Minister's definition of the technical and, possibly, substantive[33] requirements for the license, the applicant submitted documents to the ABT, which were then usually discussed in a public hearing. All applicants were given adequate opportunity for written comment on competing applications, with the ABT being required to investigate all relevant facts. The ABT then published its decision together with a comprehensive report (for details on this procedure, see Armstrong et al., 1988: 180ff.).

5.1.1. Criteria for Awarding the License

The standard for the ABT's licensing decision was nominally the "public interest."[34] Initially, the ABT had sole authority to specify this criterion. However, this situation met with opposition from the broadcasting industry, which in addition was dissatisfied with a number of ABT decisions.[35] A 1981 amendment of the Broadcasting Act of 1942, which generally led to a weakening of the ABT's position, did away with this broadly formulated power by defining the public interest criterion. It set forth the exclusive criteria to be used in making licensing decisions under the 1942 Act, which moreover were different for the various stages or conditions of license (see Armstrong et al., 1988: 181 & n. 24). But this served to limit the ABT's power to evaluate potential licensees, with the result that some concerns previously considered relevant were no longer all capable of being taken into account. For instance, the government particularly frowned upon the ABT's measures to combat concentration in urban areas. The detailed rules on media concentration had been set by Parliament, and the ABT was not allowed to alter them. Accordingly, it was permitted to consider emerging media concentration only in areas outside the state capitals. This meant that it was only possible for the ABT to take into account urban media concentration incidentally, in its review of other criteria, such as in comparing various applicants or in evaluating whether an applicant was a "fit and proper person."[36] In the aftermath of the 1981 amendment, suggestions were sometimes raised (Saunderson Report, 1988: 41ff. and Recommendation 4; more far-reaching: Communications Law Centre, 1989: 14f.) that the ABT be returned its broad discretion in awarding licensees, but these recommendations were never implemented.

In the course of granting the license and deciding on license renewal, the ABT was required to review whether this would conform with statutory requirements; whether the applicant had available the necessary financial, technical, and management capabilities needed for broadcasting; and whether an applicant requesting renewal of its license had previously complied with the norms (for details on these powers, see Armstrong et al., 1988: 181f.; Armstrong, 1987–1991: No. 6060). The ABT also had to review whether the granting of a license would threaten the "commercial viability of stations in the area." This "protectionist" respect for other broadcasters was not intended to afford protection against competition but instead to help ensure that the survival of competitors did not become threatened.

Prior to 1981, the ABT purported to exact from license applicants a special "promise of performance" in programming, which created a basis for comparing applicants and was to serve as a standard for subsequent evaluations, particularly in license renewal. In the course of curtailing the ABT's power during the 1980s, however, the legislature expressly set forth what could be expected of applicants. It was sufficient that they commit themselves to two general obligations. One was to encourage the provision of Australian programs and to use Australian creative resources in producing programs. The other was to "provide an adequate and comprehensive service." This latter obligation was a sort of written self-declaration of the desire to observe license conditions and to orient programming according to the needs of the respective audience—in particular, to take account of "the nature of the community to be served" and "the diversity of the interests of that community." In reality, licensees gave public attention to this, but legally it was enough for license applicants to reiterate the wording of section 83(1) of the Broadcasting Act of

1942 in their applications (Armstrong, 1987–1991: No. 6120). The ABT tried to clarify the situation with a policy statement (ABT, 1988c: 78), but since this had no autonomous, binding force and at best offered interpretational assistance, it was of only limited significance. Furthermore, even the ABT did not view the obligatory declaration as an occasion for evaluating future conduct in licensing decisions (Armstrong, 1987–1991: No. 6120). It is thus misleading to think that the statutory obligation to make a pledge, together with the policy statement, meant that the ABT had wide-ranging powers with regard to program-related broadcaster self-restraint.

The review of whether the applicant was a "fit and proper person to hold an interest in the license," a formulation that resulted from the 1981 legislation, was a peculiarity of the former licensing process in Australia. The review process apparently was intended to be a partial replacement for the rescinded public interest test. This allowed the review of corporate applicants to focus on the persons behind them, particularly the shareholders (Armstrong, 1987–1991: No. 6080). Although the exact meaning of this standard was also the subject of a policy statement (ABT, 1988c: 89), many doubts about its application remained.

As the High Court determined in the context of the ABT's controversial bond inquiry, this criterion also covered character issues.[37] Precipitated by a bribery accusation (see ABT, 1990c; see also Corker & Tanner, 1990: 17) involving one of the most controversial personalities in the Australian media scene, Alan Bond, the ABT's formal investigation also raised several problems regarding the application of the standard, and the repercussions from the inquiry made their way into the depths of Australia's domestic policy. Although some observers were left with the impression that the inquiry was controlled or at least overshadowed by motives of party politics, many have regarded it as a triumph of independent investigation over political and financial entanglements.

The above-mentioned review criteria applied not only for initial licensing but also—sometimes in modified form—for license renewals, as well as for decisions on modifications of the license or even its revocation. In the renewal procedure, existing licensees enjoyed de facto a relatively protected status, in that the ABT normally required substantiated accusation of specific misconduct in the past as a basis for considering denial of renewal (Armstrong et al., 1988: 183). This conforms to a statement made by a leading member of the ABT, whereby the board, despite all willingness to review, seemed to bear in mind the notion that licenses are granted for eternity (see Communications Law Centre, 1988: 17 & n. 2).

5.1.2. Payment for the License

As mentioned above (see Section 4), the broadcasting license is viewed as a privilege, the granting of which is tied to certain duties. The license to provide commercial broadcasting is thus classified as an economically valuable good, which not only entails specific obligations (e.g., programming standards) but for which payment must also be made. In addition to a one-time fee upon bestowal of the license—the establishment fee, which applied to commercial radio only—a so-called license fee must be remitted annually by commercial radio and television licensees (see, including for the relevant laws, Armstrong, 1987–1991: Nos. 6255ff.), whose amount varies depending on the annual gross earnings figures. The revenues from these fees flow into the general budget and—in contrast to some recommendations by official commissions—are not used to finance certain broadcasting sector functions, such as support for research or quality programming (for the amount of the revenues, see the data in the ABT annual reports, e.g., ABT, 1990f.: Nos. 106ff.).

5.2. Combating Concentration

Australia is considered to be the country with perhaps the greatest degree of media concentration in the world (cf. Armstrong, 1987–1991: No. 6705; see also Barr, cited in Kleinsteuber, 1989: 207 n. 2). Indeed, the term "media tycoon" appears to have been coined in Australia; it refers to an economically powerful and usually politically influential entrepreneur or corporate group that is successful on the multimedia—and usually also multinational—level and predominates in wide segments of the media market with the aid of large financial resources derived from other markets. Media concentration is also an issue in the broadcasting sector, posing a threat to regionally dispersed broadcasting. Particularly dominant are the Sydney-based broadcasters.

Norms to restrict concentration in broadcasting have been in place since 1935 in Australia. The history of the battle against concentration has been marked by the gradual decline of measures to avert this, accompanied by a weakening of the legal power of the supervisory authorities (for the history of anticoncentration efforts in television, cf. Marmont, 1990: 53ff.). Relatively clear and "reasonably effective" rules were in force during the 1960s and 1970s (this is the assessment by Armstrong et al., 1988: 184), but these were modified in 1981 and thereby considerably weakened. With the 1981 amendment, the government was reacting to changes in the media market and to pressure from influential areas of the media industry. Efforts by the ABT to stem media concentration and trim the dominant position of the Sydney and Melbourne stations in the respective networks also prompted the amendments.

In this instance, the ABT decided to take action against the interests of Rupert Murdoch, an especially influential media tycoon in possession of powerful political contacts. But after the ABT refused to consent to a takeover of station ATV 10 in Melbourne by the Murdoch-owned News Group[38] the government reacted by amending the Broadcasting Act of 1942 and significantly curtailed the ABT's powers in the area of anticoncentration measures. Particularly weighty were the amendments that limited the ability of the ABT to influence changes in ownership among licensees. It was forbidden from refusing approval of a takeover simply on the grounds that this would give some group an undesirable degree of national or interstate dominance. The requirement that the ABT give its prior consent was replaced with one of subsequent approval. In addition, generous grace periods were accorded in order to ease changes in ownership. Reference has already been made to the limitations on the ability of the ABT to counteract concentration tendencies when awarding licenses. Seen as a whole, the 1981 amendments seem to have been motivated by the desire to make the ABT aware just how strongly it was interlocked in the political system and the power relationships in place there—that it was to have very little room for shaping media policy autonomously.

Even thereafter, the media industry continued to push for reduced control of concentration. It persistently argued that only with increased concentration could sufficient performance be expected of broadcasters—in view of Australia's limited economic strength and particularly its rural regions, by all means a strong argument. It is also important to note that the applicable quantitative anticoncentration rules had become anachronistic in some respects. Even the ABT spoke out against the rules limiting the control over multiple stations, terming them "arbitrary and counterproductive for policy" (ABT, 1984: 167), especially since they failed to distinguish between large and small stations. In particular, the ABT leveled criticism against the economic power and predominating influence of the stations in Sydney and Melbourne and the resulting imbalance in Australian media provision (ABT, 1984: 167).

One government reaction to these problems was a policy of so-called equalization, which aimed at improving the supply of broadcasting in more sparsely settled regions (for equalization policy, see Armstrong, 1987–1991: Nos. 470ff.; Brown, 1987). To this end, the anticoncentration rules were relaxed and transmission territories restructured. At the same time, however, the government opted for changes that were expected to result in a strengthening of the positions of larger broadcasters. Parliament did away with the old limits on radio and television dominance, which were tied only to the number of broadcasting stations.[39] In their place, new maximum figures of permissible concentration were established. Details on the differentiations adopted need not be dealt with here. But in essence, the new rules stipulated that one and the same natural or juridical person could own as many television stations as he or she pleased as long as these did not reach more than 60% of Australia's population and member stations did not broadcast in the same service areas. In the radio sector, it was possible to have up to sixteen stations as long as several other restrictions were observed. In exceptional cases, several television channels could be operated in one and the same service area (multichannel service, or MCS), thus allowing for the acquisition of regional monopolies (see

Armstrong, 1987–1991: Nos. 475; Armstrong, 1987–1993: Nos. 6705ff.; Armstrong et al., 1988: 185f.). But in nearly all cases, cross-ownership between radio and television was strictly prohibited for the same transmission territory; only one of these media could be controlled by the same person.

The restructuring of the transmission territories aimed at the creation of larger and thus more efficient markets. This coincided with the interests of successful broadcasters in expanding their market shares, although it also threatened the local focus of programming (see Appleton, 1991: 4). Due to new concentration opportunities, the market value of established stations also increased. This led to considerable price increases in takeover bids. The prices probably also skyrocketed in the expectation that the anticoncentration rules would continue to loosen.

The new laws led to massive changes in ownership relations.[40] At the same time, the competition for programs and stars increased among televisers, thus raising prices as well. As a result, a number of broadcasters, including two of the large networks, ran into financial difficulties and in 1990 were forced to enter into receivership. The financial transactions had made it possible for companies with no previous media experience to enter the broadcasting market; these firms apparently had problems in suitably adjusting to a market with which they were unfamiliar. In addition, the hopes that the concentration laws would be relaxed further were dashed, thereby thwarting the perceived chances for developing broad national networks. Still another burden was the economic recession in 1987, which put a damper on the willingness of industry to advertise. In view of these underlying conditions, the new rules tended to have more of a dysfunctional effect. Although markets became larger, the viability of two of the three large networks suffered considerably. This was joined by a bolstered trend to centralize the broadcasting industry, thus further weakening local diversification.

As mentioned earlier, the relaxation of the restrictions on multiple ownership was coupled in 1987 with a stiffening of the limitations on cross-ownership.[41] These prohibited joint ownership of radio, television, and/or newspapers in the same area. As a result, Murdoch was forced to retreat from the broadcasting sector in order to be able to hold on to his dominant position in the press sector, where his companies controlled some 70% of the newspaper market.[42] Moreover, as a result of Murdoch's switch in citizenship, a conflict emerged with the strict Australian rules limiting media ownership to Australian citizens (see also Harrison, 1987: 16 f.).

The statutory requirements played a decisive role with regard to the ABT's decision-making discretion. In spite of enormous administrative burdens, it had only limited possibilities for exercising substantive influence, mainly as a result of the complex procedures for approving changes in media ownership (see, critically, Saunderson Report, 1988: 69ff.). As mentioned above, the ABT could not address concentration at the licensing stage because the statute itself laid down the extent of permissible holdings, from which the ABT was powerless to deviate. Subsequent supervision, which is possible only in the event of a later change in ownership relations or control, was, of course, far less effective than prior review. Statutory criteria prevented the ABT from detailed assessment of concentration impact according to the standard of public interest. Moreover, if a company violated the parameters, it was granted a period of grace (see Armstrong, 1987–1993: No. 7245) in which to undertake changes in order to avoid sanctions. Although this ensured reasonable flexibility, it also encouraged companies to accept the risk of statutory violation and simply await the ABT's reaction.[43] The risks for companies were in any case rather low, since, in the area of combating concentration, the ABT had only very limited power to pursue offenders. In particular, it was dependent on criminal prosecution by the courts, for which the written consent of the Attorney General was necessary (Armstrong, 1987–1991: No. 7285).

With the amendment of the anticoncentration rules, it was in many respects made clear to the ABT that most of the power lay elsewhere. In view of close relations that often existed between large media companies and the government, it was possible for the executive branch and industry to ally themselves against the ABT. For instance, the government thwarted "undesirable" decisions by the ABT with statutory initiatives—for example, the so-called Murdoch Amendments in 1981 (cf. Armstrong 1987–1991: No. 445)—such that the ABT could not even rely on the continued existence of statutory law.

On the one hand, the statutory rules did contribute to a limitation of concentration in the broadcasting sector: Networks would have been most interested in and certainly quite capable of extending their zones of influence across the whole of Australia had they not been prevented from doing so by law. On the other hand, the rules were only partially effective. In fact, one of the three large networks long exceeded the statutory 60% mark[44] without being forced to abandon its excess interests.

In public discussion, consideration started to be given to regulating concentration no longer—at least less—with broadcasting legislation but, instead, with general antitrust law, which would place administration in the hands of the Trade Practices Commission rather than with the ABT (for the recent discussions, see Pengilley, 1990: 8).

5.3. Programming Commitments

The Broadcasting Act of 1942 obligated the licensee to observe the programming standards set by the ABT. The standards were determined in the above-mentioned official procedure (see Section 5.1.1), particularly after an inquiry had been carried out. Since the law expressly classified advertising as programming, there were standards for advertising and for programming in both radio and television. But with regard to radio, only a limited number of (minimum) provisions were issued, on the presumption that less regulation was needed in light of the greater supply. The following remarks focus on television.

5.3.1. Programming Standards for Television: Classification and Australian Content

The interim television program standards and advertising conditions (see ABT, 1990a: 13ff., 35ff.) contained only limited obligations. They dealt with the Australian content of programs and advertising, the classification of the content of programs and advertising restrictions according to the time of day at which they could be broadcast, presentation of advertising, and the manner in which programs could be promoted.

A unique feature under the 1942 Act was so-called classification (Armstrong, 1987–1991: No. 1075; Armstrong et al., 1988: 167f.), which was introduced in 1956, primarily in the interest of protecting adolescents. Broadcasters were required to indicate with a letter visible on the television screen whether:

1. A program is especially suitable for children (C)[45]
2. A program is suitable for all viewers, including children (G)
3. It is advisable that children only view the program together with their parents (PGR: parental guidance recommended)
4. A program is suitable solely for adults (AO: adults only).

A further category referred to programs "not suitable for television." The ABT program standards also set the times of day in which restrictions for each category of programming applied, and only under certain conditions could other classes of programs be broadcast (Armstrong et al., 1988: 168).

Classification was the responsibility of the television broadcasters themselves, who normally designated a classification officer for this purpose. Advice was provided by FACTS, the Federation of Australian Commercial Television Stations. But when stations belonged to a network, the supplied program usually had already been classified at the request of the network (Armstrong et al., 1988: 167). The ABT did not undertake systematic control on its own but rather when it received complaints from viewers. Since the classification categories were outlined only roughly, the ABT restricted itself to reviewing the plausibility of the classification made by the broadcaster. Only when this was obviously incorrect was a licensee called to account. But since this merely entailed subsequent control, the consequences of such objections were rather insignificant.

A special situation applied to programs that had been classified as meeting the special needs of children (C). This classification was devised to ensure the production and broadcast of programs going further than those merely suitable for children (category G), in that they were to meet the specific personal and social needs of children in the six- to thirteen-year-old age group. Until late 1990, 5:00 to 6:00 P.M. weekdays counted as C time. During this time, children's programs had to be broadcast, and there were supplemental rules for ensuring Australian content, advertising restrictions, and so forth. In order to ensure that broadcasters met the criteria contained in these special rules, a program had to be classified prior to broadcast by the ABT in consultation with the Children's Program Committee. Although it cannot be said that this represented censorship of program contents, the procedure did function as prior control for the purpose of correct classification. Since broadcasters were interested in being certain of receiving the C classification, the ABT was able to exercise a certain degree of indirect influence on broadcasts.

In 1990 the rules on children's programming were modified and, particularly with respect to scheduling, made more flexible. However, it was decided to retain the requirements of preclassification (Armstrong, 1987–1991: No. 1570; for details, see Children's Television Standards, in ABT, 1990a: 41ff.).

Deserving of special attention are the Australian content regulations. Against the backdrop of the abovementioned geographic, historical, cultural, and economic peculiarities in Australia, a number of measures were introduced in the 1960s to ensure an Australian framework. In particular, this was intended to reduce the dominating influence of American programs and to aid in finding a means to support the search for a unique Australian identity by increasing the proportion of Australian programs (cf. Appleton, 1988: 1ff., 5ff.; Harrison, 1980: 17ff., who refers to the "confusion of objectives").

In general, the Broadcasting Act of 1942 stipulated that the licensee employ as far as possible the services of Australians in the production and presentation of programs. To encourage compliance to this duty, the former ABCB instituted a "points system" in 1973. The various categories of programming were assigned points, and the broadcaster was required to achieve a certain number of points in a given year in accordance with its annual programming hours. However, the levels were set so low that broadcasters had no difficulty in obtaining the required point figures (Harrison, 1980: 34; Appleton, 1988: 20). As a result, it was impossible to discern any relevant effects for the support of Australian productions (Harrison, 1980: 37). It is thus fair to assume that this was an example of purely symbolic policy (Harrison, 1980: 42), an appraisal that was widely accepted.

More effective than the point system were specific television quotas, particularly for "first-release Australian drama" and for "high-budget Australian specials." These were not as easily filled (Armstrong et al., 1988: 175; Harrison, 1980: 34). But through exploitation of the latitudes inherent in the drama quotas and because of concessions made in monitoring programming, the normative standard was again adjusted to match reality in order that the quotas could be deemed filled (Harrison, 1980: 35). Nevertheless, most Australian observers agree that the scheme had a generally positive impact and achieved its main objectives. A special rule was in place for the promotion of children's programming.

The attention given to ensuring Australian content focused on the category of drama in television.[46] In this area, the program production costs were especially high, and a large number of cheap import products were available. The regulation was deemed successful, because in comparison to the early stage of television, broadcasters began to make increasing use of Australian productions, and programs of Australian origin grew increasingly popular with audiences (cf. ABT, 1988a: 1, 19).

In early 1990, the rules were modified by the ABT. The new standard covered two separate requirements: a minimum level of Australian content between 6:00 P.M. and midnight, and a quota for first-release drama and for certain diversity programs.[47] There was also a quota for children's drama. The quota requirements were defined in two respects, namely, with an Australian factor determined on the basis of the amount of Australian involvement in the creative control of the program, and a quality factor based in part on purchase price per hour for the various drama formats and on the notional costs and programming risks in the case of diversity programs.

The underlying idea was to give broadcasters greater flexibility in their decisions together with incentives to broadcast programs lacking mass appeal. Thus, the requirements were an expression of the policy of enlarging broadcasters' decision-making discretion, thereby creating more room for self-regulation. Quality-related criteria applied only to external indicators; the ABT did not get involved in the difficult task of determining substantively what constituted quality.

Australian content was also demanded for commercials. Here, the rules were particularly rigorous: All commercials basically had to be produced in Australia (or New Zealand). This meant that commercials for foreign products normally had to be refilmed in Australia. Only in limited cases were exceptions possible. It is not surprising that the advertising and television industries strongly attacked these rules and called for at least greater flexibility.

The effectiveness of these Australian content requirements were augmented by supplementary measures, such as tax concessions as well as the often-cited protectionist policy in favor of established broadcasters to ensure them sufficient financial strength. Although it is certainly plausible that content regulation in this context had practical consequences, there is an apparent lack of systematic, empirical studies clarifying the causal relations.[48]

Nevertheless, Australian origin does not necessarily have any effect on program quality—regardless of how this may be defined. However, Australian institutions relied on the hypothesis used by other countries as well that fundamental participation by Australians in program production improves the chances for the coverage of specific, Australian topics, perspectives, and behavioral patterns. In addition, the place of production also seemed to have an effect on viewers with regard to program attractiveness. One indication of this is that programs produced in Sydney obtained higher ratings there than in Melbourne; the same applied the other way around for broadcasts produced in Melbourne (see ABT, 1988a: 19ff.). In general, the programs produced in Australia came to occupy mainly top standings in the ratings list (see Appleton, 1988: 23f.; ABT, 1988a: statistical references).

The policy of promoting Australian content was, however, not simply a program-oriented one; it especially aimed at developing an independent production infrastructure for Australia. In this regard, significant successes were scored, although this could be attributed mainly to tax policy (cf. Appleton, 1988: 217ff.). Melbourne and Sydney became centers for the production of television programs and commercials, as well as films, and the latter soon began to gain notice on the international market. However, concentration of production in the two metropolises Sydney and Melbourne raised concerns about furthering the cultural dominance of these two cities and thereby blocking possibilities for stronger substantive diversification (see ABT, 1988a: 39; Appleton, 1988: 232).

5.3.2. Fairness and Diversity

The Australian broadcasting order did (and does) not contain any other special requirements for the positive assurance of quality. Although broadcasters were—as mentioned above—required to commit themselves to providing adequate and comprehensive service that related to the nature of the community to be served and the diversity of the interests of that community, they were not required to take on specific duties, such as to produce special local programs or news or current affairs programs.

Above all, Australian broadcasting law did not stipulate that commercial broadcasters ensure substantive plurality or balanced consideration of all relevant interests.[49] Australia is generally suspicious of officially mandated schemes designed to guarantee fairness—or even other objectives, like morality. The bonds between broadcasting regulation and the functioning of a democratic society have thus far not been made the subject of any special normative provisions under Australian broadcasting law. One also finds no specific precautions against the risks of one-sided influence—a thoroughly remarkable situation in view of the degree of media concentration. Only prior to elections are broadcasters that take up issues of relevance to the election required to provide all parliamentary parties with sufficient broadcast opportunities. But even here, broadcasters are left with wide discretion in allocating broadcasting time (Armstrong, 1987–1991: No. 1600).

The requirement that the licensee "present news accurately, fairly and impartially" was only imposed for news programs and not, for example, current affairs programs, let alone other programs; for radio, this duty of accuracy was limited. Although defining guidelines in this area were enacted by the ABT, the duty has not been viewed as a starting point for a special fairness doctrine. However, in the fall of 1990, the ABT commenced an inquiry into the question of whether a right of reply should be introduced. This was not conceived as a mere opportunity to defend oneself on subjective legal positions, as is the case for the European right of reply relating to rights of privacy and character. Rather, the focus in Australia was a right of access to broadcasting for all persons whose interests might be affected, in order that they could present a position on controversial topics that had yet to be sufficiently elucidated (for the related difficulties of delineation and dubious results, see Bartlett, 1990: 30). This was generally viewed as an instrument to guarantee diversity of opinion, but, notably, it was not based on a programming duty of broadcasters but, rather, on the legal assurance of the viability of private and social initiatives (for details, see Bartlett, 1990: 33 n. 5; but for a broadcaster's duty, see p. 30).

The structural diversification in the different broadcasting sectors (see Sections 1, 2) seemed to be viewed as adequate for ensuring programming plurality by broadcasters. The homogeneity and level of mass-appeal programming by commercial broadcasters (which is registered with astonishment particularly by foreign observers) was, in light of the programming alternatives of ABC and SBS, apparently not seen as a problem requiring action from media policy.

The ABT subsequently extended its inquiry on the right of reply and commenced a review of whether requirements of accuracy and fairness should be created specifically for news and information programs (ABT, 1990d). In so doing, the ABT focused special attention on "talkback" programs in the radio sector, which had prompted a number of complaints. In the interest of an open, controversial public debate, however, the ABT was willing to forgo formulating requirements of impartiality (ABT, 1990d: 6). Following the inquiry, in January 1992, the ABT introduced accuracy and fairness standards for the current affairs programs on commercial television and radio. The standards included an obligation to ensure that reasonable opportunities were given to present significant viewpoints when dealing with controversial issues of public importance (Television Program Standards (TPS) 24; Radio Program Standards (RPS) 8).

5.3.3. Advertising Restrictions

With regard to advertising restrictions, one must distinguish between substance requirements and quantitative rules. The ABT Television Advertising Conditions initially set content-related rules, which also contained product-specific restrictions. In addition, the broadcasting and advertising industries have practiced widespread self-regulation since the late 1970s (see Armstrong et al., 1988: 232ff.; Media Council of Australia, 1990), and this has been approved by the Trade Practices Tribunal despite elements questionable under the law of competition.[50] The supreme body of control is the Media Council of Australia (MCA), in which umbrella organizations of commercial radio and television broadcasters (FACTS, FARB; Federation of Australian Radio Broadcasters) and of the other commercial media are represented. The MCA sets advertising codes. In conjunction with the MCA and its organs, FACTS and FARB[51] review nearly all commercials produced by advertising agencies and, when suitable, attest to their conformity with the codes and the law. The decisions can be lifted by an internal complaints authority (for the complaints procedure, see Media Council of Australia, 1990: 22). That the majority of commercials are submitted to FARB or FACTS has not least to do with the fact that nearly all stations make the broadcasting of a commercial dependent on this "clearance certificate."

Until 1987, Australian broadcasting law provided for time restrictions on advertising and for limitations on commercial interruption. In an extensive study, the ABT called the sense of some of these rules into question, particularly that of the maximum amount of advertising time (ABT, 1987). Commentary from the advertising and television industries revealed that the time restriction in force then (thirteen minutes per hour) was considered to be the maximum figure that viewers would tolerate, and this was complied even without corresponding regulation. Thus, the ABT doubted whether such time restrictions were necessary, viewing them as symbolic policy. In general, it also asked whether advertising restrictions might not have a dysfunctional side effect: The reason for making a detailed set of legal rules is usually to remove any doubt about the lawfulness of conduct in compliance with those rules. The rules have a psychological effect—they establish the idea of a licensee's free entitlement to a fixed amount of advertising effect per hour. It was believed that broadcasters would not in each case review how much advertising is advisable or tolerable for viewers but rather would simply stick to the time limits. As a result, a set of rules would have shielded the licensee from having to take public complaints seriously, for they often strip public concerns of their persuasiveness (Media Council of Australia, 1990: 13). On the other hand, the ABT deemed the rules limiting commercial interruption to be worthwhile (cf. the "criteria of success or failure" listed by the ABT in ABT, 1987: 37ff.).

Given this view of advertising rules and in conformity with the trend toward deregulation, the ABT decided to suspend the rules for two years beginning in 1987. This was intended as a regulatory experiment. Following the test period, the ABT conducted an inquiry in 1989, which indicated that there had been no serious changes in advertising conduct during this period. As a result, it was also decided not to undertake any changes, and the "experimental phase" continued in effect (see ABT, 1990e; see also Noble, 1990: 24). In this way, the ABT hoped to prompt the broadcasting industry to enforce advertising restrictions on its own as far as possible, that is, it hoped that an acceptable advertising standard would come about solely through competition among broadcasters. Following a further inquiry in 1991–1992, however, the ABT decided that the experiment of deregulating advertising time had not worked, and it reimposed advertising time restrictions, which took effect in March 1992. The new restrictions set maximum limits to the amount of "non-program material" broadcast during the day and during each hour of the day. Separate limits were imposed for prime-time and nonprime-time viewing periods (TPS 25).

Since the new restrictions on time shares of advertising were oriented to current practice, it seems fair to suspect that the new standards only constituted another form of symbolic policy, which was primarily conceived in order to give the public the impression that the ABT was keeping watch.

5.3.4. Regulation of TV Violence

The above-mentioned system of program classification also addresses the depiction of violence and pornographic or immoral programs. In view of the commercial imperative in Australian television, the depiction of violence is an especially important topic. The ABT thus felt compelled to conduct a formal investigation into the depiction of violence in Australian television, the results of which were published in four volumes (ABT, 1990b). Unfortunately, a systematic analysis of programs themselves was not conducted.[52] As is the case with such inquiries, the investigation was based mainly on submissions from the public, associations and interest groups.

The use of inquiries also serves as the main basis on which estimations of the effectiveness of regulation are based. By using this approach, the ABT's widely recognized problem of limited effectiveness has not primarily been treated as a problem of the ABT's activities. Rather, the focus has principally centered on what can be done to strengthen the population's standard of information in order to improve the use of complaints (ABT, 1990b: 86–87). For the ABT, this meant that it would only take the initiative in rare cases and in particular would refrain from systematic monitoring. It would only be able to exercise the role of guardian of the broadcasting order under the complaints procedure and the license renewal process.

In seeking to stem excessive violence in television, the ABT also appealed to the broadcasting industry that it practice greater self-regulation and augment prevailing ABT standards (ABT, 1990b: 88). In this regard, it again became clear that the ABT was attempting to stress self-regulation, in this case in the form of supervised self-regulation. The ABT would monitor industry actions in this area and where necessary intervene with its instruments. The broadcasting industry's interest association, FACTS, reacted to ABT's proposal with guidelines (FACTS, 1991). In contrast to other areas, the enactment of such a code did not encounter antitrust objections, as the effect on competition was deemed to be marginal.

6. REFORM OF REGULATION AND SUPERVISION: THE NEW REGIME UNDER THE BROADCASTING SERVICES ACT OF 1992

The Broadcasting Services Act of 1992 (hereafter, the 1992 Act), along with the Radiocommunications Act of 1992, have entailed great changes for the broadcasting sector. As already mentioned (see Section 1), there are now six different categories of broadcasting, each subject to different forms of regulation. In addition, a new regulatory authority, the Australian Broadcasting Authority has been created.

The legislation has also altered the licensing regime. Following the British model, Australian broadcasting licenses for commercial services are now generally granted on a market-based system; but in contrast to Britain, Australia places less interest on the quality of the service to be provided. Subscription radio broadcasting and television narrowcasting are less regulated under a so-called class license scheme, where the individual station requires no authorization at all. The national services (ABC and SBS) continue to be governed by the Australian Broadcasting Corporation Act of 1983 and the Special Broadcasting Service Act of 1991. However, some aspects of planning and program complaints are controlled by the 1992 Act.

This reregulation of the broadcasting order thus signals a sizable move toward deregulation. However, the level of regulation depends on the license category. As it is the ABA's task to clarify the criteria of each category (though possibly under specific directions from the government), it is endowed with far-reaching powers. The following remarks are restricted to the regulation of commercial services.

6.1. Licensing of Commercial Broadcasters

Despite some movement toward deregulation, the 1992 Act does try to protect the television industry by allowing it to adjust to new developments. No more than three licenses may be awarded for a specific transmission area, unless a special governmental review determines, among other things, that national benefits would derive from having more than three stations in the area.

Licensing itself concentrates on price as the main criterion for awarding a new license (see Broadcasting Services Act of 1992, s. 36). Since the allocation of the license is determined primarily by the market, it was a logical step to require successful applicants to pay for the license. Presumably this will be the amount contained in the bid, although the actual mechanism has yet to be determined. The amount to be paid is officially published; this payment also includes the application fee.

The allocation process is open only to those applicants that fulfill certain minimum qualifications. One such qualification is that the company must be formed in Australia. Another suitability criterion reflects the ABA's determination that allowing the licensee to hold the license would not lead to a significant risk of violation of the Broadcasting Services Act (see Broadcasting Services Act of 1992, s. 41). The earlier "fit and proper person" test was abolished, even though the suitability test makes limited reference to character criteria (see Armstrong, 1987–1993: No. 5125).

Once the license is granted, standard license conditions must be observed. These include the program standards set by the ABA; the term "program" continues to comprise advertising as well as sponsorship matters. The Act of 1992 obliges the ABA to determine standards concerning commercial television programming with respect to children and Australian content. Parliament has the right to review the standards and, if it deems necessary, to reject them (disallowance). Moreover, Parliament has the right to amend any standards determined by the ABA (Broadcasting Services Act of 1992, s. 128, as amended by Broadcasting Services [Subscription Television Broadcasting] Amendment Act of 1992). The standards may not provide that programs be subject to ABA approval prior to broadcast. There is an exception with respect to the Children's Television Standards, which permit preclassification of individual programs by the ABA or its representatives as meeting the special needs of children.

In addition, it will be expected that the major broadcaster organizations, like FACTS, develop Codes of Practice in consultation with the ABA, which will then register them. Such codes may relate to the protection of children, the promotion of accuracy and fairness in news and current affairs, the limitation of the time devoted to advertising by commercial broadcasters, the prevention of simulation of news or events in a way that misleads or alarms the audience, as well as the classification of programs by methods that reflect community standards. The list of matters to be taken into account in developing codes includes physical and psychological violence, sexual conduct and nudity, offensive language, material likely to incite or perpetuate hatred against, or vilify, any person or group on the basis of ethnicity, nationality, race, gender, sexual preference, age, religion, or physical or mental disability. Parliament reserves the right to amend any broadcasting code of practice that has been registered by the ABA (Broadcasting Services Act of 1992, s. 128).

It is evident that the 1992 Act relies even more heavily than was formerly the case on self-regulation by the broadcasting industry. However, such self-regulation is to take place in close cooperation with the ABA. The registration of a code presumes that the industry group also provides the public with adequate opportunity to comment on the code, although the public does not have any general right of participation. If the ABA finds sufficient evidence that a registered code is not operating to provide appropriate safeguards, then it is obliged to enact a standard in the code area. In this way, the old concept of a safety net has been retained.

The new rules reduce the amount of regulation while they assign greater weight to self-regulation. The ABA may vary or revoke license conditions, other than a list of standard conditions imposed by schedule 2 of the 1992 Act. The ABA may also impose additional conditions on a license (Broadcasting Services Act of 1992, ss. 43, 44). Additional conditions may include requirements to ensure that the license complies with a code of practice governing programs. In 1993 the ABA endorsed codes of practice for commercial television (developed by FACTS) and commercial radio (developed by FARB).

The licensees now enjoy a greatly protected status. While the license term generally is five years, the ABA must renew the license upon application—accompanied by a renewal fee—unless it is satisfied that the license will not be suitable (see Broadcasting Services Act of 1992, ss. 46–47). One particularly important new rule is that the ABA is not required to conduct any inquiry or investigation into a renewal. Also new is the free transferability of the license entity without prior ABA approval (and thus an inquiry into the suitability of the transferee) (see Broadcasting Services Act of 1992, s. 48).

6.2. Combating Concentration

The Broadcasting Services Act of 1992 essentially maintains the precautions to limit concentration (for details, see Armstrong, 1987–1993: Nos. 5180ff.). The Act aims to consolidate and simplify the previous control and ownership laws (see Broadcasting Services Act of 1992, part 5 and schedule 1). At the same time, however, some restrictions were lowered, such as those with respect to the total permitted television reach (now 75%, up from 60%). This change was desired by the networks. The rules on limiting cross-ownership have not changed appreciably.

Because of numerous amendments, the previous set of concentration rules under the 1942 Act was quite detailed. These rules addressed not only ownership but also diverse forms of control. Control can arise through a variety of mechanisms, including voting rights, contractual rights of direction and personal influence. The 1942 Act was criticized for the fact that media proprietors managed to find a continuous supply of loopholes, which arose in part because the 1942 Act sought to specify every means by which control could be exercised; the 1942 Act was too rigid to deal with the complex corporate structures of media undertakings. The new rules attempt to provide the ABA with the flexibility to investigate and act upon any suspected artifice or arrangement seeking to avoid the conditions of media ownership and the limits imposed on ownership concentration. Even if a merger or takeover does not contravene the concentration limits under the broadcasting law, however, it may prohibited by Australian competition law (*Austereo v. Trade Practices Commission* (1993) 115 Austl.L.R. 14).

6.3. Licensing of Subscription Television Broadcasters

The introduction of subscription television had been particularly controversial. Neighboring New Zealand, with a total population of only 3 million, had been successfully operating pay-TV for several years. However, fears were raised that pay-TV could upset the existing broadcasting system in Australia. The great political interest in this new service resulted in a set of rules that were more encompassing than originally expected. This conflicts with the regulatory policy set forth in section 4 of the Broadcasting Services Act of 1992 that the degree of regulation imposed on services should accord with their degree of influence on community views. It was broadly assumed that subscription television was less influential than general commercial television.

Parliament sought to proceed cautiously in the Broadcasting Services Amendment Act of 1992. For instance, although rules have been enacted for a transitional period (until June 30, 1997), these are considered to be dispensable in the event of expected greater competitiveness. These cover restrictions on ownership and control, the prohibition of advertising, and limitations on the number of satellite subscription services. The "laws of the market" are supposed to make such restrictions unnecessary after 1997.

But even in the period prior to this, it is expected that, under a market model, not specific broadcasting regulation but, rather, market regulation generally is indicated. Thus, provision has been made for participation by the Trade Practices Commission (TPC) in the licensing process (see Broadcasting Services Act of 1992, s. 97). Important amendments were made to the subscription television provisions in May 1993 in order to shore up the government's preference for satellites as the main delivery for pay-TV. The amendments followed alarm that microwave subscription services would become available before satellite services and a series of technical legal problems with the allocation of satellite subscription licenses. Thus, the grant of licenses for the delivery of subscription television by microwave was delayed until the commencement of satellite subscription services or December 31, 1994, whichever occurred first. Moreover, the amendment made provision for the payment of deposits for the satellite subscription services, something which had been omitted from the original rules, resulting in a number of speculative bids (see Broadcasting Services Amendment Act of 1993; Broadcasting Services Amendment Act (No. 2) of 1993).

The new rules will not be dealt with in detail here. It should be mentioned, however, that the allocation of licenses is based on price (Broadcasting Services Act of 1992, s. 93). Many of the license conditions are similar to those governing commercial broadcasting licensees. The main difference is that subscription fees must continue to be the predominant source of revenue for subscription services. In addition, subscription licensees need not contribute to the provision of an adequate and comprehensive range of broadcasting services. It is presumed that pay-TV will (and may) be used as specialized services for target groups or for certain programming categories. Special rules apply to the use of R-rated material (X-rated material is prohibited). The 1992 Act also addresses the concern of other broadcasters that subscription TV might be able to acquire exclusive rights for important events. The Minister of Transport and Communication may specify an event or kind of event, the televising of which should be available free to the general public. Subscription licensees may not acquire any rights to broadcast such events, unless a national broadcaster or commercial television licensee has acquired that right. The Minister can remove an event from this list if he or she is satisfied that free-to-air broadcasters have had a real opportunity to acquire, on a fair commercial basis, the right to broadcast that event and have not acquired the right within a reasonable time.

The Broadcasting Services Amendment Act of 1992 created new cross-ownership restrictions for subscription television that are, however, less restrictive than those in other media sectors. The restrictions prevent newspapers, commercial television broadcasters, and telecommunications carriers from having an ownership interest in—or being in a position to control—one of the initial two satellite subscription television licensees. These provisions are, however, supplemented by a competition test on the allocation of all subscription television licensees, which has been referred to previously. This test requires the Trade Practices Commission to report to the ABA on whether the allocation of a license would result in a substantial lessening of competition in the relevant market. At the time of writing, the future development of subscription broadcasting in Australia remains unclear. Although the ABA has issued two satellite pay-TV licenses, neither was issued to established broadcasting interests. One successful licensee, Australis Media Ltd., also controls a large proportion of microwave licenses in metropolitan areas. Australis is 26% owned by Telecommunications, Inc. (TCI), the large U.S. cable company. Nevertheless, attention has increasingly focused on ambitious plans by the two telecommunications carriers, Telecom and Optus, to roll out competing broadband networks. An initial alliance among Telecom, Channel Nine (Packer),

Channel Seven, and News Limited (Murdoch) dissolved after failing to acquire a satellite license. Subsequently, the commercial broadcasters (Channels Nine and Seven) have formed an alliance with Optus and the American cable company Continental Cablevision. Meanwhile, Telecom continued to pursue a strategic relationship with News Limited and was actively negotiating with TCI. At the end of 1994, however, there remained no commercial pay-TV service in operation.

7. THE ABT AND THE ABA
IN THE POLITICAL ARENA

7.1. The ABT

In the course of its existence, the ABT learned that its position was not unchallenged and above all that other powerful actors occupied the same terrain. These included influential entrepreneurs and companies and the government of the Commonwealth.

Importantly, the government retained significant powers in the area of broadcasting regulation, particularly those in frequency planning. Moreover, it was able to exercise indirect influence on the ABT's work via personnel and finance decisions. The ABT had to accept in practice the government's competence to set political target values for the broadcasting order. Therefore, the normative premises for its work were not only to be found in broadcasting laws but also resulted from the fundamentals of government policy. The ABT experienced the confines of this situation most clearly in the context of its efforts to push for greater deregulation, having been compelled to contribute to the government-supported policy of protectionist guarantees for established broadcasters. Moreover, since each administration consistently used its power to designate ABT members aligned with it politically, the control factor of "personnel" served as an additional tool in keeping the ABT in line with government objectives.

However, this did not mean that the ABT became a mere government tool. In the first place, because of the five-year term of office, personnel decisions tended to transcend phases of changing political majorities. Furthermore, members as well as staff assistants generally developed their own views and identified themselves with the ABT's special function. As a result, the ABT at various times fought strongly for its independence (cf. the data in Armstrong, 1987–1993: No. 8385). Accordingly, the relationship between the ABT and the government was usually quite tense, and the government made use of its manifold possibilities for counteracting ABT decisions with which it disapproved, particularly through initiatives for statutory amendments. This in turn probably helped to prompt the ABT to avoid conflicts with the government as far as possible.

Another important actor during the ABT's reign was the broadcasting industry. The established broadcasters and their interest association, FACTS, were by all means most appreciative that broadcasting regulation contributed to a cementing of their position, particularly in shielding them from new competition, as in the case of pay-TV. Even so, sections of the broadcasting industry often criticized regulation, especially the ABT's role. Particularly strong criticism arose when the ABT blocked expansion attempts, as with the acquisition of new stations by a licensee or with the extension of network arrangements. The ABT was often castigated as anti-industry, in particular antinetwork.

Nevertheless, the ABT maintained a remarkable level of internal and external distance from the broadcasting industry it regulated. This was probably aided by the fact that ABT members normally did not come from the broadcasting industry and in most respects also did not enter the industry following their ABT tenure.[53] They tended to be recruited from the fields of government and politics, sometimes from academia. In short, there does not seem to have been a revolving door between the ABT and the broadcasting industry.

In addition, many ABT members saw themselves as occupying a self-conscious, often critical position vis-à-vis the broadcasting industry. The distance between the two camps probably also increased as a result of the inquiries and decision-making procedures employed by the ABT, which were often marked by formality, particularly legal formality. This distance prevented the emergence of close, informal networks of cooperation with the broadcasting industry. Moreover, the broadcasting industry usually viewed the ABT as an opponent, as a controller that used avenues of communication, for example, to search for "evidence" that might be used against the interests of broadcasting companies.

Over time, the broadcasting industry came to realize that it was not dependent on arrangements with the ABT. At least in major conflicts, important sections of the broadcasting industry could count on the support of the government. For instance, the Labour government used its influence in the 1980s in a number of ways to support broadcasting companies aligned with it (e.g., the Murdoch and Packer undertakings; see Suich, 1991: 21). On occasion, laws were formulated in such a way as to allow certain companies to be in a preferred position to profit from them.

Australia's political culture had not developed adequate countermechanisms. The use of governmental power to achieve politically motivated advantages was perceived as so self-evident that it did not meet with massive public protest (e.g., with regard to personnel decisions). The press served as a public watchdog to only a limited extent.[54] Programming by the ABC—famous for its critical stance against successive governments—represented only a limited counterweight, particularly since the ABC was and still is subject to strong government influence.

Where issues of broadcasting regulation are concerned, it can hardly be expected that the public will become directly involved. Public interest groups are rather foreign to Australia,[55] and their public influence is most limited. Extensive media policy debates—similar to those experienced in Germany in the late 1970s and early 1980s—have not been conducted in Australia. Although some use was made of the possibility of public participation in the work of the ABT, especially via complaints and submissions (data can be found in the ABT's annual reports; see, e.g., ABT, 1990f: 74f.), such initiatives normally got tangled up in the ABT bureaucracy, with activities being channeled into administrative decision-making routine. It was intended with these procedures that the public would serve as a catalyst that might compensate for the scarcity of ABT self-initiative. But this form of public participation offered the ABT only limited support in controversies with the government or the broadcasting industry.

Over the course of time, the ABT's regulatory power weakened. During the early years—roughly until the Murdoch Amendments in 1981—the ABT had been able to work with much greater effectiveness than was later the case. In only a few fields—most visibly, in the area of Australian content requirements—was it able to point to notable successes. In the area of media concentration, where public criticism was most often encountered, the ABT suffered a number of defeats, and by way of its refusal to prosecute, the government made clear not only practically but also symbolically just how insignificant the ABT's powers really were. The reference to the ABT as a "toothless tiger" fed on this perception (cf. Communications Law Centre, 1988: 10). In other areas, such as that for monitoring program standards, the legal obligations on licensees were kept so low that ensuring compliance with the rules had a virtually imperceptible effect on programming conduct. Content-related stipulations, such as provisions dealing with fairness, were, until very late in its history, lacking.

As policy transition was made to deregulation policy, the ABT saw itself called upon to step down its regulatory efforts and place greater trust in self-regulation. This placed greater responsibility in the hands of the broadcasting industry itself. One sign of this trend was the trial period created for the area of quantitative advertising regulation. The broadcasting industry was supposed to prove during this period that it was capable on its own of reaching a solution satisfactory to the public. Refraining from substantial regulation, the ABT proposed maximum values oriented according to limits already practiced by the industry. The decision to forgo setting limits for commercial interruption—which had originally been considered by the ABT to be most important—represented an especially consequential retreat from the business of regulation, since the number and duration of commercial interruptions might well have been of particular importance for the public (at least this was the appraisal of the ABT, 1987: 32ff., 37f.).

In the area of advertising standards, the ABT delayed enacting a code of practice for the industry on account of anticompetition implications. Thus, self-regulation could be "controlled" in this area only by appeals to individual broadcasters, although the ABT could threaten regulatory action to strengthen the effectiveness of such requests. The catchphrase for this policy was self-regulation with a safety net.

The strategy of self-regulation was also used in dealing with violence in television, although here by using a uniform industry code (this was the recommendation by the ABT in ABT, 1990b: vol. 1, 102ff.). However, the code formulated by FACTS was characterized by many "weak" requirements, which also lacked the threat of sanction. FACTS neither created nor used an autonomous body to monitor the code. The sole sanction took the form of the ABT's safety net: Should the code fail to a large extent, the ABT could step up the regulatory possibilities it had at its disposal, a situation the ABT termed "supervised self-regulation." However, the ABT was unable to react to violations by individual broadcasters with blanket sanctions against all broadcasters, making this safety net somewhat uncertain. Nevertheless, the ABT could sanction violations of the code by reviewing broadcaster performance, particularly during license renewal.

7.2. The ABA

The government's efforts to implement deregulation in other areas as well did not shield the ABT from remaining the object of strong criticism by the government. Ultimately, the ABT was replaced by the ABA.

Although the ABA is an independent regulatory body with a wide range of powers and functions, it can only operate in a specific political setting. To this extent, it is important to note that the Broadcasting Services Act of 1992 left intact considerable regulatory powers and created some new ones (see Armstrong, 1987–1993: No. 5020). The ABA must act in accordance with policies dictated by the government, which may give general directions about the performance of any of the ABA's functions. Unlike the former ABT, the ABA is subject to governmental direction through the Minister, to whom the ABA is to report on the operation of the 1992 Act. The government even has an overriding power of direction regarding particular activities involving planning. Specific directives may be given by the Minister on matters such as the kind of price-based allocation systems to be adopted for commercial broadcasters.

This makes the ABA the entity with the greatest governmental dependence of all the regulatory bodies under study here. Although in the course of deregulation the regulatory powers over the broadcasting industry have been cut back, this authority is subject to tight governmental control.

The political pressure on the ABA can also be increased by means of Parliament. Since the ABA exercises delegated power, it is legally permissible for Parliament to reserve certain powers, such as making most ABA decisions to issue rules of widespread effect subject to disallowance (Armstrong, 1987–1993: No. 5025). Since program standards are also subject to disallowance and amendment, politicians have the power to involve themselves in what can and cannot be broadcast.

By way of the policy of greatest possible self-regulation by the broadcasting industry, the ABA's power is restricted still further: In many areas, it has only subsidiary competence. However, the interest organizations, including FACTS, have to consult the ABA within the framework of self-regulation. The ABA also possesses the power to threaten: It can set standards when it deems the code of practice to be inadequate. However, it finds itself under considerable pressure to justify its actions when it seeks to intervene in a field already subject to self-regulation.

Another important actor is the public, whose role has in some regards been strengthened. The ABA is specifically required to undertake wide public consultation in performing certain regulatory functions. This includes being required to make public its planning priorities for services in different license areas and the criteria applied. The ABA is also required to conduct research into community attitudes during the process of determining program standards. On the other hand, the ability of the public to initiate regulatory actions and to become involved in processes such as licensing, has been weakened by the new 1992 Act.

The ABA does, however, have possibilities for getting the public involved, using it as a counterweight to the power of government and industry. Since organized public opinion continues to be rather weak in Australia, the practical prospects are limited. Academically oriented media critics (cf. Wilson, 1989; Cunningham, 1992) are also concerned about issues such as diversity of media ownership, public accountability, minority rights, and portrayal of Australian culture. However, their influence is restricted to small circles.

Thus, as was the case during earlier periods, it has not been difficult for the government to exert predominant influence on developments in broadcasting. The expression applied to earlier governments—"The government is looking after its media mates"—seems to correspond also to the phase of new regulation. This, at least, was claimed with regard to the major media decisions favoring Packer (Minehan, 1993: 220). Yet, the government did not shy away from a conflict with the broadcasting industry over depictions of violence. By late 1992, the Prime Minister termed self-regulation to be insufficient and called for statutory steps (Minehan, 1993: 218). This led to the Act being amended so as to require the commercial broadcasting codes to restrict the showing of films classified as "Mature" or "Mature Adult" to certain hours of the day (see Broadcasting Services Act of 1992, s. 123 (3A)). Interestingly, the government did not care to wait and see how the Code of Practice would work and whether the ABA would itself intervene. As already mentioned (see Section 6.1), the ABA endorsed the underlying Code of Practice in 1993.

As discussed in Section 1, the Australian legal system does not contain a basic constitutional right of media freedom, but it recognizes the protection of a right to communicate on public affairs and political discussions as an essential component in the democratic process. Media freedom is a principle of political culture. But since special constitutional guarantees are missing, the media order can be made subject to the political currents of the day in a much more heavy-handed manner than can be observed in other countries with corresponding constitutional guarantees.

NOTES

1. For details on the historical development, see Green Report (Postal and Telecommunications Department, 1976): 9ff., with the chronology in Appendix 3, pp. A 21ff.

2. *Australian Capitol Television Pty Ltd. and Ors v. The Commonwealth* (1992) 66 Austl.L.Jur.R. 695 (HC).

3. In addition to public service functions, the ABC is explicitly required to give consideration in its programming to the offerings of other sectors: Armstrong et al., 1988: 165.

4. Its recommendations are reproduced in the ABC's annual reports. See, most recently, ABT, 1990f.: appendix 11. See also Armstrong et al., 1988: 165.

5. However, the Minister was formerly able to prohibit the broadcasting of a program. He or she is now only capable of demanding the broadcast of programs that lies in the "national interest."

6. For the amount of allocations, as well as further data, see Bureau of Transport and Communications Economics, 1990: 61.

7. There are currently some 150 commercial radio stations. Bureau of Transport and Communications Economics, 1990: 75.

8. Recently there have been some cross-border broadcasting issues resulting from the growth of privately run satellite systems in the Asian Pacific region, with footprints over the Australian continent.

9. For the amount of these and their allocation among the various media, see Bureau of Transport and Communications Economics, 1990: 64ff.

10. For this problem, see, for example, ABT, 1984: although lacking concrete proposals for expanding satellite broadcasting; Oswin Report, 1984.

11. For the discussion on the use of satellite technology for broadcasting, see Department of Communications, 1986; Bonney, 1987: 51f. For discussion of the use of satellite technology for broadcasting, see Department of Communications, 1986.

12. According to the Australian perspective, Sky Channel is not subject to the Broadcasting Act but, rather, to the Radio Communications Act.

13. For discussion on the introduction of pay-TV, see Department of Transport and Communications, 1989.

14. Under the Telecommunications Act of 1991, only carriers may install cable networks. There is, however, an exception to this that allows broadcasters (including pay-TV operators) to install their own networks.

15. Broadband Services Expert Group, Networking Australia's Future, Interim Report, July 1994.

16. The term "networks" (or "networking") has no roots in statute. Generally speaking, it refers to the central production and/or acquisition of programs for distribution to affiliated stations. For the various forms and structures of networking, see, for example, ABT, 1984: 33.

17. Data on use can be found, inter alia, in ABT, 1989: 50–68. The data are based on large urban regions. The annual reports by the ABC and the SBS unfortunately do not contain any information on audience ratings but, rather, only on "audience reach." By this is meant data on which share of broadcasting users has made use of the respective program offering during a certain period (e.g., within one week); see, for example, SBS Annual Report, 1989: 64f.

18. The ABC receives annual subsidies of only roughly AUS$500 million; ABT, 1990f: 78.

19. However, the responsible Minister, Kim Beazley, in 1990 assured that the ABT would continue to exist; see Beazley, 1990: 15.

20. Compare text and commentary to the substantive powers enumerated by statute, in Armstrong, 1987–1993: Nos. 8400 ff.

21. When the initiative for this stems from the Minister, these inquiries are termed "directed inquiries." For the various types, as well as the "area inquiry" not treated here, see Armstrong, 1987–1993: No. 8300.

22. The basis for these amendments was the Administrative Review Council Report, 1981: detailed in Armstrong, 1987–1993: No. 8310.

23. See Walker, 1989: 286. The Inquiries Regulation of 1986 mentioned there was, however, rescinded in 1988.

24. This professionalization (by lawyers) poses a threat to equality for the various interests groups by the related financial expense. Calls have been raised in public for financial support for legal representation of certain groups; Communications Law Centre, 1988: 30ff.

25. The degree of proceduralization of every sort of communication was demonstrated with particular clarity by the ABT's rules for direct contact with other actors; Meetings and other Contact with the Tribunal, reproduced in ABT, 1990a: 106ff. All of these contacts are listed in the annual reports with reference to the locations in the protocols. See, for example, ABT, 1989: 164f.; ABT, 1990f: 154f.

26. For such risks relating to the Australian situation, see Armstrong, 1987–1991: No. 805, who rejects the concept of agency capture for Australia as largely irrelevant.

27. One exception applies to the preclassification of children's programming; see Section 5.3.1.

28. For the restrictive criteria that have been developed for this, see, for instance, Walker, 1989: 272f.

29. The ABA does not directly impose fines. Rather, fines for a breach of the Act may be imposed following prosection, usually in the federal court. Prosection is, however, seen as a last resort. The preferred procedures are informal discussions and notices to rectify breaches of license conditions under the Broadcasting Services Act of 1992, s. 141.

30. Excerpts from official documents can be found in Armstrong, 1987–1991: No. 6010; Armstrong et al., 1988: 155.

31. Licensing of networks was also rejected in the Saunderson Report, 1988: 51ff. See also Broadcasting Review Group, 1989: 31.

32. However, special rules relating to network arrangements were permitted to a limited extent, though these were never enacted; Broadcasting Act of 1942, s. 134. See Armstrong, 1987–1991: No. 6770, who stresses that the situation of networking in Australia has increasingly come to resemble that in the United States.

33. For example, the Minister was authorized to set the categories of programming to be offered.

34. This was clearly recognized in *R. v. Australian Broadcasting Tribunal, ex parte 2HD Pty Ltd.* (1979) 27 Austl.L.R. 321.

35. This particularly applied to a decision against Rupert Murdoch; see ABT, 1980.

36. Cf. Armstrong, 1987–1991: No. 6710. Concentration control in the course of evaluating other criteria was, however, subject to the risk of being deemed an unlawful extension of ABT power.

37. *Australian Broadcasting Tribunal v. Alan Bond* (1990) 94 Austl.L.R. 11. But see Alston, 1989: 1ff.

38. ABT-Report No. 41/90 (1980), Ansett Transport Industries Ltd./Control Investments Pty Ltd.

39. Until 1987, the maximum number of stations allowed were eight radio and two television stations.

40. See the chronological survey in Chadwick, 1989: LXIff., and the table at 243ff. For the situation in 1991, see Appleton, 1991: 2ff.; see also Noble, 1990: 24ff.; Minehan, 1993: 214.

41. For the changes that took place in 1987 and then again in 1988, see Armstrong, 1987–1993: No. 6706. See also the comments on the new rules in ABT, 1989: 47ff. Space does not permit a detailed discussion here.

42. See Appleton, 1991: 4; precise figures are not known. The government has resisted detailed research, apparently out of fear of antigovernment reporting by the newspapers belonging to the Murdoch group.

43. The Saunderson Report, 1988: 72, termed the period of grace an "abuse of the legislation."

44. For some time Network Seven allegedly served an area that reached 71% of the Australian audience.

45. There is now a differentiation between C (suitable for school-aged children) and P (suitable for preschool children).

46. For the definition of this category, see ABT, 1990a: 25: "Drama means fully scripted screenplay or teleplay in which the dramatic elements of character, theme and plot are introduced and developed so as to form a narrative structure. It includes animated drama and dramatised documentary, but does not include sketches within

variety programmes, or characterisations within documentary programmes, or any other form of programme or segment within a programme which involves only the incidental use of actors."

47. The six diversity program types are variety, social documentary, arts, science, current affairs/news specials, and new concepts.

48. A good survey of the situation in Australia can be found in Appleton, 1988: 189ff.

49. Armstrong, 1987–1991: No. 1575. In the radio sector, however, nearly all stations have voluntarily signed a fairness code, which set up not only duties of care and impartiality but also the goal of adequate opportunity for depicting the viewpoints of all relevant persons and organizations in controversial matters. However, there were no mechanisms for monitoring or sanctioning these. Armstrong, 1987–1991: No. 1580; Armstrong et al., 1988: 177.

50. Re *The Herald and Weekly Times Ltd. v. The Commonwealth* (1966) 17 Austl.L.R. 281.

51. That is, the Commercial Acceptance Division of FACTS and the Commercial Clearance Office of FARB. Both of these have published detailed information.

52. The research initiated by the ABT mainly related to the impressions of viewers. See ABT, 1990b: vol. 1, 27ff.

53. However, a biography of members serving in 1990, ABT, 1990f: 3f., showed that two of the five members had formerly worked for the broadcasting industry.

54. It is here that the great influence of the Murdoch company in the press could be felt. At most, criticism came from newspapers owned by other groups, and this only when their interests were also not affected.

55. In the broadcasting sector, particularly worthy of mention is the Communications Law Centre at the University of South Wales in Sydney.

PART II

A Comparative Study of Broadcasting Regulation and Supervision

The portrayal of individual selected broadcasting systems in Part I is followed by an attempt to formulate some systematic observations on the subject of broadcasting regulation.[1] They are based mainly on the six broadcasting systems that are examined here. However, some reference will be made to other broadcasting systems,[2] partially in order to highlight additional peculiarities but also in order to illustrate that the same problems are just as likely to crop up in other broadcasting systems.

NOTES

1. The account refers to previous studies undertaken by the author (see in particular Hoffmann-Riem, 1981b, 1989a, 1989b, 1990b, 1991a, 1991b, 1992a, 1992b) but extends the range of issues dealt with considerably.

2. The other broadcasting systems cannot be dealt with here in any detail. Instead, reference should be made to studies by Head, 1985; Kleinsteuber, McQuail, & Siune, 1986; Hoffmann-Riem, 1989a; Hans-Bredow-Institut, 1994; Noam, 1991; Blumler, 1992; Euromedia Research Group, 1992.

CHAPTER 7

Justifications for the Regulation and Supervision of Broadcasting

As has been described in detail in the preceding chapters, broadcasting is extensively regulated by law throughout the world. In this regard, it is quite different from the press. At least in countries that consider themselves to be democracies, there are demonstrably fewer legal rules controlling the press and usually no licensing requirements or ongoing supervision by a special supervisory authority. In view of the universally recognized principle of freedom of communication, there arises a special need to justify the regulation of broadcasting.

The possibility to regulate broadcasting is influenced by the value accorded to media freedom in the respective society. The principle of media freedom is recognized in all Western industrialized countries, including those nations, such as Australia, where there are no corresponding constitutional guarantees, or others, such as Great Britain, where this freedom is derived from international agreements (here, the European Convention on Human Rights).

According to traditional theory, there are at least four reasons for offering special protection to free speech:

1. To facilitate individual self-fulfillment;
2. To advance knowledge and discovery of truth;
3. To promote democracy through a process of the self-governing society and check abuses of power by public officials;

4. To expedite the functioning of society, especially by assuring a proper balance between conflict and consensus, allowing social change, and fostering social integration.

Most societies value freedom of speech not as an end in itself but rather as a means to reach these normative objectives, above all, the promotion of democracy. Therefore, mass communication is deemed to have an important sociocultural dimension. Mass media render a service to society. Government, particularly the legislature, bears the responsibility of ensuring that the processes of informing the public, exchanging ideas, and thus shaping values take place in a truly free manner and are not jeopardized either by the state or private power-holders.

In the various countries, however, different conclusions are drawn from the societal importance of the media. This is related to specific historical experiences, as well as to basic attitudes toward the relationship between state and society, between liberty and obligation. In the United States, freedom of speech—as anchored in the First Amendment to the US Constitution—is rigorously protected against government intervention, even though a certain amount of regulation in the broadcasting sector is considered constitutional. In Western European countries, a strong need is perceived to balance this freedom against other values. This presupposes special regulation of broadcasting. Its sole justification is not merely spectrum scarcity; others include the prohibitive costs, the threat of a high concentration of ownership or other signs of market failure, and broadcasting's unique pervasiveness and influence on society. It has thus far been recognized that government plays an affirmative role in guaranteeing the functioning of the broadcasting order and in protecting the public interest, including the democratic quality of political decision-making—even in the age of new electronic media and expanding distribution facilities and with the advent of a merger of transmission technologies and a convergence of broadcasting and press law.

The following describes typical justifications for the regulation of broadcasting and special broadcasting supervision. The basic contours of such justification are similar in all countries studied here, even though accents peculiar to a given country are sometimes evident.

1. INTERFERENCE-FREE COMMUNICATIONS TRAFFIC

It is a common observation that the special features inherent in the transmission technology of broadcasting trigger governmental regulation. In order for broadcasting signals be transmitted with as little interference as possible, a corresponding order is recognized as necessary. Thus, broadcasting supervision was introduced in various countries to avoid a chaos in the ether and ensure protected transmission capabilities for individual broadcasters. This framework was intended to help prevent unauthorized use. Initially, the aim was only to protect against a threat to interference-free reception. As is made clear by the fight against so-called pirate broadcasters, however, this goal assumed a life of its own, since pirate broadcasters were to be closed down even when they did not disturb the reception of other programming.

The protection of interference-free communications traffic continues to be viable as a justification for regulation. The relevant national regulations are in addition supplemented by international rules, such as under the auspices of the World Administrative Radio Conference (WARC), through which individual countries are assigned certain frequency bands. Thereafter, it must be decided at the domestic level how these frequencies are to be put to use. The international body of rules became increasingly important as a result of new technologies, especially satellites. Accordingly, there were and are a number of international agreements dealing with, among other issues, satellite use.[1] These are usually limited to the regulation of the basic technical conditions for interference-free communications traffic. It is, however, perfectly clear to the participants that such technical rules have a considerable impact on the world communications order and thus also substantial political implications.

For decades, the political controversies over the way in which transmission possibilities are used substantively have revolved around the so-called world information order and the principle of free flow of information. This discussion has been and still is particularly marked by political conflicts, that is, the East–West conflict and the North–South conflict. It has not been possible to achieve international consensus on the nature of the flow of international communication and possible rules for this communication flow or to institute internationally organized supervision of the adherence to such rules (about the discussion, see, e.g., Keune, 1984; Berwanger, 1988). The following discussion therefore omits this topic.

The respective national efforts to create a system for interference-free communications traffic were, as mentioned above, initially tied to the use of terrestrial frequencies. With regard to wirebound transmission—the dominant means of transmission for radio in the 1920s—use was regulated by the owner of the transmission technology, usually the PTT. This authority to regulate was also demanded for over-the-air broadcasting in order to regulate wireless communication. However after introduction of cable systems, especially for television, there arose a fundamentally different need for regulation than with the use of exclusively over-the-air frequencies. Chaos in the use of a cable-bound medium for transmitting television or radio programming can be relatively easily avoided by the owner of the given medium by opening access to the system and then channeling use. However, it became apparent with the emergence of cable that such systems stimulate the development of new programming and that the owners of cable systems can themselves become broadcasters. This meant that cable came to be a competitor of traditional broadcasting, and new regulatory needs arose as a result (see also Section 3).

With respect to using cable or satellite as a mere transmission medium, the appropriate regulatory approach depends on who is responsible for the transmission network. In many countries, this is the PTT; however, private entrepreneurs have often been made owners, and for cable systems, communities also provide services. Governmental regulation consistently latches onto other objectives than simply ensuring interference-free transmission. It has been used as a means for influencing content in the use of media as well as for setting detailed conditions regarding media access, including use of rates. To this extent, it needs some other justification than merely the reference to interference-free transmission.

2. ACCESS TO THE MEDIUM

In the course of broadcasting development, the government's task of ensuring interference-free dissemination or reception was initially justified with the scarcity of available, that is, operable transmission possibilities. This applied particularly to terrestrial frequencies. But through technological advances—as with the introduction of new transmission technologies such as satellite and cable, and the expansion of transmission capacities, such as through digitalization and data compression—and as a result of political decisions—such as the release for private use of transmission channels originally reserved for the military—the scarcity of transmission techniques was gradually reduced. However, this was accompanied by rapid growth in the types of use, particularly an explosion in the number of broadcasters and their programming. In theory, the available means for transmission are nearly unlimited, and in practice, ever more transmission avenues are being discovered. For instance, digital compression and other technologies allow the capacity of terrestrial, cable, or satellite transmission facilities to be increased tenfold. Cable systems with up to 500 channels will be installed in the near future. The large number of channels enables the individual to develop his or her own programming tailored to his or her needs. With the expansion of satellite technology, the use of telecommunications satellites for broadcasting purposes, and the ease of reception with the aid of small, inexpensive reception antennas, a substantially augmented number of transmission paths have already been opened.

Insofar as situations of scarcity still exist with respect to transmission possibilities, governmental regulation of broadcasting will continue to be based on the scarcity problem. In such a case, scarcity serves not only to justify regulation in the interest of interference-free reception; it is also used as the occasion for exercising influence on the communications system as such. Decisions on the use of scarce resources presuppose prioritization, for which certain criteria are necessary. In broadcasting regulation, particular weight is accorded to the objective of trying to make sure with regard to access to the medium that the communications order conforms to certain target values. Thus, broadcasting's obligation to the public interest has been determinative even in broadcasting orders in which the state largely keeps itself removed from broadcasting, that is, where there is no quasi governmental broadcaster.

By way of illustration, reference can be made to the U.S. Communications Act of 1934. The allocation related to the situation of scarcity takes the form of a licensing decision. This makes it possible to ensure compliance with legal requirements and, in this way, to work toward a certain kind of broadcasting order. The same technique is used in all broadcasting systems under study here. For instance, broadcasters are licensed only if they are likely to meet governmentally set requirements, such as those with respect to plurality and quality of programming or regarding language. On occasion, preference is even given to broadcasters that have their principal place of business in the area of transmission or produce their programming there. Priority in the access to cable systems is often given to broadcasters whose programming can already be received terrestrially in the corresponding area: Users who switch to cable are not to be placed in a worse position than other broadcasting users. This also serves as a protection of local broadcasters (assurance of advertising revenues).

Because of the scarcity, but also as the result of governmental decisions as to allocation, means of transmission are viewed in most countries as a public good whose use is a kind of privilege. This makes it fair to skim off some of the "advantages" of use from those who receive the privilege. The imposition of special public service obligations is a type of consideration given for the privilege granted with the license: Broadcasters should be allowed to use the license only when the community benefits from this use as well.

In view of the opportunity to generate earnings and profits with the aid of this privilege, broadcasters are sometimes required to share economic benefits, for example, in the form of special taxes or other levies, such as a license fee. They may also be obliged to allow other broadcasters free or inexpensive use of the medium, as with so-called window programming, leased access, or must-carry obligations. When obligations call for the turning over of a certain share of revenues, these are often used to give financial support to other broadcasters, including minority broadcasters or cultural channels. In some cases, as in Australia and Great Britain, these levies also flow directly into the national budget.

The payment of economic consideration for the privilege is also taken as an approach for letting the allocation of the license be directed by the market. This concept includes provision for bidding for the license (as in Great Britain and Australia). The justification for this is seen not only in allowing the skimming off of benefits but also in the fact that the quasi-market-oriented payment rules are intended to help ensure that market mechanisms can truly be used in the selection of licensees. In no country do market mechanisms alone decide on the allocation of licenses, but they are used in conjunction with other criteria (particularly in Great Britain).

3. PERFORMANCE OF BROADCASTERS

By way of such selection decisions, indirect influence is also exerted on the activity of broadcasters. Moreover, the act of selection permits the imposition of additional legal requirements on broadcasters in the form of license conditions. This makes the structuring of access to the medium of broadcasting at once a regulation of broadcasters themselves and, indirectly, of their conduct as well. In seeking to justify such a supervisory regime, it is not enough simply to refer to the need for regulated access; rather, the actual substance of such a regulation must be justified, particularly since this influences the communicative conduct of broadcasters. If we systematize the rules underlying this manner of regulation, three main types emerge (as discussed in Sections 3.1 through 3.3).

3.1. Ensuring the Viability of the Broadcasting System

In light of the special importance of communication and its freedom for both government and society, especially for free political expression, all rules aim at ensuring a certain type or minimum amount of communication with which recipients are to be served. The government—that is, the legislature with the aid of its laws, the agent of broadcasting supervision with the aid of supervisory action—is perceived as the guarantor of the viability of the broadcasting order. Such viability is in part measured by the degree to which broadcasting is free of one-sidedness and empowerment or whether a certain quality is ensured in programming.

Government action can also enable anyone to communicate with the aid of broadcasting if he or she wishes to do so, whether in the role of broadcaster or at least on the basis of a right of access to use the medium for individual programs (access for communicators). Such intervention can instead target the role of recipients by, for example, ensuring that they have access to all available—or even all personally relevant—types of communications. The government must then ensure that such a range of contents, especially information, be truly offered and also that the financial barriers to access are kept reasonable.

In this sense, though in differing forms, the most important objective of regulation is realizing freedom of communication for communicators and recipients. One accompanying objective is diversity in the media, whether among broadcasters themselves or as regards their programming. This objective can be viewed quantitatively (e.g., in the sense of the largest possible number of competing broadcasters) or qualitatively (e.g., in the sense of quality cultural programming or of programming covering as many politically important constituencies as possible). This heading also comprises the protection of national media production, as well as efforts (as in Canada) to reflect in broadcasting the multilingual situation existing in the country or the attempt (as in Australia) to provide ethnic minorities with guaranteed means of communication. The protection of such special objectives of the communications system usually requires positive measures, that is, intentional intervention to support the production infrastructure or to ensure that different groups have access to broadcasting. But above all, such regulation serves to protect the independence of the broadcasting system from the state or other powerful interest groups, thereby safeguarding freedom of communication. Only in those countries where broadcasting has not been able to liberate itself at all from complete state domination (as is currently the case in the Commonwealth of Independent States [CIS—former Soviet Union] countries) is this objective not (yet) anchored in the media order.

Although it considers itself responsible for the viability of the broadcasting system, the government is often tempted to supervise broadcasting substantively even in those areas where it is formally independent, thereby subjecting it to specific objectives defined by the government. The history of broadcasting in France offers striking examples of this. In many countries (e.g., in Germany) the exercise of influence over broadcasting takes place in cooperation with political parties. If such threats to independence are to be alleviated, special guarantees are necessary. Some demand that independence be ensured by resorting to the economic market as the sole regulator of broadcasting. However, this can lead to new risks of market failure. In view of the massive tendencies toward concentration in the broadcasting market, the risks of accumulation of power and thus the absence of equal communicative opportunity are inherent in privately owned broadcasting orders. As a result, measures are often insisted upon not only to protect the independence of broadcasting from the state but also to check journalistic and economic power. The supervision of broadcasting is intended to help prevent all types of amassed power and at least to balance different power-holders.

State regulation to ensure the viability of the broadcasting system presumes that the requirements for such viability are defined. This means that specification of the objectives is entrusted to the political decision-making process. A glance at various broadcasting orders shows that this specification takes place in the most diverse of manners. For instance, until recently, the competition principle was recognized in Western Europe as a possible guarantor of viability. Instead, public broadcasting long held a monopoly or, as in Great Britain, there existed a duopoly of public and private broadcasting in which, in prescribed areas, commercial broadcasting was protected by a monopoly against competition for advertising revenues. Even in the United States, the supervision of broadcasting had for years tried to avoid the type of stiff competition that comes, for example, with the emergence of cable. Over an extended period, the FCC's measures served to protect local stations and networks, which used terrestrial transmission paths. Later, the FCC intentionally worked to introduce more competition among the three large networks, supporting, for example, efforts to establish a competing fourth network. The reregulation of the 1980s was marked by measures to develop a competitive structure, including making use of the wealth of transmission means offered by cable; suddenly the networks found themselves faced with threats to their existence (cf. Auletta, 1992).

These considerations make clear that every effort to ensure the viability of the broadcasting system is difficult, since a variety of interests are affected. Therefore, the regulation of broadcasting is constantly faced with the task of finding a way to balance competing interests. Nor are the interests of media companies, journalists, advertisers, and recipients by any means necessarily identical. For instance, where the interests of recipients in unmanipulated communication collide with those of the advertising industry, provision can be made for a balancing of interests in the form of advertising restrictions, nominally an effort at optimization. If certain communication needs are otherwise threatened, access rights for specified groups (e.g., minorities) can be expressly established, or a public access channel can be set up to give anyone the opportunity to articulate his or her views through broadcasting. In other words, the spectrum of conceivable regulations is very broad.

3.2. Protection of "Vulnerable" Values

In addition to measures to protect the viability of the broadcasting system, there are also safeguards to protect specific values generally defended by the legal and social order. Broadcasting poses a special threat to these values, making special precautionary measures appropriate or even necessary. Under this category fall, for example protection of personal honor and privacy, preservation of the confidentiality of business or governmental matters, assurance of the moral development of young persons, and the safeguarding of copyrights or the special interest in maintaining the integrity of works of art (for further details, see Chapter 10). Special consumer protection can be ensured with regard to advertising and teleshopping. In accomplishing all of this, regulatory bodies can generally exercise the norms of the general legal system, including all of its accompanying enforcement procedures, such as resort to the courts. But since broadcasting can present unique dangers as well, it is often subject to special rules. These can be monitored either by the bodies charged with legal supervision generally or by special authorities, such as the Broadcasting Complaints Commission or the Broadcasting Standards Council in Great Britain.

Special rules in the broadcasting sector are also created in part because the application of general norms might conversely threaten broadcasting freedom. For instance, free public discussion could suffer if the values of honor, privacy-related data, state secrets, or the moral development of young persons were protected absolutely without heed being paid to the special aspects of communication through broadcasting. As a result, from the standpoint of media policy, laws frequently strive to find a balance between a myriad of colliding values so as to ensure an optimum of exposure for both freedom of communication and for those interests in opposition to it. Various societies are clearly distinguished by whether the "optimum" value tends to lie closer to freedom of communication or to these other, opposing values.

3.3. Pursuit of Further Aims

Broadcasting regulation is sometimes used to fulfill aims with the aid of broadcasting that have only a marginal association with broadcasters and their programming. This includes the goal of generating tax revenues for the government. The extent to which this becomes a goal in itself or is instead just a byproduct of licensing (e.g., auctioning) cannot always be determined from the outset.

Insofar as broadcasting regulation is used to ensure jobs or other objectives relating to location, the situation is also ambivalent. For instance, quotas for programming of local or national (or European) origin can be justified with the goal of ensuring that appropriate attention is paid to diverse information found in the respective broadcasting area. In so doing, it is assumed that this is most easily guaranteed when at least some programming is also produced there. This is the basis on which, according to the Preamble, the quota rules of the EC television directive are premised. However, superimposed on this objective may be the goal of guaranteeing jobs in the cultural sector. Similar ambivalence can be found with measures to develop an independent production infrastructure, as with the duty to give preference to independent producers. And the same is true with other attempts to secure the viability of other media sectors, such as the press or the film industry.

4. SETTING THE CONDITIONS FOR FINANCING

From the standpoint of listeners and viewers, the fact that they are offered programming that they find satisfactory does not suffice in itself; consumers are also interested in the financial conditions that provide access to such programming. When financing takes place by means of advertising, costs are borne by all those who acquire the goods and services being advertised. However, the costs involved in advertising comprise not only the payments made to those broadcasters who make use of advertising but also the considerable compensation paid for production of the commercials. For this reason, advertising is an expensive way to finance broadcast programming. Costs are defrayed not just by citizens in their role as broadcasting consumers but indeed by all citizens in their role as consumers of goods and services—regardless of whether they are broadcasting recipients. Because of this indirect method of financing, broadcasting regulation normally does not deal with these types of payments.

In the area of pay-TV, in contrast, broadcasting consumers bear the costs themselves. Even the listener fee for public broadcasting—for instance in Germany—is a form of pay-TV. In Germany, this fee must be paid by those who possess reception facilities regardless of whether they use public service programming. Comparable to this is per-channel pay-TV, although the costs are borne only by those who order the programming of the relevant channel. Pay-per-view offerings are paid for only by those who order or access the specific broadcast.

If programming is broadcast via special technologies, particular rates are often calculated for the use of the transmission medium. This is most often the case for cable. The rates usually cover both the financing of the special transmission technology and particular cable programming. To the extent that the companies responsible for dissemination of programming occupy monopoly-like positions in their areas of coverage (as is particularly the case with cable), there are distinct risks of abuse. Attempts are made to alleviate these in part through rate regulation. Rate regulation has a significant impact on both the financial viability of broadcasters and barriers to access for consumers. Governmental rate regulation is to this extent a consumer protection measure.

5. STRUCTURING A UNIFIED EUROPEAN BROADCASTING MARKET

Broadcasting also continues to be regulated in order to structure transborder transmission. The unification of Europe is a good example: With the abolition of trade barriers among member states, there arose a special need for regulation at the EC level.

The EC intentionally uses broadcasting regulation to achieve the harmonization of laws necessary to create a unified broadcasting market. In so doing, it focuses special attention on monitoring observance of the rules of the EC television directive. There is (as yet) no serious talk of creating a special European broadcasting authority, and national supervisory bodies continue to be responsible for licensing and supervision of broadcasting in their respective countries or regions. The same essentially holds true for the observance of rules by member states of the Council of Europe under the European Convention on Transfrontier Television.[2]

The states, however, are obligated in relation to one another and in relation to the EC to provide suitable ways for implementing the norms of the EC television directive and the European convention. If they violate this duty, EC member states can be penalized with the set of sanction instruments under the EEC Treaty. Corresponding authority to sanction is, though, not available within the scope of the Council of Europe. For this reason, the Council's convention provides for new mechanisms of conflict management, which include not only the commitment to cooperate but also the establishment of a standing committee and conciliation and arbitration procedures.

The special sanctioning and conflict-resolution mechanisms become important only when the member states are not able to ensure observance of the legal duties within their sphere of responsibility. In order to accomplish the latter, not only must suitable norms of broadcasting law be enacted; they must also be monitored on an ongoing basis. To an extent, both international instruments are based on the presumption that supervision of the provisions of broadcasting law is initially a duty solely in the province of the authorities of the transmitting nation. Even when the supervisory organs of the receiving state suspect that the directive has been violated, they may not prevent transmission of the broadcast program. Only in cases of particularly serious, repeated violations of norms can retransmission be suspended. The situation is similar under the Council of Europe's convention: Violations must first be tolerated and thereafter lead to the above-mentioned conciliatory procedures. Retransmission may only be interrupted in exceptional cases. This means that the responsibility of the transmitting member state for the effectiveness of broadcasting supervision becomes especially important.

It would conform to the basic idea of the European Common Market to structure the supervision of broadcasting by the member states as uniformly as possible. However experience within the EC, and not just in the broadcasting sector, shows that a harmonization of laws is not an adequate guarantee of the universality of their observance. Citizens (and governmental institutions) in the various member states have differing attitudes about interpreting the meaning of the law and the role of obedience to it. Measures taken by authorities and courts to implement laws also vary considerably. This can be observed in the area of broadcasting supervision. For instance, the chances for implementing norms in Italy are fundamentally different from those in France or Germany. But when the EC nevertheless refuses to enact uniform broadcasting supervision, it ultimately must satisfy itself with the fiction of equally intensive supervision in the various member states. Apparently, such a fiction is still viewed as an means of regulation for attaining a common broadcasting market.

6. CONTINUED REGULATION
OF BROADCASTING

The extent to which the above-described objectives can justify distinct broadcasting regulation in the future is the subject of debate in all countries. The most likely consensus is that regulation and supervision are justified when the issue concerns the protection of those values that are generally safeguarded by the legal system (such as privacy or copyrights). Such protection may be tailored to broadcasting, but it must also take account of the special qualities of freedom of communication. Regulations to protect the viability of the broadcasting system are, in contrast, disputed in media policy and, to some degree, in constitutional law. Some allege that such regulations are legitimate only in a period of frequency scarcity and that this has now ended or will shortly. Government is thus claimed to have no business acting as regulator in the broadcasting sector and must refrain from regulation because of the constitutional guarantees of freedom of communication.

Freedom of broadcasting is viewed as a subcase of general freedom of expression, which may be limited only to protect values generally recognized in the legal system.

However, this is only one of several aspects of modern basic rights doctrine. For instance, it is acknowledged in most legal systems that to be effective, basic rights depend on positive measures, particularly those in the form of laws. The well-known U.S. Supreme Court decision in the Red Lion case (*Red Lion Broadcasting Co. v. FCC,* 395 U.S. 367, 390, 89 S.Ct. 179 (1969)) was based on this concept. The German Federal Constitutional Court has developed this position quite persistently (see the portrayal by Hoffmann-Riem, 1990a: 49ff.). Likewise, the position was long recognized under article 10 of the European Convention on Human Rights (ECHR); the licensing proviso regarding broadcasting found in art. 10 (section 1) was understood as justification for traditional broadcasting regulation in Western Europe, which always aimed at structuring the broadcasting order with positive measures (cf. Hoffmann-Riem, 1991b: 184ff.).

However, a reversal of position can now be noted in Western Europe. Beginning with the EC Commission's green paper, "Television without Frontiers" (European Community, 1984) and continuing with recent decisions of the European Court of Human Rights (see in particular European Court of Human Rights, Judgment of 28 March 1990, Nr. 14/1988/158/214, Ser. A 173; EuGRZ 1990, 255—*Groppera Radio AG et al. v. Suisse*), ever more credence has been given to the view that the licensing proviso in article 10 (section 1) may only be enforced within the limitations set forth in article 10 (section 2) (for details, see Engel, 1993a, and Petersen, 1994). This reinterpretation of article 10 is taking place parallel to the manifold deregulatory efforts in the European Community, which in turn fit into a larger deregulation trend occurring in the United States and Australia. Both the EC Commission and the EC Council regard reregulation in order to develop market freedom as a way of creating a common internal market in Europe.

It is, however, also apparent that in spite of the trend toward liberalization evident in all countries, there continues to be extensive regulation of broadcasting. A good example is the British Broadcasting Act of 1990, which, though marked from an economic standpoint by the deregulatory philosophy of the Thatcher government, also contains detailed, wide-ranging rules to ensure programming quality, limit competition, and protect "vulnerable" values. Other reregulations during the 1990s—the Radio Communications Act of 1992 and the Broadcasting Services Act of 1992 in Australia, or the German Interstate Broadcasting Treaty of 1991—likewise adhere to extensive regulation, which applies both to the viability of the broadcasting order and to the safeguarding of generally protected values. Even the EC television directive uses the tool of broadcasting regulation to achieve a concensus of broadcasting law in Europe, which is the legal prerequisite to an internal market in the broadcasting sector.

This continued commitment to broadcasting regulation is supported by broad parts of the general public. Such support can be seen in the criticism of media concentration or in the concern about the effects of depictions of violence on television, which have both increased with growing competition. The political underpinnings for justification of broadcasting regulation have, to all appearances, not yet evaporated; in the legal systems as well, it continues to be possible to find legal justifications, as with constitutional guarantees. But this does not exclude the possibility that not all of the usual aims of regulation can be justified or that new requests for regulation will not be forthcoming.

It is particularly difficult to justify special broadcasting regulations when the regulatory instruments of the general legal system are sufficient for the task. When, for instance, the choice is made in favor of the market broadcasting model, special broadcasting supervision does not have to be very rigorous as long as there are other types of effective regulatory measures, such as those in the area of antitrust law for safeguarding economic competition. Within the framework of the market model, aspects peculiar to broadcasting can be taken care of with sector-specific antitrust rules similar to the precautionary measures otherwise employed to ensure the ability of markets to function. If, in contrast, special, broadcasting-specific forms are created outside of the market model—as, for example, with group broadcasting in the Netherlands or Germany's internally pluralistic integration broadcasting—then parallels to other economic sectors will be most unlikely, making special broadcasting regulations unavoidable.

In a number of broadcasting orders, it can be observed that provision is made for an interlocking of broadcasting supervision with general economic regulation. For example, with regard to subscription TV, the Australian Broadcasting Services Act provides that a broadcasting license be granted only after the Trade Practices Commission has given its authorization under the Trade Practices Act of 1974. Some German media laws require clearance from the monopoly authorities prior to granting a license. In the United States, the FCC does not deal with advertising that is misleading or deceptive; supervision of this area is the province of the Federal Trade Commission.

Nevertheless, market broadcasting does not completely dispense with specific broadcasting regulation. In fact, no country is convinced that the market alone as regulator can suffice to achieve the desired positive effects and avoid dysfunctional side effects. The British Broadcasting Act of 1990 may serve as prototype here. In spite of the stipulation that the market be used as the principal instrument of control, the act still contains far-reaching rules regarding, for example, a "quality threshold," the number and type of competing broadcasters, special programming duties, and precautions for combating concentration. Formulated with a similar penchant for regulation are the French law on freedom of communication and Germany's state media laws. The weakest regulations are to be found in the Italian act on the regulation of the public and private broadcasting system (Act No. 223 of 6 August 1990; see Mazzoleni, 1992; Barile & Rao, 1992: 261ff.).

NOTES

1. This aspect is not addressed in detail. See instead Magiera, 1981; Fisher, 1990; Smith, 1990.

2. The European Convention was also signed by European countries that are not members of the EC; however, it is also valid in the EC. Its rules were made consistent with those of the EC television directive.

CHAPTER 8

Approaches to Broadcasting Regulation

T he task imposed upon the bodies responsible for supervision of broadcasting and the instruments they employ depend on the framework in which the laws are set. In this regard, it can be observed that various countries follow different regulatory philosophies. In the following, an effort is made to systematize the various approaches that have been selected.

From an analytical standpoint, two particular types of regulation and control may be distinguished. The one model may be termed imperative regulation with conduct control (command-and-control regulation). This type seeks to steer the conduct of the object of control directly by way of guidelines, requirements, orders and prohibitions; violations can normally be sanctioned, for instance, negatively, with fines, revocation of license, or other penalties, or positively, with subsidies. The other regulatory model may be called structural regulation. Here, the government sets up a general structural framework for broadcasting that is to influence indirectly the conduct of broadcasting companies and other entities that are active in the broadcasting sector. In this manner, the basic economic structure, including the type of financing, can be stipulated, or special procedural or organizational rules can be created. Such structural requirements are intended to ensure, indirectly, that the actors are steered toward the realization of the desired goals, so that development does not proceed in a rampant, uncharted fashion. Imperative and structural regulation may of course be combined with one another.

The following sections attempt to describe these two types of control in somewhat greater detail.

1. IMPERATIVE REGULATION FOR SAFEGUARDING GENERALLY PROTECTED VALUES THREATENED BY BROADCASTING ACTIVITIES

Imperative control is employed, primarily with prohibitions, when limits are to be placed on communicative activities in order to ensure that these do not harm values afforded protection by the legal system generally. By this is meant, for example, the protection of the rights of other private parties or of the interests of the state or society when these might be especially endangered by communication. Examples include prohibitions on violating human dignity, privacy, and religious feelings, advertising for drugs or alcohol, attacking state symbols, and interfering with copyrights. As mentioned above, precautionary measures of protection may be found not only in laws with general validity but also in norms formulated specifically for broadcasting. The latter are illustrated by the British Broadcasting Standards Council's code of practice, the juvenile protection rules established by the German Interstate Treaty on Broadcasting and by related state media authorities, and the advertising restrictions contained in Articles 11 et seq. of the EC television directive.

2. IMPERATIVE REGULATION FOR SAFEGUARDING THE COMMUNICATIONS ORDER OR OTHER VALUES RELATED TO COMMUNICATION

Just as widespread are guidelines, requirements, orders, and prohibitions aimed at protecting rights directly associated with freedom of communication or the communications order. These particularly include measures designed to ensure the capacity of the broadcasting order to function adequately by such regulatory measures as the ordering of the technical communications infra-structure and by prohibiting broadcasting without a license. The prototype for programming-related norms is the original U.S. fairness doctrine. In the form of the prohibition of one-sided reporting or the requirement of balanced coverage, it has also been adopted and modified by European legislatures, as in France, Great Britain, and Germany. When the legal system contains the requirement that a broadcaster serve all categories of programming or provide special educational programs tailored to minors, that the broadcasting time allotted in the license be used in full, or that advertising be broadcast in blocks, we have further examples of imperative control. Other examples include quotas relating to informational or cultural programs (as in the licensing procedures of the British Independent Television Commission) or aimed at ensuring that national or regional productions comprise a certain share of overall programming (as in France, Canada, or Australia). With regard to the latter example, the goal aspired to is not merely in the area of communications policy; rather, objectives of siting and labor-market policy also play a role.

3. STRUCTURAL SAFEGUARDS FOR EXTERNALLY PLURALISTIC COMPETITION

Of special importance are regulations aimed at setting basic structural conditions for the operation of broadcasting. For instance, ever since the introduction of private broadcasting—as in the United States and Australia from the very outset—governments resorted to a regulatory concept also employed in other sectors of an order based on private enterprise: the economic market. Economic competition serves to ensure journalistic competition and thus diversity. This model for the broadcasting order is often termed "external pluralism." Under this externally pluralistic form of broadcasting, an effort is made to provide diversity in programming by enabling various interest groups to have access to it and thus by giving coverage to the greatest possible variety of information, positions, and opinions in the overall programming of all broadcasters.

This market model is oriented on the archetype of the private press. In principle, everyone has a right under this model to become a broadcaster. He or she must raise the enormous financial resources needed and may use broadcasting for profit-making purposes, whether through advertising financing, pay-TV, or sponsoring. Even though market access is limited on account of scarcity of available transmission means, the government remains largely uninvolved. In particular, it leaves the operations of licensed broadcasters subject to the latter's self-control, which is mainly determined by economic factors and motives. To the extent that norms of broadcasting law are created, these constitute a framework intended to help ensure that self-control functions. Accordingly, rules on the allocation of scarce frequencies are thereby just as legitimate as precautionary measures aimed at securing the manner in which economic competition functions, such as measures to combat concentration or to prevent unfair competition.

Externally pluralistic broadcasting is not merely to be found in the form of the market model; another type is group broadcasting in the Netherlands.[1]

The guarantors of diversity in programming are various social groups competing with each other for influence. They are granted a privileged status—such as the sole right of broadcast transmission access—and they are endowed with responsibility for programming. The pluralistic group contest is to assist in ensuring diversity in the various programming contents. In the process, the state respects the groups' programming autonomy while it establishes a framework for group competition, such as with an abstract, general definition of group privilege and rules on allocation of frequencies.

4. STRUCTURAL SAFEGUARDS FOR INTERNALLY PLURALISTIC INTEGRATION PROGRAMMING

External pluralism relies on content-based competition among various broadcasters with various interests. It is left up to the recipient as to whether he or she wishes to select the programming of one particular broadcaster and accordingly perhaps receive one-sided information or instead prepare his or her own individual communications "menu" from the program offerings of various broadcasters. A different approach is taken under the so-called internally pluralistic model, which primarily used to be followed in European public broadcasting. In this model, each broadcaster's programming palette gives coverage to the broad spectrum of positions and information important to society, that is, integrates these positions and information into its overall programming. The recipient is thereby offered a pluralistically balanced communications menu. In theory, then, he or she has access to all relevant facts, opinions and positions when selecting from a given palette of programming.

Since such internal pluralism in programming content does not come about by itself, the government must lay down framework data. This has taken place to a particularly far-reaching extent in German public broadcasting, where internal pluralism is also reflected in the broadcaster's organization through the inclusion of pluralistically composed decision-making organs, the broadcasting councils (Rundfunkräte). But it is also possible to rely instead, or additionally, on other control factors, such as the (autonomous) professional inclinations of those responsible for programming. The BBC, for example, is obligated to provide pluralistic integration programming, even though it is not mandated that the board of governors be composed pluralistically. It is nevertheless ensured on an informal level that the interests of various groups will be taken into account, and at the same time, there is the hope that journalists will see themselves committed to upholding the BBC's public-service concept.

The financing of pluralistic integration programming normally does not take place over the market. One reason is that such programming is also intended to include broadcasts that are not, or are only insufficiently, amenable to market-based funding, such as minority programs. Therefore, other types of financing are typical, for example, user fees or government subsidies (as are used in Australia). Still, some broadcasting orders supplement this in part with market financing, as in Germany and France. Additional legal commitments are to ensure that partial advertising financing does not dictate programming conduct.

5. STRUCTURAL DIVERSIFICATION

Experience with the various broadcasting systems demonstrates that the models that are described above may also have deficits. For instance, public broadcasting in many countries is marked by a proximity to the federal government (as with public broadcasting in France) or to political parties (as in Italy and Germany), and this can pose a threat to its independence. Although private, market-based broadcasting is more independent in this regard, it can easily become vulnerable to powerful economic interests (as can be seen everywhere, but especially in Italy). Moreover, private broadcasting often does not strive to satisfy all of the citizens' communications interests, concentrating instead on, for example, entertainment programs with mass appeal. Since the advantages and disadvantages of the various broadcasting models are located in different sectors, most Western European countries have created the dual broadcasting system so that both types—public and private—complement one another. This principle of reciprocal compensation may be termed "structural diversification" (Hoffmann-Riem, 1990b: 38ff.). The private and public pillars of the broadcasting system are each structured differently in order to combine the various benefits, thereby compensating as much as possible for the detriments in each subsystem. This principle of reciprocal compensation was formulated with singular clarity by the German Federal Constitutional Court: It holds "basic provision" by public broadcasting to be indispensable, and it tolerates the imposition of milder programming duties on private broadcasters than those for public broadcasters only to the extent that the latter's continued existence is ensured (BVerfGE 73, 118, 157ff.; 74, 297, 325ff.). In essence, dual broadcasting systems are based on the principle that, in the area of private broadcasting, the market may very well facilitate freedom of communication, but it can also trigger dysfunctional side effects. Dual broadcasting systems close a kind of reinsurance policy with public broadcasting, which traditionally is under stronger obligations. The Italian RAI, the French France Television 2, the British BBC, the German stations ARD and ZDF, and the Australian ABC are all examples of public broadcasters obligated to offer integration programming.

In reality, not only are differently structured broadcasting systems set up alongside one another; mixed forms are also selected. For example, market broadcasting is normally not allowed to pursue whatever programming policies it pleases. Even in the United States, which has the longest tradition of commercial broadcasting, the Communications Act recognizes certain programming commitments, pointing to broadcasting's trustee role. The British ITC program companies operating under its auspices are reliant on private financing from advertising, but for a long time private broadcasters were protected against competition from other private companies through statutory transmission monopolies. Group broadcasting in the Netherlands makes the groups in part dependent on quasi market financing, that is, on the dues payments by their members as ascertained by the subscription to program guides. Internally pluralistic public broadcasting in Germany is also based on partial financing via the market, namely, through advertising.

6. COMBINATION OF STRUCTURAL CONTROL WITH IMPERATIVE CONTROL

Broadcasting systems typically display a combination of structural requirements and imperative dictates and prohibitions regarding conduct. Imperative approaches are normally employed in this regard in order to compensate for possible deficits in structural safeguards. One prototype is reflected in the programming commitments placed on private broadcasters. It is, however, well known that orders and prohibitions are less likely to be observed the more these requirements conflict with the broadcasters' economic (or editorial) interests. This has led to a situation where the commitments placed on private broadcasting are structured more weakly the more economic competition is relied upon as regulator. It is thus no mere coincidence that in connection with the licensing of new competitors, the British Broadcasting Act of 1990 lifted the strict commitments imposed on private broadcasting under the IBA act and subjected it instead to "light-handed" control by the ITC. German legislatures have, in contrast, proceeded quite cautiously. Most of them have mandated that private broadcasters also be obligated to offer integration programming when there are not a sufficient number of competing broadcasters (the minimum figure is usually set at three) or when it becomes clear that, in spite of competing broadcasters, diversity is not ensured.

NOTE

1. Jehoram, 1981; Nieuwenhuis, 1992; Brants and McQuail, 1992. The Netherlands is currently in a phase of transition from group broadcasting to market-oriented commercial broadcasting.

CHAPTER 9

Organization and Procedures of Licensing and Supervision

I nsofar as special regulation of broadcasting is justified, there is normally also justification for the establishment of a special body responsible for the supervision of broadcasting. None of the countries studied here neglects to transfer to a special authority the task of ensuring whether the specific norms of broadcasting law are being observed. The organizational framework of such authorities and their powers are addressed in the sections that follow.

1. BODIES RESPONSIBLE FOR LICENSING AND SUPERVISION

In some countries, for example in the United States, Canada, and France, there is a sole authority responsible for both public broadcasting and commercial broadcasting. In other countries, such as in Great Britain, however, a separate supervisory body is set up for private broadcasting (formerly the IBA and now the ITC). The supervisory authorities are always empowered to decide on questions of broadcasting. In some cases, they may also rule on which frequencies are to be used for broadcasting generally or for which specific broadcasting services. Usually however, the government also plays a part in making basic decisions as to how frequencies are allotted, or it may even have sole power to do so. Whereas some supervisory authorities—for example the

German state media authorities or the British ITC—are responsible only for broadcasting (as a mass medium), others—such as the American FCC or the French CSA—also have powers in other telecommunications sectors. This sort of organizational double role seems also to have an influence on the way in which responsibilities are carried out. For instance, in the United States, this will likely promote the convergence of broadcasting and telecommunications services, particularly if companies are allowed to operate in various sectors simultaneously.

Most supervisory bodies are central authorities whose competence covers the entire national territory. Only Germany provides for separate state media authorities in each of its sixteen (formerly, eleven) states. But even in systems with centralized broadcasting supervision, one often finds decentralized subauthorities, as in the form of regional offices.

Supervisory bodies are more or less autonomous organizations usually severed from the governmental structure. Legal autonomy does not, however, necessarily ensure independence in day-to-day work. For instance, most governments and/or parliaments monitor supervisory activities, and they are often given the right to participate in the selection of top personnel and, in some cases, financial decisions. The Australian ABA is more closely tied to the government than any other institution studied here. The ABA must perform its functions in accordance with government policies, and it is subject to government direction.

In addition to powers of intervention, there are a variety of informal ways to exert influence. The legal form is little indication of the intensity of the influence exercised. Thus, the former IBA in Great Britain was in fact quite independent, although the power of influence granted to the government by law was particularly far-reaching. Nevertheless, the government long refrained from making use of its possibilities for exerting influence, a situation apparently to be attributed to British political culture. As a result, the number of program-related interventions was quite low. The German state media authorities, in contrast, are in some ways de facto relatively closely tied to the government, despite statutory guarantees of independence. The situation is similar for the various French supervisory bodies, even though the CSA has been endowed with greater practical independence than its predecessors.

Most supervisory bodies have at their disposal a collegially organized directorial organ (e.g., a board) and an administrative staff, which takes care of day-to-day work and prepares the activities of the directorial organ. The directorial organs are composed of a small number of members (usually five to thirteen). Following the model of the broadcasting councils for public broadcasting, most of the main organs in Germany are constituted as considerably larger, collegial organs, which because of their size and the diversity of their members, often operate quite lethargically; consequently, true authority usually rests with the director, a monocratic position (Wagner, 1990a: 131).

In addition to the supervisory bodies responsible for licensing and general supervision, one also finds on occasion special organs with specific duties. Particularly worthy of mention here are the British Broadcasting Standards Council, whose mandate is the "maintenance of standards in matters of violence, sex, taste and decency," and the British Broadcasting Complaints Commission, which can take action in order to protect private rights.

The special broadcasting supervisory bodies are confined to the licensing and monitoring of broadcasting in the respective national or regional areas. Their powers usually extend much farther with respect to the operation of broadcasting than the mere retransmission of programs via cable or satellite. In spite of the increasing internationalization of broadcasting (see Chapter 7, Section 5), supervision of broadcasting remains a national affair. Thus, as already mentioned, neither the EC television directive nor the television convention of the Council of Europe provides for pan-European supervision.

2. LAW AS A MEANS OF REGULATION

In all of the countries studied here, legal norms set forth the scope of duties to be exercised by supervisory bodies and the regulatory instruments at their disposal. In some countries (such as Canada and United States), statutory powers are kept very broad, so that supervisory bodies are able to utilize broad delegated power. In other countries (such as Great Britain and Germany), parliamentary laws go into considerable detail; however they usually also leave scope for further specification.

A basic legal framework is generally a requirement for broadcast regulation in liberal democracies. Moreover, in the area of freedom of communication, requirements can usually be found in constitutional norms (and in article 10 of the European Convention on Human Rights), which ensure that interference with freedom of broadcasting is permissible to only a limited extent. In a number of legal systems, such interference requires the form of statutes.

On a general, abstract level, the most important supervisory instruments are the capacity to influence the structure of the broadcasting order and the power to set forth additional conduct duties in legal norms, as well as in guidelines, codes, and so forth, that specify those duties. The specification of corresponding requirements in the individual conditions of license or in additional, possibly subsequent terms and conditions represents an attempt to detail the amount of latitude in each broadcaster's conduct and thereby to determine more precisely the mutual rights and duties in its relationship with the supervisory body.

Regulatory norms usually contain a number of undefined—namely, prognostic and value-assessing—legal terms and allow a certain amount of discretion in their application. Furthermore, there are often a variety of decision-making criteria, which are all taken into consideration in a given conflict without there being need to define only one possible result. This process likewise allows multiple options.

In all broadcasting systems, regulatory norms are characterized by relatively vague and usually broad delegations of power (cf. Russel, 1993: 185ff.; Ziethen, 1989; Hoffmann-Riem, 1981b: 278ff.). The latitude with regard to licensing and supervising is particularly large: In selecting certain applicants and specifying duties as well as supervising these obligations, there is room for exercising political influence. The transfer of certain aspects of selection decisions to the market—one example of this is the cash bid in Great Britain— is an indication that efforts are being made to curtail the decision-making authority of the licensing bodies; but as may be seen by exceptions in Great Britain, the latter's decision-making and structuring power has remained, at least in part, intact.

Internationally, however, the legal requirements regarding the level of regulation vary to a great extent. In countries like the United States, where freedom of speech is broadly protected, permissible legal restrictions are drawn quite narrowly. Vagueness and overbreadth must be avoided as far as possible, and restrictions must be proportional, and not excessive, to the end sought. Such legal requirements may, however, prove to be a barrier to effective supervision. They limit flexibility in reacting to unforeseen or new questions and make it possible for the broadcasting industry to find loopholes. For this reason, the reverse path was taken in Australia in 1992. Earlier regulations to combat concentration, which had often been very detailed, were replaced in the interest of greater effectiveness with more general, and thus more flexible, requirements.

The conflict between strict, normative specification of supervisory powers and their effectiveness is present in all media orders. In many cases, supervisory bodies have consequently begun to work with "soft" regulations as far as possible. Typical examples are published policy statements (e.g., in Canada and the United States), which are not binding effect but nevertheless have a great effect on the orientations of broadcasters and their supervisory activities. Examples of this usage are the (former) FCC guidelines that related to program content and advertising. They did not contain binding requirements but, rather, rules of thumb that the FCC was to use in deciding whether the public interest was safeguarded. In this way, broadcasters knew the direction being taken, but this did not mean that the FCC could not deviate from the guidelines in a given case. De facto, however, the guidelines often worked like quotas. In the area of equal employment opportunity, the FCC continues to use the same regulatory technique. Although it has not adopted a required system of quotas, goals, or timetables with respect to hiring and advancement, the Commission has intimated that a renewal application might be denied where the licensee indicates the percentage of minority or female employment to be "outside the zone of reasonableness." These guidelines are neither quotas nor safe harbors. In the area of protection of minors, on the other hand, broadcasters do have safe harbors with regard to the limitation on the time of broadcast of certain programs.

One special regulatory technique consists of the supervisory body not setting limits itself but instead using guidelines created by the broadcasting industry in the course of self-regulation in clarifying undefined legal terms. This was the path long taken by the FCC with regard to the NAB code. In addition, the Australian ABA uses the codes of practice developed by the broadcasting industry as an orientational aid in applying the law. This policy was expressly provided by the Australian legislature, and as a result it required that the ABA be included in the process of creating codes, particularly through the imposition of consultation duties. Moreover, the ability of the ABA alternatively to set standards itself provides it with some threatening power, which is meant to have repercussions on the conduct of the broadcasting industry. The magic formula set forth here is self-regulation of broadcasting with a regulatory safety net.

Rules of thumb, unofficial guidelines, safe harbors, safety nets, and so forth, are catchwords for identifying a type of broadcasting regulation unsure of itself. Soft, flexible formulas and correspondingly soft regulatory practice are intended to help make supervisory measures less easily contested. As a result, regulatory intervention recedes and is replaced with an expanded scope of responsibility on the part of broadcasters. The risks of such a regulatory technique are plain. As one American observer noted:

> The deregulatory trend . . . has shifted the balance of programming discretion significantly in favor of the broadcaster. However, it has also made more difficult the determination of whether a broadcaster is operating "in the public interest." If the public interest becomes equated solely with broadcaster discretion, the standard loses effective meaning. (Zuckman et al., 1988: 409)

3. AWARDING OF LICENSES

The point of departure for broadcasting regulation is the allocation of means of transmission. In addition to the effect of policing frequencies, the granting of the license is the starting point both for checking basic qualifications (e.g., suitability, financial viability, citizenship, technical capabilities) as well as—in a kind of piggyback procedure—for detailing conditions for the use of the means of transmission. The granting of the license brings with it, first, the obligation to furnish broadcasting material and, in addition, the commitment to observe the laws or the supplementary directives, codes, and regulations that may exist. Moreover, further obligations may be imposed on licensees. This occurred to a great extent, for example, with the granting of franchises by the IBA, which was preceded by extensive negotiations on the details of the franchise. The license was thereafter to be awarded unilaterally, and although it also permitted closer specification, it did not need to be as far-reaching in this regard as the franchise. In France, the conditions are additionally set down in the individually negotiated "conventions" following the awarding of the license. Licenses in Germany can be tied to additional terms and conditions. These must of course remain within the framework of the broadcasting laws, which nevertheless permit detailed commitments.

Since several applicants normally compete for one license and since licenses are generally awarded in comparative allocation procedures, the licensing authority is basically able to select the applicant with the most attractive offer and to bind it legally to its promise. Although in Great Britain this possibility has likely been reduced with the transition to the cash-bid system, it has not been completely ruled out, since the ITC must review whether the offer satisfies the statutory requirements, which in turn leave room for interpretation. Furthermore, under exceptional circumstances, it is also possible to select a "superior" applicant even when that applicant has not submitted the highest cash bid.

Licenses are usually awarded for relatively long periods. The applicant enjoys a protected status during the term of the license. Even following expiration of the license, the last holder has a good chance of securing a renewal, particularly since it is by no means certain that a different applicant would offer better programming. The renewal applicant can rely on a sort of "renewal expectancy." The Australian Broadcasting Services Act of 1992 provides for a right of renewal unless, under very limited criteria, the ABA is satisfied that the licensee would not be suitable. In the United States, the Communications Act authorizes renewals only on the same terms as initial grants. Nevertheless, the FCC has designed the procedure and evaluation in favor of the incumbent. If the incumbent is found to have rendered substantial service, its renewal expectancy will be recognized, and it will be given a comparative preference. Because of deregulatory measures with regard to past performance—for example, the elimination of the requirement of program logs in the United States—it is difficult to prove that the incumbent has not met the public interest.

There are, however, rare cases in which licenses have not been renewed. Most remarkable are the nonrenewal decisions by the IBA and the ITC in Great Britain. The legal and/or factual renewal expectancy makes it difficult for the supervisory authority to enforce commitments or to set new obligations. But under changed circumstances, it is normally possible, although only within narrowly defined terms, to supplement the conditions in the license, and in the event of failure to observe the duty, the license may be suspended or revoked.

4. MONITORING

The effectiveness of broadcasting regulation also depends on enforce-ment, one important aspect of which is monitoring. Supervisory authorities generally are able to monitor conduct in a number of ways. Basically, super-visory bodies are able to act on their own initiative in order to establish and punish any violations of the rules from outside. A typical occasion for checking compliance occurs when the license is being renewed. Complaints by com-petitors or members of the public may cause special checks. In the course of the renewal process, license-holders are obliged to render information on their former conduct. During the course of the license period, supervision is usually not so intensive. As soon as there is no duty on the part of the broadcaster to report on a continual basis, the supervisory bodies usually only act in response to complaints.

The most sensitive area of monitoring is that of broadcast content. In some broadcasting systems (e.g., in Germany), broadcasters are obliged to tape their programming and then store it for a certain length of time; if necessary, the tapes are to be provided to the supervisory authority. However, systematic, on-going control of, for example, programming conduct (which the IBA was empowered to perform), is nowhere to be found. German media authorities may monitor a program after it has been broadcast, and some perform this task on a random basis. In most cases, supervisory bodies take action only in response to specific complaints; in others, they themselves may also take the initiative. On some occasions, action originates with the broadcasters directly, as when they request the supervisory body to review a program in advance for compliance with regulations. In Australia and Canada, the broadcasting indus-try has developed its own mechanisms of control for advance review of clearance, especially with respect to advertising regulations. Prior clearance in the form of classification by the ABA takes place in Australia for children's programs. In all other areas, it has been taken for granted that prior control would constitute undesirable—and unconstitutional—state intervention in an area sensitive for reasons of freedom of speech and creativity.

In other fields, the supervisory authorities use a wide range of monitoring techniques. For instance, broadcasters may be subject to notification duties, usually in the event of a change in ownership relationships. In Australia, though, the transfer of a commercial broadcasting license does not even require prior notice. The American FCC monitors equal employment opportunity performance by requiring stations with five or more full-time employees to file yearly employment profiles.

The former U.S. requirement of community ascertainment studies called for a highly formalized ascertainment process to determine the needs and interests of the communities served. These procedures were abolished during the deregulatory phase. Now the broadcaster need only list in the station's public file of programs that it has presented programs that meet the issues confronting the community. The manner of ascertainment is neither a basic nor a comparative criterion at renewal time.

The renewal of licenses usually is the best occasion for reacting to complaints by the public or competitors and thus for "monitoring" previous performance. Over the course of time, the regulatory provisions and the agencies' policies have reduced the need for systematic monitoring by softening the renewal requirements. Again, the renewal applicant can rely on some sort of renewal expectancy (see Section 3).

5. SANCTIONS

Should the supervisory body ascertain a violation of any specified duty, it then has at its disposal a set of sanctions that is usually graduated. The first step generally is comprised of notifications, thereafter warnings. If these prove ineffective, then administrative or even criminal fines may be imposed. Some supervisory bodies are able to impose such sanctions themselves (e.g., the German media authorities impose administrative fines); others are reliant upon the intervention of the courts (as is the case in France) or the Minister (as the Minister of Transport and Communications in Australia). The imposition of such sanctions suffers from the fact that the supervisory authority normally bears the burden of proving that norms have been violated. All the same, the situation is improved when broadcasters are obligated to deposit a bond as a guarantee that they will observe programming commitments (as in Great Britain). A similar construction has been introduced in Canada: Broadcasters pay a license fee that is refunded if they fulfill programming duties.

In the event that rules are violated, most broadcasting orders permit action to be taken against the license, for example, by suspending it for a certain amount of time, by shortening its length, or, when all else fails, by revoking it. The possibility of refusing to renew a license when it expires is an especially important supervisory tool with great practical utility, although it is not expressly mentioned in the law as a sanctioning instrument. Procedural requirements may also have the character of a sanction. For instance, the U.S. broadcasting industry generally views the costly procedure of a formal hearing before the FCC as a severe sanction, which is to be avoided at all costs in the interest of saving time and other expense but also to avoid potentially adverse public discussion.

Practice has shown that a subtle, graduated set of sanctions offers greater chances for effective control than if the supervisory body were limited to only a few, possibly quite drastic instruments (such as revocation of license), which it would then ultimately be reluctant to use. With this awareness, most legislatures and supervisory bodies have, through time, refined the initially somewhat crude instruments provided by law.

6. PROTECTION AND SUBSIDIZATION

The sanctions discussed above constitute a form of negative control. However supervisory bodies also make an effort to support broadcasters positively in their endeavors to observe the duties placed upon them. Thus, supervisory bodies have practiced restraint, particularly during the start-up phase in the launching of a private broadcasting order or with the initial licensing of a broadcaster, and have respected the difficulties encountered in getting operations under way. In addition, they have often seen themselves as protectors of licensed broadcasters, such as by according them special, protected status in comparative licensing decisions. Repeatedly cited also is the prolonged effort by the FCC to shield the networks against emerging competition from cable. The delays in the introduction of subscription TV in Australia likewise have served to protect established broadcasters.

Regulations that provide that certain programs have to be carried over cable, or at least are to have priority, may as well constitute an effort to protect the profitability of the corresponding broadcasters. Such priority rules may particularly be found in those areas where there is still a scarcity in the means of transmission, as is the case in many European cable systems. Must-carry rules were also long used in the United States to protect programming that served community needs, but these were in part found to be unconstitutional (*Century Communications Corp. v. FCC*, 835 F.2d 292 (D.C. Cir. 1987)). The Cable Television Consumer Protection and Competition Act of 1992 nevertheless also contains must-carry rules (cf. *Turner Broadcasting System Inc. v. FCC*, 114 S.Ct. 2445 (1994)).

Corresponding concerns about protection are behind the U.S. network nonduplication rules and the Canadian substitution rules, which ensure that local or regional broadcasters have priority in the broadcasting of a certain program by cable when several competing broadcasters are airing it at the same time. This particularly has to do with the protection of the advertising revenues of local or regional broadcasters. Aimed at the protection of certain desirable broadcasters are rules setting up monopolies, as was earlier the case with the British ITV companies or can currently be found in the German state of North Rhine–Westphalia to protect local radio broadcasters. With broadcasters that encountered or seemed on the verge of encountering economic difficulties, the supervisory authorities were often willing to permit exceptions to anticoncentration rules or quota obligations. However, if it better suited the scheme of things, a bankruptcy would also be accepted, as with La Cinq in France.

There are also a number of special measures regarding subsidization. For instance, broadcasting laws often provide that certain broadcasters that cannot acquire adequate financing on the market are to be supported by funds stemming from the revenues of other broadcasters (as is the alternate plan for the financing of Channel 4 in Great Britain). Canada provided for subsidies to protect Canadian productions, which were administered by the Canadian Broadcast Development Fund. The expansion of the technical infrastructure for the transmission of the programming of private broadcasters is subsidized in Germany with funds siphoned off from the broadcasting fees paid to support public broadcasting. In the German state of Bavaria, there are special subsidies to support certain minority broadcasts.

CHAPTER 10

Fields of Supervisory Action

This chapter offers a description of the most relevant areas of broadcasting regulation, focusing primarily on the commercial television sector. No attempt is made here to further systematize the sections or to form main categories, as this would presuppose the application of uniform criteria when creating broadcasting systems. However in view of the heterogeneous nature of the various broadcasting systems, this approach would be extremely difficult to ascertain.

1. PLURALISM, DIVERSITY, FAIRNESS, AND IMPARTIALITY

In all countries studied here, broadcasting is seen as a means of communicative self-development and as important to the functioning of a democratic society. There are, however, considerable differences as to whether democracy is essentially automatically supported when freedom of broadcasting is guaranteed or whether, instead, additional assurances are necessary in the broadcasting order. When broadcasting is subjected to public service obligations, this stricture is based on the assumption that special precautions are required in the area of broadcasting. This has been expressly confirmed by, above all, European courts. In particular, the German Constitutional Court has recognized the fundamental importance of mass communication for the free formation of public opinion, and it has called for positive (affirmative) assurances in the broadcasting order that all citizens be provided with such types of communication as are important for their political, social, and cultural orientation. Other courts as well—such as the French Conseil Constitutionnel and the U.S. Supreme Court—have stressed the relationship between broadcasting and democracy.

Broadcasting is one of the media for the responsible airing of civil affairs and political debate. The demand for pluralism in broadcast programming is a direct reflection of the Western social orders that are anchored in the concept of democratic pluralism. In the pluralistic process, one finds growing awareness that not simply informational programs are important for the formation of public opinion but also—or, perhaps, more importantly—entertainment programs. This obligation applies particularly to television, which has grown to become the entertainment medium par excellence. Accordingly, the political calls for pluralism are by no means limited to informational diversity in special news and current affairs programs but, rather, relate as well to the operation of broadcasting as a whole. However, most content-related regulations are drawn quite narrowly and largely ignore entertainment programming.

As regards news and current affairs programs, one finds special precautions ensure pluralism. Prototypical in this respect was the U.S. "fairness doctrine," which has since been almost abolished, although attempts are being made to reestablish it. Substantively speaking, the fairness doctrine covered two main duties. First, it was expected of each broadcaster that it allocate a reasonable amount of its programming time to the treatment of controversial issues. Second, each broadcaster was obligated to undertake such treatment in a fair manner, that is, also to afford a reasonable opportunity for presentation of contrary positions. Broadcasters in other countries (a good example is Germany) are expected to observe a minimum of substantive balance, objectivity, and mutual respect. Furthermore, individual programs must not unduly exert one-sided influence on the formation of public opinion. The British Broadcasting Act of 1990 goes even further, requiring that "due impartiality [be] preserved on the part of the person providing the service as respects matters of political or industrial controversy or relating to current public policy."

Programming duties such as these are alien to a broadcasting system that is based on competition and therefore adheres to the concept of external pluralism. If—so the theory goes—economic competition ensures journalistic competition virtually automatically and thus satisfies the communication interests of the public, then there is no need for additional guarantees: Therefore, it is consistent that they be reduced or even abolished in the course of regulation. According to this theory, every broadcaster in every program broadcast should be allowed to be as one-sided as it sees fit, as supposedly other competitors will compensate for the one-sidedness by producing a program that will ensure a balance is achieved. However, it is noticeable that most broadcasting laws shy away from this step. Clearly, faith in market forces is blanketed not only by fear of market failure but also by the view that too much is being asked of viewers if they are expected to consume all competing programs in order to achieve a balanced supply of information. It is even more difficult for the recipient to attempt to consume through his or her choice of entertainment programs a varied diet of values, stereotypes, and so on that are conveyed via programs of this kind. Because of the practical difficulties in ensuring variety through the market, most broadcasting rules contain a kind of safety net by expecting a minimum of fairness and impartiality from each program—which means, however, limiting one-sidedness.

These types of programming duties are particularly difficult to administer, as has long been made clear by the U.S. fairness doctrine. Only to a limited extent do mass-appeal broadcasters tend to treat controversial topics and depict controversial opinions (see, e.g., Barrow, 1975: 673f.; Mallamud, 1973: 115ff.; Malone, 1972: 216ff.). They apparently go to great lengths to avoid affronting even segments of their audience or excessively straining viewers'/listeners' tolerance for minority views. As a result, every attempt to exercise control represents a nearly insurmountable task when it involves the requirement of comprehensive treatment of political or other public controversies.

Similar difficulties are encountered with enforcement of the duty to give pertinent parties the opportunity to present contrary positions. This consumes television time, and according to current opinion, it is often less attractive for the majority of viewers than the simplified expression of only one position. Moreover, the advertising industry is hardly interested in a programming atmosphere characterized by disharmony. It therefore comes as no surprise that even during the period of full validity of the fairness doctrine, the large U.S. networks made an effort to confine the spectrum of controversies. They also specifically decided, for example, not to accept any commercial on controversial issues, since they would then be obligated to broadcast advertising with the contrary message. While taking pains to enforce the fairness doctrine, the FCC nevertheless practiced a considerable amount of moderation, as reflected in the low number of successful fairness complaints.

In other countries, there are and have been in some respects better possibilities for enforcing programming commitments. For instance, on account of its position as broadcaster and the accompanying self-responsibility for programming, the British IBA was able to exert lasting influence on the structure of programming and, in particular, to enforce the duty of impartiality. Since the ITV companies did not compete with one another for advertising revenues, they usually could afford economically to uphold or comply with programming commitments even when their observance might meet with opposition from audience segments. In this respect, however, the basic conditions in Great Britain have since changed. In France and Germany, the supervisory bodies have less enforcement authority than was the case with the

IBA, and there have been only very few known instances of action taken to implement programming obligations. At the same time, however, almost no intervention was taken with regard to the prohibition on one-sided programs: Commercial broadcasters tend to avoid obvious bias in order not to irritate their viewers.

Nevertheless, inherent in their programming is the risk of subtle one-sidedness, such as when advertising clients are catered to in such a way as to offer them a programming setting that increases viewers' willingness to purchase. Programming commitments established thus far have not tried or at least they have been powerless to combat such subtle bias.

2. EQUAL OPPORTUNITY IN POLITICAL BROADCASTING

Closely related to the requirement of impartiality are the rules dealing with political broadcasting. There is extensive experience with these in the United States where Section 315 of the Communications Act sets forth equal opportunity requirements for political candidates. A broadcaster may not improperly influence an election by affording only one candidate the opportunity to reach its audience. Equal opportunity relates not only to the time of coverage but also to the rates charged for advertisements. There are also provisions dealing with political editorializing by broadcasters. Other broadcasting orders (e.g., in Canada), likewise provide that all political parties and candidates are treated equally and fairly in election broadcasts as well as in informational programming. These regulations normally do not dictate that candidates be given access to broadcasting but only that they be treated evenhandedly. In Germany, however, there are special norms guaranteeing political parties free access to broadcasting time during election campaigns; paid political advertisements are prohibited.

Although such rules deal with the content of programming, they have been enacted without hesitation even by those legislatures that are otherwise reluctant to stipulate content. These rules help to foster genuine interests of politicians. But at the same time, they serve to guard against one-sided influence, particularly when exercised by those with sufficient capital to use broadcasting to support their political agendas. The U.S. requirement that a station charge a candidate no more than the "lowest unit charge for the same class and amount of time of commercial advertisements" operates to keep costs down, as does the German prohibition on paid political advertising in broadcasting, since this is coupled with a limited right of free access during election campaigns.

Supervision of the observance of these rules is relatively easy when they are highly formalized, as is the case in the United States. Nevertheless, there remain considerable problems of delineation.

3. PUBLIC RESPONSIBILITY IN AIRING DIFFERENT INTERESTS AND COUNTERING OF STEREOTYPES

Associated with the above-described programming commitments is the duty to provide socially relevant groups and forces, especially minorities, with a reasonable opportunity to air their concerns in broadcast programming. This might be accomplished by giving such groups preference in licensing (such as by setting up minority channels), by providing for their partial ownership of certain broadcasters (e.g., preferential acquisition of shares), or by ensuring their access to programming (as in the form of window programming with other broadcasters); free, public access channels in cable television are another possibility. Such rules pay regard to a group's interest in the role of communicator. But they cannot prevent these groups from being relegated to the niches of the programming spectrum, and above all they cannot ensure that the depicted interests and attitudes will actually be seen by those recipients who prefer to concentrate on mass-appeal programming.

In the interest of social integration and in order to help ensure that diverse perspectives reach the public eye, broadcasters are sometimes required to allow all relevant groups an appropriate chance to be heard in their programming, or at least to address their concerns. This is the concept behind internally pluralistic programming. But such precautions conflict with the basic concept of market-based broadcasting, since under this model broadcasters must be able to establish themselves in the market by being one-sided, if necessary. In Germany, however, the principle of internal pluralism applies in weakened form to commercial broadcasters, although they actually operate under the external pluralism model. The enforcement of such duties is somewhat difficult, not least because there are no recognized standards for what is socially relevant. Moreover, these duties may lead to a situation where (as could be observed in Germany with public broadcasting) groups and interest associations tend to measure balance in programming with a stopwatch, repeatedly complaining that they are at a disadvantage. Public broadcasting in Germany was often paralyzed during the 1960s and 1970s by such debates about balance, though these debates have waned, at least in private broadcasting. The more the broadcasting order approaches the external pluralism model, the less can various broadcasters be obligated to take into account all interests. And it was for this reason that a duty of this sort was not set up under Dutch group broadcasting; however, the public broadcaster, NOS, served as a sort of balance reserve (Nieuwenhuis, 1992).

Regulatory efforts are aimed at a special form of public responsibility to ensure that various ethnic, religious, and other groups are portrayed in a reasonable, sociopolitically desirable manner and that gender roles are not shown in a biased or stereotyped fashion. Such requirements seek to achieve prosocial effects, including the furtherance of social integration. This is based on the repeatedly confirmed empirical finding that broadcasting is a means of constructing reality, that its programming influences not just the recipients' level of information and knowledge but also the social climate, and that broadcasting tends to follow existing biases and reinforce stereotypes. As a result of these and other insights or presumptions, broadcasting's potential is required to be used in a sociopolitically desirable manner. However, broadcasting laws rarely provide for rules to address such findings. One example of such rules can be found in Canada, where broadcasters are required not only to take steps to counter gender stereotypes but also pay regard to the lifestyles of ethnic and other minorities. The observance of relevant guidelines falls under the province of the Canadian Broadcast Standards Council.

Such duties are programming-related obligations. In the United States, a different approach is taken with the long-established duties to adopt and file an affirmative action equal opportunity program. Steps by broadcasters to ensure equal opportunity are a prerequisite to licensing. The main objective pursued by these FCC regulations is not necessarily to ensure programming that takes into account minorities, although they may achieve this indirectly. Their central intention is, as in other areas, to ensure equal employment opportunities in broadcasting. Therefore, from a systematic viewpoint, these measures are not considered to be programming-related duties.

4. DUE ACCURACY OF NEWS

In some broadcasting systems, informational programs are subject to special duties relating to the traditional professional ethics of journalism. For instance, prior to broadcast, news reports must be scrutinized with requisite care with respect to their truth and origin; they must be objective and independent. It is often also required that commentaries be clearly separated from news reporting. These obligations may be dealt with in statutes—as in Germany and Great Britain—or in industry codes of practice.

However, as concerns supervision, it is difficult to enforce such duties. Debate often revolves around the normative requirements, such as objectivity and care. Notice can be taken of the conflict surrounding investigative reporting. Moreover, strict requirements threaten to have chilling effects. A rigid administration of such rules can dampen free communication and robust public debate, as well as the media's watchdog role. In the interest of lively democracy, but also out of fear of the conflicts that supervisory action can trigger, supervisory bodies generally tend to practice restraint here.

5. MAINTENANCE OF CULTURAL AND LINGUISTIC IDENTITY

Ensuring pluralism includes the efforts to safeguard a society's cultural identity, or that of its important groups. Programming-related orders and prohibitions have, however, little effect when it comes to enforcing such goals. Of greater importance is the creation of structures in the broadcasting system that provide a society's local or regional, as well as social, subsystems with access to broadcast communication. When regulations ensure that national broadcasting is accompanied by regional or local broadcasting, this is an attempt to promote cultural identity.

Special problems arise in countries with active cultural minorities, such as in Australia or Spain, particularly when these groups resort to their own languages. The pressure exerted by minorities for access to broadcasting was acknowledged in some cases by the creation of special broadcasters—as in Australia through the Special Broadcasting Service, in Spain for the Basques and the Catelonians, and in Great Britain for the Welsh. Similarly, Belgium's linguistic dispute is reflected in its television order, with special broadcasters licensed for the Walloons and the Flemish. However, this additional licensing was also motivated by efforts to control the alienation resulting from the programming of foreign broadcasters. Since Belgium has a high rate of cable penetration, foreign programs have long been accessible. Belgium attempted to tie the retransmission of foreign programming in cable networks to the payment of levies designed to benefit the financing of national audiovisual productions. This has met with opposition from the EC Commission.

The protection of multilingualism often presupposes financial support for minority broadcasters. For instance, the public broadcasting system in Switzerland provides for a type of internal compensation in order to ensure that linguistic minorities, such as the Italo-Swiss in Tessin, have their own programming.

Legal guarantees of multilingualism typically relate primarily to public broadcasting, unless the linguistic groups are large enough to finance of commercial programs. Commercial broadcasters are most likely to satisfy the members of linguistic minorities when their language is also spoken in other countries. It is then at least possible to have access to programs in the native tongue via cable or satellite retransmission of relevant foreign programming. But even then, the substantive basis of such programming fails to do sufficient justice to the problems of the respective minority. The concern for identity is thus satisfied only to a limited extent.

Commercial broadcasting is dependent on sufficiently large viewer or listener audiences and therefore looks for the largest possible dissemination territories. Rules that seek to address local or regional television therefore usually meet with opposition from broadcasters and very often have almost no effect—unless the broadcasters learn that some types of local or regional programming can be profitable. Networks are in most cases created to supply additional programming from other regions, meaning that local or regional emphasis often remains marginal at best. The opposition undertaken to prevent a regionalization of private television in Germany may serve as one good example here. The state media authorities sought to ensure regional television offerings by obligating licensed television broadcasters to provide at least so-called regional window programming. The nationally televised, although regionally licensed, broadcasters RTL and SAT 1 were adamantly opposed to such obligations, but they had to give way, since otherwise they would not have received a license to broadcast their programming terrestrially.

6. PROMOTION OF INTERNATIONAL UNDERSTANDING

One of the most controversial political issues concerning the media, particularly on the international stage, is the question as to how foreign societies or countries are represented in the media. The discussion about the new world information order and the imbalance of the flow of information, especially from north to south, or about interference in internal affairs demonstrates just how sensitive the flow of both trans- and international communication is.

This problem primarily comes under the rubric of national politics. No international supervisory bodies have been set up to deal with it, and international broadcasting regulations touch upon the issues outlined here only to a very limited extent. For example, according to the German broadcasting rules, broadcasters are obligated to give the recipients an objective overall view of world events and to serve international understanding. Yet there is no evidence of the supervisory boards' undertaking any action to enforce these programming duties. Such programming duties are also difficult to put into operation and should they actually be realized, they run a severe risk of being politically misused.

7. MAINTENANCE OF HIGH-QUALITY PROGRAMMING

The idea that broadcasters should offer high-quality programs is commonplace; it crops up again and again in speeches. Just what is meant by high-quality programming however, is debatable (see NHK, 1991, 1992, 1993). Proponents of the market model usually maintain that the appropriate level is reached when a significant number of recipients accept the programming. In this case, ratings are considered to be a reasonable indicator of quality. But in most instances, quality is perceived to be a more complex category, and this generally triggers considerable controversies as to criteria and how these are operationalized.

When a broadcasting order is expressly structured according to the objective of plurality, as in Germany, a requirement of quality is considered to be part of the duty in achieving such diversity. With the transition from internal to external pluralism—and the associated right of broadcasters to be one-sided and not obliged to take all interests into account—this requirement of quality (diversity) can no longer be directed at the programming of each broadcaster but, rather, at best directed at the total programming of all broadcasters put together. But this means that the requirement of quality is no longer a viable standard for evaluating the individual programming.

In view of the problems related to operationalizing a pertinent standard, most broadcasting laws refrain from setting rules on quality; at most, they make appeals in this direction without any legally binding force. One exception is the British Broadcasting Act of 1990, which requires certain broadcasters to provide "a sufficient amount of programs which are of high quality," a term that obviously relates to more dimensions of quality than just diversity. Programming quality is also intended to be a factor in the licensing decision, when an exception can be made from the rule that the license be awarded to the highest bidder. Apparently, the British Parliament believes that the ITC can define quality without overstepping its authority, that is, by becoming a taste dictator. One observer writes: "Anyone adopting a relativist position ('Who is to say what is quality?') will be dissatisfied with these terms, but since they are firmly embedded in British consciousness about broadcasting and in the Act, they cannot be wished away nor should they be. It is better to aim high and fail than to aim low and hit your target" (Hearst, 1992: 73).

8. MULTIPLICITY OF PROGRAM FORMATS

Supervision of broadcasting also ensures that transmission capacities are actually used; this might take the form of duties to make real use of means of transmission. In addition, supervisory bodies often make an effort to ensure a certain amount of breadth in program offerings. For instance, German broadcasting laws provide that certain broadcasters must air so-called full programming, by which is meant programming covering not only entertainment but also information, education, and advice. This is intended to prevent broadcasting from becoming a medium of pure entertainment. But when broadcasters are entitled to air special or target-audience programming, then authorities tolerate or even purposely advocate that the focus be placed on certain program categories.

Broadcasting laws often provide that a specific minimum amount of news and current affairs programs or local programs be broadcast. Some laws stipulate only generally worded duties, whereas others set forth exact percentages for the various aspects of programming. The FCC, with years of experience with such requirements, had enacted corresponding guidelines, but it abolished these during the 1980s in the course of deregulation. Since it is difficult to define such categories precisely, and since the gray area is necessarily wide, broadcasters have repeatedly found ways to show they are meeting requirements by redefining their programming and getting it classified in the desired category, even though this misses the actual objective. Particularly evident are strategies whereby the legally required program categories that are not very attractive for the advertising industry are aired at a time of day that is unprofitable anyway. Sometimes supervisory bodies react to this by stipulating the time at which the programming duty is to be fulfilled.

Broadcasters have on some occasions learned that the programming duties imposed upon them against their will ultimately prove lucrative for them. For instance, although some broadcasters were at first reluctant to air local or regional information programs in the United States, they realized over the course of time that there was a significant interest for these among recipients and that these program categories could be structured profitably as well.

9. COVERAGE OF IMPORTANT EVENTS

Because of rising competition between broadcasters, concerns have arisen as to appropriate coverage of various programming categories. Commercial broadcasters, particularly in the pay-TV field, are able to earn large profits for selected, especially popular broadcasts, and for this reason, they go to great lengths to acquire exclusive broadcast rights. As a result, organizers of important events attempt to sell such broadcasting rights on an exclusive basis. In view of the increased importance of sports for the attractiveness of broadcast programming, there has emerged a market for exclusive television rights to sports events. But the marketing of exclusive rights for other events, including entertainment, disasters or spectacular political events, is also possible.

In German media law, precautions are in place to protect against this in the form of rights to cost-free short reporting, which is available to all broadcasters. Likewise, the Australian Broadcasting Services Act limits the ability of subscription TV to acquire exclusive rights. And the U.S. Cable Television Consumer Protection and Competition Act of 1992 obligates the FCC to undertake corresponding investigations as a preliminary to possible regulation.

10. MAINTENANCE OF COMPETITION: PROTECTION AGAINST ABUSE OF MARKET POWER

Under the market model, broadcasting's ability to function presupposes a variety of broadcasters in competition with one another. But as can easily be evidenced with economic analyses of the media, there are considerable incentives in market-oriented broadcasting in favor of concentration. Media concentration is therefore one of the greatest problems associated with private broadcasting. Italy may serve as an example of the way in which concentration processes can operate in the absence of government regulation: The successful Fininvest industrial group (headed by Berlusconi) was able within a short period of time to gain influence over all of Italy's national television networks. The group is also strongly represented in the press sector, and this has given rise to cross-ownership problems (Barile & Rao, 1992). The Italian broadcasting law (1990) has largely accepted this situation and called for only marginal changes. Berlusconi used his media power unscrupulously in order to become Prime Minister in 1994—although it did not help him to stay in power for long.

Even though it is debatable whether a certain minimum amount of concentration might not in fact enhance efficiency in the broadcasting field, all broadcasting laws and supervisory authorities nevertheless go to great lengths to stem concentration. Typical instruments include precautionary measures designed to restrict both multiple and cross-ownership. The history of the efforts to combat concentration is, however, in all respects marked by the limited success these have had. In the first place, the anticoncentration rules are usually full of loopholes and exclude relevant fields, such as the area of program supply. Moreover, the rules also provide for a diverse number of exceptions, including the "grandfathering" of existing concentration. Most of them are, in addition, relatively unwieldy in cases of aversion, as when ownership is masked through figureheads or entangled, interlocked corporate structures, although there have been attempts to tackle the problem (e.g., by the detailed regulations in Australia). In many "control" instances, supervisory authorities lack suitable instruments for gaining a reliable overview of the respective degree of concentration and integration. The failures experienced in combating concentration are also evident from the fact that countries often react to rising concentration with statutory amendments, at times even relaxing requirements, or by making even greater use of the power to grant exceptions. The norms and their administration are adjusted to match economic reality— and not the other way around. At the same time, however, the will to regulate continues to be vigorous in several countries, and defensive action has at least been able to slow the rotation of the concentration spiral.

Supervisory bodies in the field of commercial broadcasting seem to be caught in a dilemma: They generally operate on the presumption that ultimately only the economically powerful will be successful (this applies at least to the cost-intensive television sector) and that, with regard to dissemination, an interlocked multimedia system may be able to promote efficiency in production. Accordingly, economic and journalistic efficiency often become licensing criteria. It is most certainly also significant that the supervisory authority will take the credit for an economically sound media industry.

It is no coincidence, for example, that the IBA encouraged publishers in the 1950s to invest in the television sector. The U.S. FCC urged economically and journalistically productive companies, such as that of Murdoch, to open a fourth network, and in the interest of global economic expansion, it has promoted cooperation among U.S. media companies at the international level that it would not have tolerated in the domestic market. The periodic relaxation in the FCC's ownership rules reflects the belief in the substantial "benefits resulting from group ownership, such as the proposition of program service diversity and the development of new broadcast services."[1] The German state media authorities have usually given preference in the awarding of television licenses to companies with strong participation in press undertakings. In other respects as well, they have defined the requirements going to productive efficiency in such a way that, in close license cases, media companies have had a practical advantage over firms foreign to the sector. In awarding the television licenses for the fifth and sixth channels, the French CNCL intentionally supported multimedia concentration, giving preference also to media companies with international operations.

Now that private broadcasting has developed into quite a profitable enterprise, some supervisory authorities have tried to do an about-face. The German state media authorities, for instance, began in 1992–1993 to study the process of concentration more closely and primarily to place limits on multiple ownership in those areas where influence over several other broadcasters is exercised indirectly, such as via shell corporations or figureheads. Strong attention is also given to program supply. Yet these authorities met the same experience as others before them: once it has become formidable, concentration can hardly be reversed. The history of broadcasting supervision in the United States, for instance, shows that over time, anticoncentration rules were always adjusted (i.e., relaxed) to accommodate the respective concentration developments. The area of media concentration is thus a consummate example of the law's limited ability to control when economic factors point in the opposite direction.

Over the course of the history of anticoncentration measures, at least two, contrary philosophies have fought for recognition. One viewed any kind of concentration as being a retreat from the ideal model of the greatest possible number of diverse, mutually independent, competing broadcasters. This was based on considerations stemming from the classic model of the marketplace of ideas. The other philosophy saw the emergence of economically and journalistically strong companies as being a guarantee for viable broadcasting. Under this view, it was sufficient when several (if only very few) such companies competed with one another and kept any one particular firm from gaining market dominance. However, these two doctrines, which are also common in general competition theory, were subjected to additional considerations. Cross-ownership restrictions, for example, have been justified not just with the objective of limiting economic power but also and primarily with the desire to facilitate diverse influence in a variety of media sectors. Entirely unrelated to notions of economic competition are strategies pursued in Germany whereby several competing broadcasters are made owners of one and the same broadcaster so that the latter can offer the most balanced programming possible. It is hoped that the internally pluralistic programming concept of public broadcasting can be approximated with such a broadcaster structure in private broadcasting without setting a duty to offer balanced, internally pluralistic programming.

In no other aspect of regulation have broadcasting supervisory authorities been so helpless and at the same time so unsuccessful as in combating concentration. Resourceful companies have not only found ever newer avenues for growth and integration; they were also able to hide their level of concentration relatively well from supervisory authorities. Primarily, however, supervisory authorities lacked clear concepts for combating concentration. In view of multimedia and multinational ties and the large, usually international, dimensions of markets, broadcasting supervision was usually too regional and thematically limited to have a significant impact. Compounding the problem was the fact that the media sector had branched out into a number of markets, such as production, acquisition, and retransmission rights, advertising as well as the actual broadcasting of programming. It proved to be a particular handicap that broadcasting supervision was largely restricted to the market for the broadcasting of programming itself. For example, it was the case in the United States for decades that the networks were essentially not covered by the supervisory regime, which targeted licensed broadcasters. Although the FCC eventually did enact special network rules, these were largely abolished in the early 1990s.

In all countries studied here, it can be seen that economic developments in production and marketing fields have had a massive effect on the type of programming broadcast and the manner in which this is accomplished. Since broadcasting supervision is unable to regulate conduct in these markets (or at best, can do so only indirectly or in peripheral areas), it simply cannot reach far enough to have any noticeable regulatory impact. Since production and marketing are moreover not limited just to use by broadcasting but extend also to other media (such as cinema, video, and recorded music) that are not regulated or are not monitored by the broadcasting supervisory authorities, relevant areas remain unsupervised.

Anyone who still presumes the market to be the best regulator will undoubtedly be able to shrug his or her shoulders at this development. However, broadcasting laws continue to be marked by the notion that governmental regulation is necessary, particularly since the economic market can lead to the amassing of journalistic power, which could threaten journalistic diversity and interfere with recipients' freedom of communication. All the same, the possibilities for counteracting this trend via broadcasting supervision are clearly limited.

Anticoncentration measures aim at restricting power and thus at ensuring competition. As has been repeatedly mentioned, the general competition law is available for the purpose as well. But all broadcasting orders have concluded that additional, broadcasting-specific precautions are necessary.

The traditional licensing of broadcasters is related to access to a medium of transmission, particularly terrestrial transmission. Licenses can of course also be awarded for the use of new technologies, such as cable or satellite. In most countries awarded licenses are limited to specific transmission media. If no special program-related license is necessary for transmission via cable—as is the case in the United States—the limitation of the license to terrestrial broadcasting does not prevent transmission via cable. In some instances, broadcasters with a license for only one transmission medium also have the privilege of using other media. This is the case, for example, with the German media laws for the cable transmission of certain types of programming.

When a transmission medium is privately owned—as with cable in a number of countries—access is made available to this medium not through a governmental license but in other ways, as with a contractual agreement with the cable operator. At most, relevant laws set forth minimum conditions to prevent the abuse of market power.

Even when a natural monopoly is indiscernible for cable or other new technologies, monopolies or monopoly-like situations are often provided for certain means of transmission. As a result, broadcasting-specific rules may be needed to prevent the abuse of market power. One example is the U.S. Cable Television Consumer Protection and Competition Act. It regulates, broadcasters' access to cable through, on the one hand, must-carry obligations on the

part of cable operators and, on the other, rules on contractual relations (including fees) between broadcasters and cable operators.

On another level are rules regarding the awarding of franchises to cable
operators. These are designed, for instance, to protect the cable operator
against excessively high payments to local franchising authorities or to keep
open the possibilities for competitive franchises.

11. VIABILITY OF THE DIFFERENT MEDIA SECTORS

The problem of media concentration is not limited to the broadcasting
sector. As the rules on cross-ownership especially demonstrate, various sectors
of the media and consequently the media markets themselves are affected by
concentration. The more differentiated the media markets become, the more
attention must be paid to the dynamics of the different markets and also to the
fact that the same companies or conglomerates are often active in different
markets.

The broadcasting system's ability to function—particularly with regard
to the principle of competition—can be affected by events in other media
markets. However developments in the field of broadcasting are equally likely
to affect other areas of the media, as well. For example the decision to introduce
advertising-based broadcasting has always evoked fear of negative repercussions upon the financing of the press. Measures introduced to ease the access
of publishing houses to broadcasting were sought to help protect the viability
of the press by giving it access to earnings from broadcasting; it was not simply
a matter of making use of the specific know-how of those working in the press
for the benefit of those in broadcasting. The cinematic field is particularly
endangered by television. Its ability to survive is also threatened by the
increasing popularity of videos. For this reason some broadcasting laws (e.g.,
article 7 of the EC television directive) acknowledge temporal protection
zones for cinema films, in which selling of videos and screening on television
is forbidden.

A further field of regulation consists of measures introduced to protect
the national audiovisual industry, particularly the independent producers (see
Section 12). Those bodies authorized to supervise the broadcasting industry
are not authorized to monitor the entire media sector. For most media markets,
there are no specific regulations whatsoever or even supervisory bodies (e.g.,
for press, film, and production). In any case the general rules of the legal system
apply to them, for example, the antitrust laws or the norms for warding off
unfair competition. Broadcasting supervision covers these branches of the
media only indirectly, namely, insofar as the effects of these markets upon the

markets in the broadcasting sector can be taken as a starting point for regulation (e.g., cross-ownership rules). Thus, there is a notable lack of symmetry in the regulation of the various media markets.

12. STRENGTHENING INDEPENDENT PRODUCERS

In several European countries, special quota rules have been created with the goal of protecting cultural and political identity. These apply, on the one hand, to shares of programming stemming from, for example, European, national, or regional producers (see Section 13). In some cases, quotas are also used to give preference to programs from independent producers in order to help them develop and to give them a chance to survive in an ever more concentrated field.

In contrast to those in the United States, broadcasters in Western Europe were initially also the producers of most of their programs. Public broadcasters (one example is the BBC) and in early days private broadcasters as well (e.g., the ITV companies in Great Britain) had widely built up their own production facilities. On the other hand, U.S. broadcasters, and in particular the networks, relied heavily on outside production companies and only rarely produced their own programs (such as news broadcasts). The FCC bolstered this structure for the broadcasting industry with specific network rules, although these were relaxed considerably over the course of time. During the 1980s a similar production structure was gradually adopted in Western Europe as well.

These changes were supported by attempts observed in other industry sectors to create "lean production" linked with "lean management." Many broadcasters started to realize that they could produce at less cost and, above all, were more flexible when they commissioned other producers or purchased programming elsewhere. The industry tried to increase both mobility and flexibility of employment practices and conditions to make use of freelance and contracted staff. They especially made growing use of externally supplied contractual service. Many broadcasters separated the management of their broadcasting, program production, and resource activities to ensure appropriate managerial focus and function more cost effectively in each activity. This structural change also affected public broadcasting (for information on the BBC, see BBC, 1992: 51ff.). Commercial broadcasters (and, of those, in particular the newly formed broadcasters) were able in this way to keep start-up investments low and spread economic risks to other companies. Production companies were, however, sometimes connected with the broadcasters, mainly through common parent companies but also through direct capital investment. The trend toward the use of external producers was often reinforced with normative requirements, especially in Great Britain. The object of these rules chiefly was to create purchase guarantees for so-called independent producers and in this manner to develop a production infrastructure that was self-reliant.

The shifting of production out of the broadcasting area—a move suggested by the economic advantages involved and only reinforced by broadcasting supervision—has led to considerable structural changes in the media industry in Western Europe. Broadcasters gradually have become pure publishers. This also has had an impact on the effectiveness of broadcasting supervision. When broadcasters are not integrated into the production process, their ability to have an effect on program producers is either nonexistent or very limited. But such integration does not exist if ready productions are bought on the market and is only possible to a limited extent with coproductions. Although supervisory authorities can continue to hold broadcasters legally responsible for programming, the assumed responsibility comes close to being a fiction. This necessarily leads to a loss of effectiveness in supervision when supervisory measures are still based on broadcaster duties and sanctions can be addressed only to broadcasters.

13. STRENGTHENING NATIONAL AND REGIONAL PRODUCTION RESOURCES

Many broadcasting laws and supervisory measures seek to ensure that as many broadcasting programs as possible are produced within the country. The United States does not have such rules: The American broadcasting industry is the worldwide leader, and at the same time, foreign productions are not popular with American audiences. This is also evident from experiences in the area of public television, where the British-produced programming aired by PBS has nowhere near the viewer ratings that it does in its home country. The production of nearly all television programs in the United States, or at least that by American companies, is, moreover, not ensured by any legal rules. The virtually unilingual American market is also so large that costs for U.S. productions are usually covered through domestic distribution alone.

These advantages are not open to small countries. Because of its multiplicity of languages and diverse national traditions, e̶v̶ Western Europe fails to possess a corresponding "edge" in the market. The same is true for other English-speaking countries such as Canada and Australia, which consequently see themselves forced to adopt special protective measures. They must witness how domestic broadcasters acquire a large amount of American programming. Even when these aren't as successful as domestic productions—and this is clearly evident in Australia—they usually promise a better cost–revenue relationship on account of their lower purchase price. Since American productions usually recoup costs in the U.S. market while also being sold abroad, they are considerably less expensive than original productions.

The regulatory steps undertaken on account of this situation often pursue different purposes. Operating on the assumptions that foreign programs cannot respond to the cultural aspects unique to the territory of reception and that there tends to be a high amount of internationally oriented commercial programming, precautionary measures in this area aim to maintain a nation's or region's cultural distinguishing characteristics. At the same time, however, they also seek to preserve jobs in the domestic production sector—and presumably also the tax-paying power of national firms. This makes sense not only from an economic standpoint but also for the purpose of ensuring cultural identity, since know-how and other aspects of the production infrastructure that a country needs in order to produce its own, culturally distinct programming would otherwise be lacking.

The experiences with quota rules are ambivalent. While Canada has been unable to prevent an "Americanization" of programming due to broadcasts originating in the United States, it has succeeded in bolstering domestic production, which has since become so proficient that even U.S. broadcasters often have their programs produced in Canada (Kleinsteuber & Wiesner, 1988). The relevant quota rules in Australia have been deemed relatively successful by most Australian observers, including taking into account the development of a domestic production infrastructure.

Enforcing quota rules by way of supervision is difficult, particularly since the terms employed are not easily operationalized, and there are a number of ways to circumvent them. It is no coincidence that the EC's television directive deals extensively with definitions, and this despite the fact that the quota obligations it contains are formulated rather weakly.

The Canadian and Australian examples demonstrate that quotas are most likely to achieve the desired results when they are coupled with other measures that create positive incentives for compliance. In this regard, Canada is attempting to augment quotas with subsidies for desired programs or otherwise to take into account broadcasters' economic calculations. In particular, an effort has been made to make use of cross-subsidization measures. Australia has introduced tax breaks.

14. PROTECTION OF JUVENILES AND FOSTERING OF EDUCATIONAL PROGRAMMING FOR CHILDREN

Children and juveniles are also television consumers, and here one can observe, particularly in problem groups, intensive television consumption among the so-called "TV addicts." There is broad consensus that television's likely socialization effects warrant special precautions to ward off dangers for children and juveniles. The corresponding measures enacted can be split into two groups: Negative juvenile protection is mainly undertaken by way of precautions against depictions of violence, pornography, or other morally offensive programs. But also conceivable is positive action to protect juveniles, such as through obligations to broadcast child- and juvenile-oriented or other educational programming.

The area of negative juvenile protection is usually comprised of two types of precautions. On the one hand, certain program content is absolutely prohibited—with the aim of protecting adults as well. This may include broadcasts that incite to racial hatred, that glorify war or minimize its horrors, or that fall under the heading of pornography. On the other hand, there are rules that, while not completely forbidding the category of programs, seek to ensure with special precautions that children and juveniles are largely unable to view them. The main problem here is violence in television and averting possible dangers for children and juveniles. A well-trodden path to this end is the so-called family viewing policy, which stipulates that programs broadcast during prime time be kept free of excessive violence and that they accordingly be suitable for the "entire family."

Such rules are sometimes contained in statutes (as in Germany), sometimes in the codes of supervisory authorities (as in Great Britain), and sometimes in the regulations of the broadcasting industry itself (as was earlier the case in the United States in the form of the NAB code). The enforcement of these rules is difficult, since here as well the terms employed are not easily operationalized, and moreover there is a well-founded risk that they might be abused to suppress content that is for some reason undesirable. Legal systems that grant freedom of communication virtually absolute protection (as in the United States) are more moderate here than those in which the ability to restrict freedom of communication in the interest of other values is taken for granted (as in Great Britain and Germany).

The enforcement of corresponding programming duties has met with vigorous opposition, particularly in commercial broadcasting. It seems that programs steeped in violence are more appealing to viewers than other broadcasts; especially advertising-financed programming, which is dependent on high viewer ratings, makes particularly intensive use of depictions of violence as a dramaturgic tool. The more heated the battle for ratings, the smaller the chances for voluntary compliance with the corresponding duties (cf. Hoffmann-Riem, 1981b: 168f.; Barnett, 1988: 117). Whereas larger, mass-oriented commercial broadcasters, which target all segments of the population, tend to exercise a certain degree of restraint (at least with morally offensive broadcasts), some cable broadcasters have virtually specialized in filling a self-created gap in the market with "adult channels" and "cable porn."

Operating under the somewhat fictitious assumption that children and juveniles do not have access to such programs or that parents can prevent this access, the supervisory authorities tend to practice restraint here. At the same time, however, one can observe incremental differences in tolerance; the most consistent supervisory bodies in the area of juvenile protection seem to be those in Great Britain and Germany.

Positive juvenile protection is justified with the concern of taking the often extensive television consumption by children—which then drops off somewhat among juveniles—as the occasion for offering them educational programs tailored to their age group and using television to achieve "prosocial effects" (cf. National Institute of Mental Health, 1982). Notable programs of this variety have especially been produced by the public broadcasting system. Efforts to encourage private commercial broadcasters to televise such programs, however, have met with only limited success. The relatively disappointing results achieved by the FCC may serve as a paradigm for the difficulties in creating and then enforcing duties to broadcast programs suitable for children. Weakly structured from a legal standpoint, the duty to televise special children's programming was observed by the networks only at the outset; over the course of time, these programs came to be shifted to useless time slots and were then dropped altogether. Special children's broadcasts are of limited interest to the advertising industry; it is more attractive (and lucrative) to gain children as the recipients of advertising messages by way of programs that also appeal to adults. Certainly children and juveniles can be of interest to specific sections of the advertising industry, in particular to the toy and confectionery industries. These advertisers are, however, primarily interested in using the program content to enhance the effect of the advertisement; they are not interested in utilizing children's programs for educational purposes as well.

U.S. networks often fulfilled their duty to broadcast special children's programs by declaring shows originally produced for adults to be henceforth considered children's broadcasts. However, a notable number of special offerings for children have come about in the area of cable television. This was not the product of supervisory action but, rather, of economic and journalistic considerations: Cable networks used children's programs as an advertising tool for gaining cable subscribers. This niche left open by the terrestrial networks was intentionally filled by cable television. But there are also clear indications that children's programs are also curtailed in cable television when broadcasting time is needed for other programs (see Le Duc, 1987: 65).

In Germany there was at one time an apparently contrary trend to what was taking place in the United States. With the exception of SAT 1, terrestrial television broadcasters originally included a remarkable variety of children's broadcasts in their programming (Schmidt, 1990). They were apparently interested not simply in making a good impression both on the public and the supervisory bodies but also in steering the viewing habits of children and juveniles toward commercial programming at an early stage in order to expand future viewer potential. It has furthermore become evident that children and juveniles possess significant purchasing power, and a corresponding advertising market exists for certain products. There is, however, no evidence that these juvenile-oriented programs met the educational criterion of achieving prosocial effects. Over time, commercial television broadcasters in Germany began to devote less and less attention to children's programming. Nevertheless, they continue to target young viewers, since juveniles and young adults apparently are the most receptive to advertising. Programs aimed at juveniles also garner sufficiently high viewer ratings with adults. Such programs have created problems, particularly when they are marked by depictions of violence.

15. MAINTENANCE OF STANDARDS IN MATTERS OF VIOLENCE, SEX, TASTE, AND DECENCY

In the above description of the measures regarding negative juvenile protection, we made reference to measures for maintaining standards in matters of violence and sex. Protection of adults may in some respects be sought as well. But with regard to regulation of viewing for adults, broadcasting orders practice great restraint in regulating such matters. For instance, apart from the above-mentioned prohibitions on certain types of broadcast content, most broadcasting orders do not recognize any additional measures for the protection of adult viewers (e.g., Germany, the United States). In contrast, the protection of such values in Great Britain has been increased with the creation of the Broadcasting Standards Council and its normative reinforcement in the Broadcasting Act of 1990.

The British Conservative government was particularly interested in going beyond the policy practiced by the IBA and the BBC (Home Office, 1988). However, the specifications that followed in the Broadcasting Standard Council's code of practice (in particular, the exemplary explanations) demonstrate how difficult it is to operationalize the standards in detail. Supervision must be conducted with great sensitivity if it hopes to avoid risking the suppression of opinion. Particularly in issues of taste and decency, one can observe rapid swings in a society's values, but it cannot be the task of a supervisory authority to thwart or otherwise steer social change from a moral standpoint. One can, in addition, observe that political minorities tend to go to the limits of what a society tolerates—not least with the goal of attracting attention otherwise denied to them in the media order. Difficult questions arise here with respect to the delineation between the legitimate goal of having a forum for a concern and the equally legitimate goal of upholding minimum standards for all.

A special role in the discussion about violence has been played by so-called reality TV. Broadcasts that show actual accidents, catastrophes, or crimes, mostly using footage, achieve a special dramatic effect by offering a realistic depiction of uncommon events, including violence. Since this cannot be classified as fiction, broadcasters have sought to justify their conduct by citing their duty to offer information, and they have claimed to have a specially protected basic right: The depiction of true events cannot be prohibited by the state. Although this may seem hypocritical to some, it does point out a conflict: Measures to limit depictions of violence in television easily come into conflict with freedom of communication and broadcasting's duty to provide information. Even when only the "excessive" use of violence is forbidden and the prohibition does not apply to informational programming whose depiction of violence is within the scope of the public interest, the delineation still remains difficult. As long as broadcasters feel that the depiction of violence is profitable (i.e., ensures high ratings in the battle for viewers), they will always find new ways to circumvent corresponding restrictions. Moreover, broadcasting supervision usually does not go far enough, particularly when it tries to avoid interfering with the content of communication in the interest of the communicator's freedom to communicate. If this boundary is not respected, there is a risk of suppressing or even abusing freedom of communication.

16. PERSONAL INTEGRITY

It has long been known that the various media, through their subject matter, can threaten personal interests. Due to the speed of the medium and its wide reach as well as the considerable value of its pictures and other documents as "evidence," broadcasting, particularly television, has at its disposal a broad range of possibilities for consistently violating individual rights, at times even worldwide. General laws—for example, criminal law or civil damages law—serve to protect one's reputation and good name, the right to one's own image, privacy, and private life. In most broadcasting regulations there are supplementary protective norms that deal specifically with the media.

One typical protective measure, albeit very limited in its scope, is the right of reply to which everyone is entitled if legitimate integrity interests have been damaged by an assertion of incorrect facts in a program. However, this tool does not protect against value judgments and most certainly not against violations of privacy. Nor is it applicable in the case of inaccuracies and falsehoods that do not violate individual rights. For the most part, the courts are entrusted with supervising the right of reply, not the broadcasting supervisory bodies. As far as these also share responsibility for ensuring that broadcasters comply with the general rules, they may also take additional action. In this respect, they usually wait until a complaint has been made by the injured party before they take action. For exceptional cases—one example would be the Broadcasting Complaints Commission in Great Britain—a special complaints authority is set up, outside of the control of the general broadcasting authorities. This body can deal specifically with asserted violations of rights of third parties, particularly rights of privacy.

17. RESPONSIBLE ADVERTISING

Advertising forms the most important basis of financing for private broadcasters; a number of public broadcasters also finance their operations with advertising revenues, although to a subordinate extent. It is feared that financing through advertising may have an impact on programming. Accordingly, quantitative rules on advertising time seek to ensure that commercials do not comprise an excessive amount of programming. In order to assure that the reception of programming is as undisturbed as possible and to stem the indirect influence of advertising on programming (e.g., on dramaturgy and content), commercial interruptions are often prohibited or restricted, and commercial spots may be required to be televised in blocks. Moreover, some broadcasting orders require that advertising be identified as such so that the recipient is able to recognize it clearly. Broadly worded prohibitions may expressly ban advertising from having a direct influence on programming. In order to protect program sources from extraneous, especially commercial, influences, such precautions are usually also adopted for programs supported by a sponsor, although they are structured differently.

In the United States, however, rules restricting advertising were quite limited from the outset and have since been nearly completely abolished; similarly, Canada and Australia rely mainly, but not entirely, on industry self-regulation. But in the EC's television directive and in all Western European broadcasting orders, advertising is subject to a variety of restrictions. Since most rules run counter to the interests of the advertising industry and thus to those of commercial broadcasters, violations are frequent. No other field is so often the target of informal and formal action by European supervisory authorities. Nevertheless, it has become apparent that supervisory authorities have difficulty operationalizing the rules.

It is also evident that the advertising industry's ingenuity in developing new forms of advertisement is considerably greater than that of the legislature or supervisory body in regulating these. Correspondingly, supervisory authorities are relatively helpless in the face of ever newer forms of product placement, concealed advertising, sponsoring, bartering, and other methods of combining programming with advertising. Sometimes it even seems the advertising industry is merely interested in testing the extent of the supervisory authority's tolerance. For instance, new forms of advertising are often devised just after protests have been raised against a given method. At the same time, the limits of tolerated advertising are in constant flux. This is compounded by the internationalization of the media landscape. If, for example, one country imports programs from another, the commercials can certainly be edited out—but not the influence they have had on the dramaturgy, or, above all, the subliminal comingling of program and advertising interests.

Most easily enforced are rules of a formal nature, such as time restrictions, the requirement of commercial blocks, interruption limitations. Nevertheless, dysfunctional side effects cannot be ruled out entirely. For example, supervisory bodies are largely powerless when broadcasters try to circumvent the interruption prohibition by artificially segmenting programs into several broadcasts or when they intentionally create "natural breaks," insofar as norms permit advertisements during these. Moreover, special advertising restrictions, as for children's broadcasts, may also operate dysfunctionally when they result in the wide omission of children's programs.

The pressure of the advertising industry's innovative practices and the awareness of some dysfunctional effects have meant that the practical administration of advertising rules has grown increasingly weak, with an accompanying reduction in normative commitments as well. Teleshopping has made these problems even more apparent. The classification of teleshopping as advertising—as under the EC's television directive and the Council of Europe's convention—has led to a relaxation of restrictions on advertising time. There is no effective method of counteracting the intensive, substantive interweaving of advertisement and entertainment in teleshopping broadcasts, in which the viewers' entertainment needs are exploited for in part subliminal, in part obvious, but by all means very effective advertising ends. Precautions for consumer protection are difficult to administer.

Other advertising restrictions seek to limit certain content. Widespread rules include those to restrict or prohibit advertisements for pharmaceutical, alcohol, and tobacco products. It is relatively simple to monitor these rules, and infractions are rare. Furthermore there are rules about political advertising, especially in relation to election campaigning (see Section 2).

18. PROTECTION OF THE INTEGRITY OF WORKS OF ART

The prohibition of commercial interruption, addressed in the preceding section, attains a special dimension in the Italian discussion of broadcasting. Renowned directors and other filmmakers have rallied broad public support behind their assertion that the interruption of cinematic films with commercials impairs them as works of art. Because regulations for private broadcasting were lacking for many years in Italy, commercial interests have been able to prevail massively, particularly in the area of television advertising. It was therefore all the more surprising that a certain amount of political support was able to be garnered for the filmmakers' desire to reduce commercial interruptions during cinematic films. The Italian Broadcasting Act of 1990 has taken this concern into account to a limited extent. The EC's television directive also sets forth special limits on the interruption of feature and made-for-television films; these do not apply to series, serials, light entertainment programs, or documentaries. The latter are governed by the less restrictive general rules on interruption.

19. CONSUMER PROTECTION

The foregoing review has attempted to demonstrate first, that the areas of broadcasting supervision are still quite diverse and second, that supervision is also linked to program content. However, the worldwide trend toward de- and reregulation has also brought about a reduction in the requirements to which broadcasters are subject and in the tools of the supervisory authorities. At the same time, however, new fields of regulation have been uncovered, such as the limitation of exclusive broadcast rights (see Section 9).

Further changes have been wrought by technological expansion and the accompanying development of new services, as well as by the growing reliance on market forces that can be witnessed everywhere. A typical example of this are characteristic aspects of regulation in the area of cable broadcasting. The more government refrains from the regulation of content, the more new regulatory problems crop up. For instance, in the United States, the financial barriers to cable access piqued regulatory interest. As is illustrated by the U.S. Cable Television Consumer Protection and Competition Act, the focus was fixed on the contractual relationships between the cable operators and the broadcasters or program providers. In addition, the government felt compelled to regulate the relationship between cable operators and consumers. These are rules not so much to protect the viability of the broadcasting order but, rather, for consumer protection. It is no coincidence that the U.S. act contains this concern in its very title. Consumer protection relates, among other fields, to consumer electronic equipment compatibility, but it mainly applies to rates. The level of the rates, including their gradation when particular program packets are subscribed to, may be a central steering element with regard to the type of available programming. However it cannot be ruled out that rate regulations have less to do with such repercussions for program content than with general considerations of fair pricing.

20. SKIMMING THE ECONOMIC VALUE OF TRANSMISSION PRIVILEGES

Broadcasters as well as cable operators generally enjoy a privilege on account of governmental regulation, mainly through licensing. They may use this privilege for profit-making activities. As far as governmental regulation protects against competition, however, the possibilities for profit-making are not restrained by market forces. It follows from the basic idea behind broadcasting regulation that some of these privilege-based profits should be "skimmed off" for the benefit of the general public (see Chapter 7, Section 2). Broadcasting's traditional public service obligations are a kind of noncommercial counterperformance in exchange for an economically valuable commodity. However, economic counterperformance is also conceivable. The greater the reduction in public service obligations, the more justifiable it is to demand payment in return for the license.

For instance, American communities have long sought to tie the granting of a cable franchise to payment commitments on the part of cable operators. German states have attempted to couple the awarding of a terrestrial broadcasting frequency to pledges that production capacities will be expanded in the state. Australia and Canada, as well as a number of European countries and the EC itself, have used their regulatory competence as a means of industrial policy, seeking to help in the development of independent production infrastructures—even though they have done this in the guise of objectives of cultural policy (production quotas). Obligations to maintain quotas for productions of specific origin are a form of capturing the economic benefits of broadcasting.

The universal curtailment of the public service obligations placed on broadcasters in many countries has led to the charging of licensees with fees, either on a one-time or recurring basis. The more the market principle is used in broadcasting regulation, the fairer it seems to determine the amount of compensation with reference to the market, for example, through auctioning or special taxation (e.g., by a percentage of the advertising and pay-TV revenues levied on all broadcasters or cable and satellite companies). This is one way of being able to finance those kinds of programs that the market does not adequately finance but that are regarded as indispensable in terms of public interest.

The corresponding Australian and British auctioning model will likely continue to find further imitators. Since auctioning means the conscious choice of a market-oriented criterion for licensing, it primarily creates the opportunity for generating revenues. This permits the government to use broadcasting to relieve the strain on the national budget. But this can also lead to a change in the accent of broadcasting regulation: The state may succumb to the temptation of viewing the revenue-producing power of broadcasting as a primary goal of its regulation. Broadcasting policy then becomes fiscal policy. There are, of course, alternatives, such as using receipts to finance broadcasts not or only insufficiently addressed by the market mechanism (e.g., programs for minorities or educational broadcasts).[2] It must, however, be noted that governments generally prefer to acquire revenues for the general budget.

When the supervisory authority is involved with the auctioning process or is otherwise charged with determining the amount of the license fee, its functions take on a new—a fiscal—dimension. However, broadcasting laws usually couple this aspect with the duties of monitoring certain quality requirements and of combining these during the licensing process with financial aspects. This has been accomplished quite clearly under the British Broadcasting Act, which emphasizes the importance of quality requirements; under the Australian Broadcasting Services Act, in contrast, the financial aspect has been accorded greater weight. In auctioning the licenses for Channel 3 in Great Britain, the ITC considered the viability of broadcasters to be more important than the securing of large revenues. In so doing, it clearly acted in the tradition of the IBA.

NOTES

1. Report and Order, In re Revision of Radio Rules and Policies, 7 F.C.C.Rcd. 2755 (1992).

2. Compare the models dicussed by Noam (1994b) in relation to telecommunication.

CHAPTER 11

Patterns of Practical Supervisory Action

1. SOFT REGULATION AND INFORMAL COOPERATION

Not only in the area of broadcasting supervision but also in the area of economic regulation generally has practice shown that supervisory bodies by no means restrict themselves to the formal instruments provided in norms but also seek to supplement or even replace these by resorting to a number of further measures and special supervisory strategies. The broadcasting authorities are particularly interested in "soft" regulation, as they fear that regulation in the sensitive field of freedom of speech could be interpreted as unconstitutional abridgement. The usefulness of informal networks of communication and cooperation is quite obvious in the media sector in view of the sensitivity of the material and, in particular, the broadcasters' desire for independence, which is legitimated through basic constitutional rights as well as in light of the usually quite vaguely formulated description of the regulatory task. As a result, efforts are made to ensure a maximum level of flexibility and to act in cooperation with the companies concerned as much as possible.

For instance, a certain caution is demonstrated when it comes to composing rules. The FCC redefined its older, content-related rules as mere unofficial guidelines to separate the unquestionably acceptable from the troublesome. As long as the broadcasters adhered to these guidelines, the staff alone had the power to decide (for instance, on a renewal application). In case of noncompliance, the commission itself had to decide, but it was not bound by the guidelines.

It is particularly evident in the field of day-to-day activities that supervisory bodies make an effort to avoid employing their formal sanctioning instruments, and if so, only as a last resort. They prefer informal paths for resolving conflicts (Guillou & Padioleau, 1988; Krasnow et al., 1982; Hoffmann-Riem, 1989b: 239ff.). The tools employed range from the "policy of raised eyebrows" and informing the public to threatening formal supervisory action. The supervisory authority usually seeks to deal with occasions for supervisory action and other problems in informal coordination and cooperation with those supervised, thereby hoping to keep irritation as well as the risk of subsequent court battles to a minimum. Networks of informal and formal communication have been developed for this purpose. These can be seen in, for instance, special commissions (as in Great Britain), as well as in periodic meetings or the use of informal avenues of contact. Supervisory reality is not characterized mainly by orders and license revocation but typified as well by telephone requests and meaningful remarks during cocktail parties, and so forth.

Cooperative solutions to problems not only avoid the expense and risks associated with obtaining legal relief but also, substantively speaking, may be deemed advantageous by both sides. In the context of constructive cooperation, the supervisory authority receives certain information that it would not, or only with difficulty, obtain when acting in a sovereign capacity. It can react flexibly to the desires or problems of those supervised, and it can agree to tolerate a violation for a short period of time in exchange for a promise to remedy this situation voluntarily. Through intensive informal communication, those supervised are given the opportunity to present their version of the problem and influence the way in which the supervisory authority views it and thereby to mitigate sanctions otherwise likely.

Cooperation becomes problematic when the supervisory authority in this manner enters into close camaraderie with those it is to supervise or when it decides not to employ the pertinent supervisory instruments. The legal authority to supervise thereby turns into bartering power in working out the substance of—or refraining from the use of—conceivable supervisory action. One thus finds in this area as well evidence of what is termed "agency capture" in the field of economic regulation (see generally Bernstein, 1955; Peltzmann, 1976).

Cooperative relationships may also be cemented on a personal level. For example, a continuous interchange of personnel between the broadcasting industry and the supervisory authority is nothing unusual for FCC employees, the situation often being justified as merely the exploitation of professional knowledge. This "revolving door" in personnel policy also ensures an uninhibited transfer of the respective approaches to problems. Although in European countries such a long tradition of mutual exchange has not yet developed, the revolving door is nevertheless beginning to swing. This is accompanied in Germany by a close exchange between supervisory authority and ministerial bureaucracy or the political machinery in general; for example, the directors of many state media authorities are recruited from the chancellery offices of the state minister–presidents, where they had previously taken part in the drafting of the corresponding broadcasting laws.

2. PRIORITY FOR LIMITED SELF-REGULATION

The difficulties associated with broadcasting supervision, as well as the respect for entrepreneurial autonomy, have led to a situation where supervisory authorities often seek to leave much to the self-responsibility of broadcasters. For instance, German state media authorities initially entrusted competing license applicants with finding a way to allocate the scarce transmission frequencies, that is, to reach agreement among themselves as far as possible. In so doing, the supervisory body functioned as moderator among the parties to this agreement.

Another example of the respect given to self-administration was the FCC's effort to depart from its own programming codes, such as in the case of the limitation of depictions of violence, and encourage the broadcasting industry's interest representative—the National Association of Broadcasters— to enact a code. The FCC used the NAB code for many years as an orientational aid in specifying the statutory obligations of broadcasters. Only as a consequence of antitrust objections was it necessary to rescind this code. In Australia as well, self-regulation has been urged by the ABT and the ABA. It leaves regulation to the broadcasting industry, although making clear that the authority will take action if certain minimum elements of regulation are not upheld (a kind of self-regulation with a state safety net). Self-regulation is also used in Australia in the form of prior control, notably regarding classification in connection with protection of children and young people. Voluntary rating and classification schemes are sometimes adopted by trade organizations in order to stop legislative actions in this field (e.g., in the United States by the motion picture association voluntary rating scheme for violent motion pictures).

3. IDENTITY OF INTERESTS WITH REGARD TO THE SUCCESS AND PROTECTED STATUS OF BROADCASTERS

As time passes, supervisory authorities increasingly tend to align themselves with the broadcasting industry's interests in the success of their operations. They also contribute to maintaining the status quo—or deviate from it only gradually. As a result, once broadcasters have been licensed, their interest in a renewal of the license is usually satisfied. However, Great Britain has in this respect often—most recently, in 1991—experienced considerable surprises: Licenses were not renewed on a number of occasions. Although the claim to license renewal normally in place affords protection against surprises—no one can ensure that a different broadcaster will provide better programming—it also confines the latitude of supervisory action and may reduce tactical options. The prospect of certain license renewal as well as the awarding of long-term licenses can deflate the effectiveness of other supervisory instruments.

If a broadcaster is successful in getting its programming firmly implanted among its recipients, this also opens special possibilities for initiating political pressure in order to bolster its interest in protected status. Worthy of mention here is the experience of the French Haute Autorité in its efforts to prohibit radio broadcasters from unlawfully increasing their transmission strength. The affected radio stations were able to mobilize their listeners and organize a public protest, which ended in a demonstration of some 100,000 listeners, mostly teen-agers, and ultimately led the Haute Autorité to back down. Since television broadcasters, in their guise as employers as well as vehicles of economic power, can also activate interests, they are often able to count on the support of lobbies to advocate for protected status.

Thus, it is in no way surprising that the supervisory bodies, when in doubt, play their part in supporting the incumbent rather than a competing applicant. Nor do they usually raise any basic objections when laws are passed that facilitate license renewal. The obvious strengthening of licensed broadcasters' long-term protection, both in legal and in real terms, is a side effect of the establishment of the broadcasting order. The supervisory bodies are often even involved in protecting the profitability of the broadcasters; examples of this are the U.S. and Canadian signal carriage rules to protect local companies from competitors in the cable system.

4. THE BROADCASTING INDUSTRY AND REGULATORS AS ALLIES

Burdensome broadcasting supervision generally meets with strong criticism from the broadcasting industry. It often objects to petty, bureaucratic controls, alleging that supervision has dysfunctional effects. It is, however, apparent that the industry also seems to favor regulation. In the United States, one could observe during the 1960s and 1970s that a large section of the broadcasting industry attached great importance to the maintenance of a regulatory framework because this apparently also conformed to its interest in status protection, especially with the prospect of cable television (Le Duc, 1973; Cole & Oettinger, 1978). The delayed introduction of subscription TV in Australia was also in the interests of most broadcasters and networks who wanted to be protected against a fragmentation of the market through further competition.

A phenomenon becomes evident here that also appears in other instances of economic regulation: Prosperous economic entities tend not to reject regulation as such but, rather, only that which runs counter to their own interests. In phases of expansion, successful companies usually push for the dismantling of barriers to expansion; during periods of consolidation, they gratefully rely on a regulatory framework as a buffer to campaigns by newcomers against their positions. For instance, the U.S. FCC for many years protected the networks against the expansionist wishes of cable television. The FCC abandoned its policy during periods in which traditional broadcasters likewise saw new opportunities for expansion and were looking for opportunities to invest their money.

The broadcasting industry can quickly become a proponent of regulation if it is a matter of fending off or diluting a different regulation with more powerful effects. This was the case with the U.S. cable legislation that was aimed at preempting the state and local authorities.

The broadcasting industry can also ally itself to the supervisory authorities in other ways. For example the growing influence of citizens upon decisions of the FCC about renewals of licenses was eventually considered by many broadcasters to be annoying. In many cases the broadcasting industry looked for the support of the FCC and often got it. The commission made skilled use of constant procedural delay to deny the citizens' groups the hearings they sought (Teeter & Le Duc, 1992: 377). The broadcasting industry's protection was also further increased by measures introduced later by the FCC explicitly to ease the process of license renewal.

However not only the citizen' groups but also other affected broadcasters make sure that the rules are upheld when it comes to granting licenses. Competitors particularly tend to pay attention to the observance of governmental regulations, as, for example, when a broadcaster transgresses the legal framework and attempts to increase its market share at the expense of established firms. Affected broadcasters have in some cases (e.g., in the United States) the option of court intervention. Squabbles in court, brought about perhaps by an applicant being rejected, also help to ensure that the rules are followed. In Germany and the United States, the courts have a considerable say in enforcing rules.

Although broadcasting regulation finds approval among those being regulated, their interests vary considerably. In terms of the future, it is quite unclear what effect the process of concentration that is taking place worldwide will have upon the broadcasting industry's support of regulation. Media companies that operate at the international level have ways and means of dodging supervisory actions by national supervisory bodies. Even at the national level they are less and less dependent upon the support of government supervisory authorities. Their continued existence is safeguarded in any case; instead regulation can prevent their desire to expand. Regulation as a means of protecting weaker companies will no doubt be resisted by successful companies.

5. INCREMENTALISM

Supervisory authorities are generally reluctant to undertake radical structural changes in the media order or abrupt breaks in their policy. If they occur at all, the changes implemented are gradual and incrementally structured. In the process, formal amendments in the rules are usually prepared during preliminary phases, in which the old rules are no longer applied in full. An example of this may be found in the way in which the fairness doctrine was treated, which the FCC had ceased to enforce before it was officially abandoned.

Far-reaching structural changes—such as the transition from broadcasting regulation to widespread deregulation—tend to be introduced only incrementally and after lengthy preparations. Regulators usually wait with changes until the political climate is favorable to them, giving broadcasters time to get adjusted to deregulation and to ensure the political support of the legislature and the executive branch, as well as wide sections of the general public. Deregulatory measures have consistently been implemented during periods in which they were called for by the broadcasting industry; in this way, the authorities ensured unanimity of interests with the majority of those regulated and, at the same time, could expect that most of the media would publicize their support.

Supervisory bodies try to keep transitional problems to a minimum with other types of changes in the rules. Provision is made for long transitional phases, as in the British Broadcasting Act of 1990. In addition, there are a number of ways to ensure that new regulations (e.g., in the area of anticoncentration measures) do not interfere with accumulated holdings (grandfathering). Furthermore, in the transition from illegal to legal conditions, supervisory authorities often tend to base their activities on the situation that has already developed—this can be seen particularly in France and Italy—that is, they eliminate an unlawful situation by making it subsequently legal. Maintenance of the status quo—and this more according to practice than legal norms—is also observed with regard to the enforcement of advertising rules. When it becomes evident that a norm is being systematically disobeyed and that the supervisory authority's power is insufficient to rectify the situation, then the strategy resorted to in many countries is to resolve the contradiction between the norm and reality by modifying the norm to match the reality. Examples include the statutory amendments in France entailed by the Haute Autorité's unsuccessful efforts to enforce integration restraints and advertising limitations or the relaxing of the advertising rules in Germany's Interstate Broadcasting Treaty of 1987 and 1991.

6. SAFEGUARDING THE SUPERVISORY AUTHORITY'S ORGANIZATIONAL INTERESTS

Cooperation and guaranteeing protected status fit in with a phenomenon well known to organizational sociology: Public authorities develop an interest in their own protected status, which is, depending on the situation, directed either at mere survival or at expansion—and this even when the original mandate is no longer applicable. Whereas norms may come and go, public authorities withstand nearly every blow. For instance, the FCC had little difficulty surviving the reregulation of the broadcasting order. Although the British IBA could not hope to survive in the face of the government's reform plans, it did meet with success in getting the ITC to take on as many of its employees as possible. The Australian ABA more or less took over the staff of the ABT.

The further deregulation progresses, the greater is the risk that the supervisory body will become superfluous. If, however, it manages to extend its field of duties to new areas of regulation, then its existence is not under threat in the same way. It cannot be ruled out that the positive attitude of the FCC toward the deregulation of broadcasting was influenced by the fact that the FCC is responsible for the entire telecommunications field. The emergence of new technologies and services in the wake of deregulation has opened up new areas of responsibility, for instance, in the field of cable and other up-and-coming technologies. Even as the industry to some extent pressed for federal rules in order to preempt state and community regulatory power, this also provided the FCC with new duties. The U.S. cable legislation of 1992 provides a considerable amount of material on the subject.

7. INCORPORATION OF BROADCASTING SUPERVISION IN THE POLITICAL ARENA

Supervisory authorities in all countries must take into account that they are locked into the respective political arenas of their countries. The French Haute Autorité and the CNCL got a taste of this when they were being dismantled. Of course, the political surroundings also have an effect on day-to-day work. Inroads for political access include not only personnel decisions—these are used with particular openness in Australia—but also isolated criticism or specific directives, and of course it is also possible to exert influence by amending legal norms or the form of organization. Subjecting ABA decisions to make rules of widespread effect subject to nullification by the Australian Parliament provides another example.

The attempts of the U.S. Congress to reintroduce the fairness doctrine despite its being annulled by the FCC illustrate as well that there are different actors involved. The veto of President Reagan against the reintroduction of the fairness doctrine focused on another actor in the political arena. In Australia, the Minister has retained several powers in the broadcasting era, thus subordinating the ABA under the government's policies.

The intensity of state influence is also characterized by the constitutional framework. For example, the German Constitutional Court acknowledges that broadcasting supervision must be free of government influence in all matters pertaining to broadcasters' programming. In Australia, on the other hand, there is only restricted constitutional protection of independence. In Great Britain, such legal protection is also lacking; however, the special characteristics of political culture in Britain create an equivalent. Independence is by and large respected despite the government's having legal possibilities to exert influence.

Various interest groups also have an effect on supervision. Their chances for success apparently depend on the political, legal, and social position of their representatives. In some countries—for example, in the United States and Australia—citizens' groups have expressly recognized participation rights in some supervisory procedures. This has strengthened them, especially in the United States, and has facilitated their formation. However in most countries, it is not yet possible for the public to assume a significant voice in the form of citizen participation. In Germany, the special, pluralistically composed supervisory organs of the state media authorities offer an official form of interest advocacy by "socially relevant groups."

The extent of the supervisory bodies' independence is directly related to the assessed importance of their activities. It is thus certainly no coincidence that in Germany, political parties have made a considerably greater effort to gain influence in public broadcasting organs than in the supervisory bodies for private broadcasting: As external supervisory bodies, the state media authorities are substantially weaker than the broadcasting councils of public broadcasters, which are a part of the broadcasters' organization and share responsibility for their programming.

8. INTERORGANIZATIONAL COOPERATION

Where different authorities are responsible for broadcasting, there is a need for mutual coordination. One example is the allocation of duties between the FCC and the FTC. The absence of any clear boundary between the FCC's authority over advertising and commercial broadcasting and the FTC's general authority over broadcasting led to an agreement between the agencies.

As a result of the federalist structure of Germany, there is a need for coordination among the different supervisory authorities. Their representatives meet regularly in conferences and joint committees. The supervisory bodies have agreed on coordinated standards for assessing similar situations, and they coordinate their activities, for instance, in the field of anticoncentration measures, advertising restrictions, and the protection of juveniles.

There is a need for similar coordination among the West European supervisory bodies in implementation of the EC television directive and the TV convention of the European Council. This need has not yet been met.

9. PROSPECT OF JUDICIAL RESTRAINT

Successful broadcasting regulation requires a multitude of value judgments, the balancing of interests, and predictions of future developments. Thus the legislative branch has immense scope in shaping the broadcasting order.

This also has an effect upon the supervisory bodies. They are mostly formed with the aim of functioning as a kind of expert committee. By appointing boards of individuals from diverse backgrounds, Germany also tries to achieve an additional organizational legitimacy for value judgments. As a result, the supervisory bodies maintain mostly an assessing and predicting prerogative that has a particular effect upon the judicial control of their actions. The courts respect their expert knowledge, and consequently they exercise judicial restraint. Just how far this goes depends upon the respective legal frameworks. In the United States, inspection is mainly limited to ascertaining whether an agency is acting arbitrarily or capriciously.[1] In Germany the courts demonstrate a certain leeway when passing judgment (OVG Lüneburg, DVBl, 1986: 1112).

The supervisory bodies, therefore, need not fear that their actions will be easily altered by the courts. In spite of this, the supervisory bodies have often lost in court, especially in Germany and the United States. Consequently and understandably, they take great care that the risks are kept as low as possible. As the broadcasters are obviously taking increasing advantage of legal protection, estimating the risk of a procedure is a particularly important deciding factor for the supervisory bodies.

10. GRANTING POLITICAL LEGITIMACY

The supervisory authorities or acting persons realize that their activities are of limited effect in the shaping of the broadcasting order. In private conversations, they usually admit this. The majority of actors seem to be convinced of the necessity of regulation and as a result experience frustration in their jobs. This can be seen quite clearly in the German state media authorities. A series of disappointments and annoyances were one of the motivating forces behind their attempt in 1992–1993 to take more decisive action, particularly against media concentration.

In view of the difficulties associated with effective broadcasting supervision, one may raise the question whether the main task of supervision lies more in the area of the political legitimizing of the present broadcasting system and less in the influencing of broadcasters' conduct. As early as 1950, M. Edelman, author of the well-known book *The Symbolic Uses of Politics* (Edelman, 1964), developed theories in this direction with regard to the U.S. FCC (Edelman, 1950). In more recent literature, emphasis has also been placed on the symbolic–ritualistic character inherent in some aspects of broadcasting supervision (Streeter, 1983; Hoffmann-Riem, 1981a). A look at the new Italian broadcasting act shows, for example, that it is not set up in such a way as to have a fundamental influence on the broadcasting order. Nevertheless, its enactment pretends to satisfy the corresponding demands of the Italian Constitutional Court and presumably the expectations of the general public as well.

When one looks at many European regulations, it is quite evident that legislatures have introduced instruments of broadcasting supervision that, in light of foreign, particularly U.S. experiences, could not be expected to be effective. But in view of, for instance, the heated controversies over media policy in Germany, it was both politically and legally impossible to introduce private broadcasting without wrapping it in a set of regulations. It may very well be that the German state legislatures were primarily interested in the significance of supervisory instruments for political legitimacy, seeking to placate the general public with the norms. It is not unusual for norms to be enacted or retained even though it is known that they have only limited effectiveness. Their political value may lie in the proof of political activity or in the deflating of public opposition.

However, such analytical approaches should not be incorrectly taken to mean that broadcasting norms and supervisory tools are necessarily ineffective and only justifiable under legitimacy aspects. At the same time, however, it is doubtful whether supervision in a system of economic and disseminational competition among broadcasters is capable of fully attaining the objectives set by the legislatures.

But even when this is not the case, the norms may still be of significance in ensuring legitimacy. To all appearances the supervisory authorities accept such a role. In any case, almost no fundamental criticism of how the job is done is openly expressed nor is used as a reason to discuss the need for a new orientation. Interest in the success of the broadcasting order they themselves helped form includes defending the broadcasting order against fundamental criticism.

Commercial broadcasting has firmly established itself in all countries analyzed here. Whereas in the past there were heated debates on media policy, these have largely abated. However, again and again the media is denounced— mostly by outsiders—as having failed or as posing a danger to society; occasionally, the role of the supervisory authorities is also criticized. Those critics who have the greatest effect upon the public—in the United States, notably Neil Postman (1986) and Noam Chomsky (1989; Chomsky & Herman, 1988)—aim their attacks directly at the media, not at the supervisory bodies. An increasing amount of criticism—voiced by conservatives whose supporters very often are the most vocal proponents of deregulation in general—is expressed on the depiction of violence and on other subject matter deemed liable to harm the development of young people. Criticism is also made (mainly by those people who are critical of cultural policy), about the reduction in quality in programs, often resulting in demands for more action on the part of the supervisory authorities. The legislatures also expect an increased level of activity from the supervisory authorities, especially in the areas of fairness as well as protection of children and young people. Generally, the supervisory authorities are assuming this mantle, albeit hesitantly in some cases. On the other hand as the limited effect of their actions is patently obvious to everyone anyway, this call for more supervisory measures seems rather to be aimed at using these supervisory measures as a means of legitimization and exoneration.

NOTE

1. For example, for judicial restraint, *Office of Communication, United Church of Christ v. FCC,* 707 F.2d 1413 (D.C.Cir. 1983).

CHAPTER 12

Assessment

1. OBJECTIVES OF BROADCASTING REGULATION

An evaluation of the way in which broadcasting orders and particularly their supervision function naturally depends on which objectives are looked at. The observable changes in broadcasting regulation go hand in hand with changes in the objectives.

From the very outset, broadcasting regulation was not charged simply with the pragmatic role of "traffic police" but in addition by continual reference to a special public interest in freedom of communication. In Chapter 7, we noted the connection between the justification for freedom of communication and the regulation of broadcasting as based on this consideration. In this regard, all countries consider reference to the functioning of democracy to be a canon in the context of such justifications. Even in phases of deregulation and increased reliance on market forces, this justification has never been renounced in official documents.

However, it would be a mistake to study primarily the rhetorical surface. In particular, the term "democracy" is flexible and is susceptible to a wide variety of meanings; the term "freedom" is likewise multifaceted. A rough distinction reveals two primary poles. At one pole the concepts of democracy and freedom are oriented according to certain material objectives and require that a democracy be measured against them. Accordingly, it is not enough simply to anchor freedoms and democratic procedures in the legal system. They must also be truly implementable and aid in achieving qualitatively defined objectives for the common good, such as true equality and the emancipation of citizens. In this case, democracy and liberty are usually tied to the personal development of citizens. It is not enough to guarantee a legal opportunity to realize the goals; rather, the viability of the political order is measured against whether citizens are in actuality able to use this chance. If this is not or only insufficiently the case, additional guarantees are needed from the government. The way in which the order functions thus actually determines the concept's viability. When functioning is disturbed, the state takes on a duty of active structuring.

At the other pole can be found formally oriented concepts of democracy and freedom. These are concerned with rights and procedures that ensure chances of personal development. Democracy requires a process for reaching decisions, one that builds on the premise of legal equality. The proponents of this concept of democracy are, however, distrustful of materially defined objectives; in particular, they believe the possibility of abuse is likely when the state assumes the task of acting to realize these goals. They assume that the creation of a legal order for the exercise of freedom, especially for the exchange of services on the market, suffices as guarantor for ensuring the common good—everything else will take care of itself in the course of society's self-regulation.

Such classifications should not blur the fact that in reality mixed and transitional forms usually exist. Nevertheless, distinguishing between these concepts can help in evaluating the situation and analyzing changes. In this sense, the history of broadcasting regulation seems to be that of a movement from one pole to the other.

Special importance is accorded in the European discussion to the state's duty to build up a framework of broadcasting that responds to the functioning of democracy. The public service obligation imposed by government on broadcasting was justified, particularly in the period following the World War II, with the argument that broadcasting must carry out special public tasks, in particular, that of helping to perfect democracy. It was during this time that the British and American occupying armies in Germany made this an important objective, viewing the broadcasting order as a building block in their efforts to reeducate the German public. But even in the absence of this special historical situation, the trustee and proxy concept was developed for broadcasting. The concept also helped to establish special programming duties in commercial broadcasting as well, especially in the United States, Canada, and Australia. Examples include obligatory program time for informational, educational, local, and other programming; duties of balance and fairness imposed on broadcasters; or must-carry obligations on the part of cable operators.

In the course of liberalization, the orientation toward the public interest was never formally abandoned, and the promotion of democracy continues to serve as the formula for justifying remaining regulations. But deregulation was accompanied by diminished governmental responsibility and increased reliance on market forces. The economic market came to be viewed not merely as the process for satisfying commercial interests in the best way possible but generally as a procedure for determining and satisfying the wants and needs of the public. One example of such reasoning can be found in the following excerpt from an appendix of the FCC report on the deregulation of radio:[1]

> Although the public interest has many elements, the primary objective is broadcast service responsive to the wants and needs of the public—what economists call consumer satisfaction. In this regard, one of the major responsibilities of the FCC is to determine what kind of regulatory framework would yield a broadcast system most responsive to the public's diverse wants and needs. The Commission, of course, can only act as a catalyst in the provision of such service; it is the licensees that actually offer the programming. The Commission's limited role in this regard is to determine what actions it should—or should not—take to foster responsive programming. Accordingly, the key issues to consider are who is best able to determine the wants and needs of radio audiences and, once these wants and needs are recognized, what forces are most likely to lead licensees to be responsive to them. The economic model suggests that, given the status of radio broadcasting today, the marketplace and competitive forces are more likely to attain these public interest objectives than are regulatory guidelines and procedures. Universally applied rules or guide-lines cannot take into account differences among communities; therefore, they often will be unresponsive to the wants and needs of the public in individual markets. Also, the administrative procedures that must be followed to change rules in light of changes in circumstances are often cumbersome.
>
> In general, competitive markets are responsive to wants because there are natural forces (i.e., the profit motive) that induce the entrepreneur to discover what consumers want and that penalize him if he fails to respond to these wants. While advertiser-supported systems utilize different adjustment mechanisms than direct-pay systems, the same basic forces are at work in each. Commercial broadcasters, like all businessmen, seek to maximize profits.

Even though the FCC does not want to see this economic model put into place in its pure form,[2] it typically relies for its deregulatory action on the market mechanism as doing the best job of satisfying interests and thus for achieving the set objectives. To this extent, this concept proposes to be empirically oriented as well. However, empirical data, which are used to deflect criticism, seem to offer a somewhat different picture. Examples include the argumentation used in the two reports on the deregulation of radio and TV[3] and, more recently, that in the "Report on the Revision of Radio Rules and Policies" in the ownership rules (1992)[4]: These particularly discuss quantitative data—such as the number of available stations and alternative media or the specialization of program formats—that indicate diversity. Insofar as wants and needs are addressed, this argumentation takes place mainly in terms of consumer satisfaction. From this standpoint, it makes sense not to differentiate between consumers and citizens and to exclude a qualitative assessment of possible wants and needs. However, this view means that the functioning of democracy is not the starting point of deliberations. And such a concept all but eliminates an inquiry as to the extent to which commercial media offerings influence (or manipulate) wants and needs in a certain manner and enable (or prevent) citizens from pursuing certain interests (e.g., the emancipation interest).

This view is particularly evident in the Statement of Policy attached to the American Cable Television Consumer Protection and Competition Act of 1992. The focus of the law, which is embodied in the act's very title, is described in the following objectives (mentioned in Chapter 1, Section 6):

It is the policy of the Congress in the act to

1. promote the availability to the public of a diversity of views and information through cable television and other video distribution media;
2. rely on the marketplace, to the maximum extent feasible, to achieve that availability;
3. ensure that cable operators continue to expand, where economically justified, their capacity and the programs offered over their cable systems;
4. where cable television systems are not subject to effective competition, ensure that consumer interests are protected in receipt of cable service; and
5. ensure that cable television operators do not have undue market power vis-à-vis programmers and consumers.

These are market-oriented goals. Although diversity of views and information is also mentioned as an objective, it becomes clear that the path to this end is via the market and the expansion of offerings. It is therefore sensible to consider measures to check undue market power and to protect consumer interests. As the act itself shows, such protection takes place through, for example, rate regulations or protection of consumer service, as well as must-carry rules. But at most, this ensures that available programming is also carried over cable. The question of what type of programming is actually available remains, however, beyond the scope of regulation, such that measures to support this objective are absent. Market performance is described exclusively in quantitative categories, and categories of substantive quality are, of course, not dealt with.

With respect to the abolishment of programming guidelines, the FCC (1984) expressly mentioned its considerations:

Our decision to eliminate the processing guide-lines is based on two fundamental considerations. First, our view of the record and study of station performance persuades us that licensees will continue to supply informational, local and non-entertainment programming in response to existing as well as future marketplace incentives, thus obviating the need for the existing guide-lines. . . . Second, our re-examination of the current regulatory scheme reveals several inherent disadvantages, including: potential conflicts with Congressional policies expressed in the Regulatory Flexibility Act and the Paperwork Reduction Act, imposition of burdensome compliance costs, possibly unnecessary infringement on the editorial discretion of broadcasters, and distortion of the Commission's traditional policy goals in promulgating and monitoring programming responsibilities.[5]

Here the line of argument in deregulation becomes clear: It is asserted that regulation leads to undesirable side effects, particularly to regulatory effort and expense and to restrictions on broadcasters' decision-making freedom. However, these burdens are not specified in detail, and the possible disadvantages of regulation are not weighed against the possible advantages. Since some advantages *are* seen, the reference to burdens more or less serves as supplemental emphasis for doing away with regulation.

The reasoning behind reregulation in other countries studied here, although it usually attaches less importance to economic considerations and makes greater use of the public service formula, is still consistent with the trend toward increased reliance on market forces (see, e.g., Home Office, 1988). With increasing use of market-oriented instruments, a redefinition of the objectives simultaneously occurs. Public interest can and should (only) be pursued in the manner in which the market is able to do so. The economic market organizes processes of exchange. These may of course exhibit discrepancies because of the various forms of financing broadcasting. However, it is true of all of them that communicative services are rendered only in accordance with the economic returns to be expected. To avoid failures, the objectives of such a model must therefore be reduced to those services that can be stimulated through market incentives. The most the governmental side can do is to see to it that the functioning of the market has additional safeguards, as through the intervention of monopoly laws.

However, when a purely market-based model is not employed, as is everywhere the case, further objectives can no longer be ruled out. But the pursuit of other goals presumes intervention that may not be in conformity with the market and thus necessarily collides with the structural dictates of the market model. This gives rise to a conflict of goals and instruments, and the broadcasters react differently to these.

For instance, the British broadcasting order presupposes that divergent goals can be pursued parallel to one another. Even though broadcaster conduct is tied to market processes, the relationship is not permitted to threaten the quality of programs. In order to achieve quality, some aspects of control are withdrawn from the market. This approach is particularly evident with the granting of licenses for Channel 3: The quality threshold becomes important during licensing, and within their regions, licensees are exposed to only limited competition from other private broadcasters. In other words, complete deregulation has not taken place.

In general, it can be said of the various broadcasting orders that reregulation is normally accompanied by a relinquishing of the claim to pursue those goals that cannot be ensured via the market. Similarly, broadcasting legislation limits the pursuit of non-market-conforming objectives to narrowly defined fields, especially the protection of juveniles, which in many countries is accomplished with special emphasis on suppressing indecent speech. Although it is conceded that a democracy has the important task of ensuring plurality in communicative content and especially giving coverage to all of a society's interests, this duty is reduced to rigid fairness or impartiality obligations in informational programming that are very hard to implement. The area of entertainment, which appears to play a predominate role in creating value judgments, attitudes, and stereotypes, is, apart from a few exceptions,[6] spared content-related rules.

However, a number of broadcasting orders continue to follow traditional public service concepts in other ways: While commercial broadcasters may not be required to pursue objectives not in conformity with the market, this duty is to be balanced by retaining an alternative in the dual broadcasting system, namely, public service broadcasting (see also Section 4). Behind this is the concept of structural diversification: Potential disadvantages associated with the commercial control of broadcasting are to be compensated for by other broadcasters that are not financed via the market, or, if so, only to a minor degree. They would receive revenues from fees or subsidies out of the general budget, and in return they would be confronted with the expectation that they satisfy those interests not addressed in commercial broadcasting, for instance, those of minorities. Since the way such entities are financed itself raises risks—such as the hazard that the government might influence programming—commercial broadcasters can compensate for the deficits of public broadcasting. The two "pillars" of the broadcasting order are to complement one another.

2. THE PARADIGM SHIFT IN REGULATORY POLICY

The following discussion attempts to characterize the changes in regulatory policy by describing the typical trends evident in various countries.

2.1. From the Trustee to the Market Model

The above-described changes in the regulation of broadcasting can be portrayed as a shift from the trustee to the market model of broadcasting.

According to the trustee model, broadcasting is legally organized in trusteeship for the whole of society. Broadcasters are obliged to take into account different social interests; unbalanced and biased programming is not permitted. Under the trustee model, broadcasting is not geared primarily toward profit making: It draws its financial support from license fees, fixed by the state, or in some cases by tax expenditure.

At the opposite pole stands the market model. Broadcasting under this model is financed from the revenues of free enterprise, in particular, from advertising income or direct viewer payments. Its proprietors may strive to realize profit and orient programming toward the goal of profit maximization—or toward any other goal that they themselves may choose. Unbalanced and biased programming is not out of the question. Under the market model, there exists no obligation to foster the wider communication interests of all citizens; market forces decide which communication needs are satisfied.

In fact, broadcasting reality has long reflected contrasting mixes of these models. Thus, the market model has traditionally been embedded in some kind of regulatory framework involving, for instance, license procedures and programming codes of conduct. A well-ordered minimum of fairness and balance has been imposed upon free-enterprise broadcasters. In turn, the trustee model can be mingled with elements of free-enterprise and market orientation. In this way, many public service broadcasters benefit from regulatory arrangements that allow them to finance their activities partially from the market, for instance, from advertising revenues.

Even today, this mix of models has not been abandoned entirely. At the same time, however, there is a tendency to eliminate the trustee orientation in commercial broadcasting. On the other hand, in its competitive battle with private broadcasting, public broadcasting increasingly sees itself compelled to act as if it were controlled by market forces. In fact, in the procurement and advertising markets, as well as that for broadcasting rights, there is total commercial competition, which has an indirect effect on programming. For public broadcasting, the battle for audiences may not be one of economic competition (or, in the event of partial advertising financing, only a limited one). But in the interest of long-term survival—that is, for the purpose of ensuring political support in decisions on the continued existence of public broadcasting—even public broadcasters feel obliged to garner the largest possible audiences.

2.2. From Cultural to Economic Legitimization of the Broadcasting System

The turn toward the market model is accompanied by a change in the aims of state regulation of broadcasting. In Western Europe, government responsibility for broadcasting has long been legitimized politically in terms of the priority given to the political and cultural significance of broadcasting for state and society. Thus, broadcasting must fulfill a responsibility for public information and orientation. Insofar as broadcasters are now compelled to think in terms of maintaining a competitive hold on the economic market, these traditional duties cannot be upheld. The change of legitimization occurs through the widespread social and political acceptance of the subordination of mass communication to inherent economic laws. In this case, any degree of programming diversity that might be attained by the market's operation would be deemed sufficient.

2.3. From Freedom of Communication to Freedom of Broadcasting Entrepreneurship and Supply Services

The change of emphasis in the political legitimization of the broadcasting system has repercussions at various levels. One of these is the understanding of the fundamental right to freedom of communication. In the United States, the commercial basis of the media system led at an early stage to the integration of the principle of economic freedom—within the field of communication—into the protection of the First Amendment. Insofar as it concerns the spectrum of the media system in general, and the structure of individual media organizations in particular, the proprietor of the broadcasting station or the cable operator is regarded as the principal beneficiary of protection of constitutional guarantees to freedom. Consequently, it is not fundamentally questioned in U.S. constitutional commentary and debate that, analogous to freedom of the press, broadcasting freedom embraces the right to deploy one's wealth to communicate one's views. The right to use broadcasting vectors according to such an understanding of free communication, on the one hand, and the property rights of broadcasting entrepreneurs, on the other, are thus closely intertwined in the interpretation of the First Amendment and in the application of statutory provisions. A similar development can now be detected in the constitutional debate over the new media in other countries.

From a historical point of view, the transformation of the principle of freedom of communication into the protection of a commercial business could be considered surprising. The history of freedom of communication in Western nations is rooted in the early liberal tradition of constitutional law, which sought to prevent government from restricting the content of communication and which was oriented toward promoting individual self-fulfillment through communication and facilitating the quest for truth. Since then, however, the right to communicate and the right to run a commercial business have become merged. In principle, this new complex freedom extends to all media enterprises, broadcasters as well as cable operators, including multimedia concerns and multinational conglomerates.

A parallel development can be observed in the European Commission's activities, including its interpretation of article 10 of the European Convention on Human Rights (ECHR), the basic human right to freedom of opinion (Commission of the European Community, 1984: 127ff.). The basic rights contained in the EEC treaty with respect to freedom to supply goods and services and to establish businesses are, however, geared toward economic activities and do not aim at the right to freedom of communication vis-à-vis a legal definition of cultural undertakings. Therefore, the organs of the European Community have combined article 10 of the ECHR with the EEC Treaty in such a manner that the entrepreneurial activities of broadcasters may be covered by a unitary (if complex) basic right. The combination of those freedoms that are provided by the EEC Treaty and those basic rights designated by the ECHR applies: " . . . also to the freedom, guaranteed under Community law, to supply services within the Community as manifested in the freedom of cross-frontier broadcasting, on the one hand, and to the fundamental right, enshrined in the Convention on Human Rights, to the freedom of expression regardless of frontiers as manifested in the free flow of broadcasts, on the other" (Commission of the European Community, 1984: 128).

By combining these two basic rights, each of which has very different legal qualities, two objectives are to be achieved. Article 10 of the ECHR— which focuses on the free expression and dissemination of communicated messages, but not the economic activity of employing communication for commercial purposes—will be enriched with the additional legal ingredient of freedom to supply services and establish businesses. Conversely, the pre-dominantly economic-oriented freedoms contained in the EEC Treaty will receive the cultural-legal dimension of article 10 of the ECHR. The intention is to endow the European Commission with liberty to maneuver within a framework of EC economic liberalism.

2.4. From Primacy of the Communicator and Recipient of Information to Primacy of the Entrepreneur

This change goes hand in hand with a shift in the emphasis of state regulation itself. The basic right of freedom of communication protects the communicator and the recipient of information in equal measure. This right has been expressly established by several courts, especially both the U.S. Supreme Court and the German Federal Constitutional Court. The latter reasons that freedom of the media refers to communication as a process; protection should be assured for both the freedom to express and disseminate opinions and the freedom to receive full knowledge about, and to inform oneself by, opinions that have been expressed (BVerfGE 57, 295, 319). The U.S. Supreme Court reached a similar conclusion in establishing that "It is the right of the viewers and listeners, not the right of the broadcasters, which is paramount."[7]

In the transition to the market model, this orientation toward the accommodation of wider public interests in communications is threatened with oblivion. The citizen is left with a discretionary role as a potential communicator "free" to establish her- or himself as a broadcaster. If she or he is not in a position to do so, the result is not counted as particularly unjust (or in need of change), since in reality this situation applies in other areas of market activity as well. Not all markets are equally accessible. As a recipient, however, the citizen figures in the enjoyment of broadcasting freedom but has to change roles, that is, to act as a consumer. The significance of this change is especially clear in the case of broadcasting financed by advertising revenues. The true partners of the broadcasters are the advertisers, not the recipients (consumers). The communication or information interests of the recipients (consumers) in the pure market model are only relevant to the extent that the satisfaction of any particular group of consumers correlates with an increase in advertising revenues or the attainment of a favorable relationship between program expenditure and advertising revenues. Media economists have often demonstrated that under this state of affairs, not all consumers' interests are satisfied in the same measure.

This problem of inadequate satisfaction of communication interests loses its relevance for law when the legal protection of basic rights is no longer understood in its traditional sense. If broadcasting freedom comes to be conceived of as primarily entrepreneurial freedom, then deficits of any broader character are no longer open to diagnosis and remedy. The interests of the consumer are still considered relevant under the premise that the market indirectly satisfied consumer interests and, moreover, that it still does so, in spite of its many shortcomings, better than any other regulatory instrument.

2.5. From Special Culturally Based Broadcasting Regulation to General Economic Regulation

These tendencies can also be seen in the legal instruments used to aid in the regulation of broadcasting. Particularly characteristic is traditional broadcasting law being subjected to, and increasingly replaced by, rules of economic law. Traditional broadcasting law is being supplemented by instruments that have generally been used for economic regulation but are tailored specifically to broadcasting. One example is special rules of broadcasting law designed to combat concentration or for rate regulation. In addition, broadcasting law is supplemented by generally applicable economic law, such as resort to general rules of antitrust or the law of unfair competition. Some broadcasting laws expressly provide for the combined application of broadcasting law and economic law (e.g., in Australia). An additional step has to do with largely forgoing specific broadcasting law and relying primarily or even exclusively on generally applicable economic law. Although this step has been called for in policy discussions, it has not yet taken place on a broad basis in practice.

This shift of the legal basis of regulation is not just an expression of a changed perspective. There has also been a qualitative change in regulation. For instance, economic law is, if at all, able only to a limited extent to take into account the special sociocultural features of broadcasting; instead, it must treat the activities of broadcasting companies just like those of any other commercial enterprise. In so doing, it may become apparent that peculiarities of broadcasting markets are not given appropriate treatment. Thus, concentration processes may often not be countered at the local or regional level when the markets are not defined in the case of small areas. Internal growth will not give rise to combating concentration if merger control, which reacts to external growth, is the only tool available. The problems of bonding between different markets—particularly diagonal and vertical integration—can be treated merely from general economic aspects, often ignoring the specific substantive consequences for programming and the exercise of journalistic power. Further, the transition to economic regulation is commonly accompanied by a change of bureaucratic competence: Antitrust authorities replace broadcasting authorities. This organizational change usually also leads to modifications in attention and in the interests at the forefront of the activities.

2.6. From Comprehensive to Limited Regulatory Responsibility

It corresponded with the earlier trustee model that broadcasting had to fulfill public service obligations in all fields of activity. The governmental duty to guarantee this was accordingly broad. In the course of new regulatory techniques and accompanying deregulation, the fields in which broadcasting is subject to duties become narrower. It is apparent that programming requirements are being stepped down, and in particular, these requirements are no longer being derived from the democratic principle. The abolishment or relaxing of fairness duties is one example of this trend. Another is that German state media authorities no longer address duties of balanced programming to specific broadcasters but, rather, simply require that the programming of all private broadcasters together is balanced. This conforms to the reliance all market models place in the creation of external diversity.

All the same, the area of programming is not totally free of regulative requirements. The maintenance of morality and good taste and the protection of the development of juveniles continue to be recognized regulatory goals with the understanding that the market alone cannot ensure these. Although this protection is particularly strong in Great Britain, it is recognized in essence in every country studied here. At the height of its deregulatory efforts in the United States, the FCC placed increased attention on the elimination of indecent speech—apparently in reaction to public opinion in favor of this and to corresponding policy by the Reagan and Bush administrations.

During the 1990s, the U.S. Congress has required the FCC to take strong action to protect juveniles. German state media authorities made increased efforts toward juvenile protection the more it became clear that supervision of broadcasting generally encountered considerable difficulties. In Australia there is a well-developed system of classification of programs, whose goal is protection of juveniles. Considerable activity, although few successes, can be observed in all countries in the area of juvenile protection. Experience shows that measures of positive juvenile protection are much more difficult to enforce than negative ones (for this distinction, see Chapter 10, Section 12).

The fact that the activities of parliaments and supervisory bodies continue to center on protection of juveniles seems to demonstrate that the paradigmatic shift to market broadcasting has not yet been fully completed. Supervisory authorities retain certain duties; however, these no longer have to do with the functioning of a democratic order but rather with the protection of selected other vulnerable values.

2.7. From Programming-Oriented to Allocation-Oriented Broadcasting Regulation

Traditional broadcasting regulation dealt with the organization and financing of broadcasters and with programming duties. Since broadcasters largely produced programming themselves (although this was less the case in the United States), this type of regulation also covered the area of production. In the course of its development, the broadcasting sector has become increasingly differentiated. Production, transmission and dissemination, residual use, and advertising are all unique fields of activity with different markets. The significant differentiation among activities leads to intensive division of labor within the media industry. However, the companies operating in the various markets are often interlocked with one another. They attempt to achieve synergistic effects through horizontal, vertical, and diagonal concentration and thereby use the possibilities of cross-subsidization and cross-promotion. Traditional broadcasting regulation does not, however, address this diversity of activities. Rather, it is largely limited to broadcasting per se, that is, the legal responsibility to put together and then transmit broadcast programming. Only in exceptional cases—for example, the U.S. network rules—does broadcasting regulation extend to the phases of the production or purchase of programs; normally, regulation cannot adequately take account of the influence these areas have on broadcaster conduct, thereby making many measures ineffectual.

To only a limited degree does broadcasting supervision react to the extensive interlocking of companies. Common tools are restrictions on multiple and cross-ownership. But these usually cover only one section of true media integration: generally (but not everywhere) the relationship between broadcasting and the press or between broadcasters and cable operators; occasionally also the dependence of broadcasters on program suppliers or advertising companies.

Current regulations are marked less by the effort to deal with the diversity of media activities than by the effort to uncouple broadcasting regulation from the level of operations and to shift it to that of programming transmission. New, market-oriented regulation accepts that a market is engaged in the allocation of goods and services, forcing regulation to address the level of distribution. The functioning of the market also has indirect effects on production. Market growth means growth in means of allocation, which is normally accompanied by a growth in the means of production. New broadcasting technologies, such as satellite and cable, and the increased performance of both old and new technologies, by way of digitalization, for example, expand the allocation potential and stimulate both new services and the provision of new capital.

The new regulatory philosophy is based on the assumption that the expansion of this potential in the means of allocation is a value in itself, since it can be expected that the potential will be used to satisfy needs and wants in a more satisfactory fashion. This philosophy is well expressed in the earlier-cited policy statement in the U.S. Cable Television Consumer Protection and Competition Act 1992 (see Section 1). In all countries, however, the transmission of programming via terrestrial frequencies continues to require a license, for which a potential broadcaster must apply. Licensing is still often required for transmission with other media, particularly cable and satellite, although regulations in these areas are less dense. As a consequence of expanded means of transmission—such as the future chances for digital compressing—it is increasingly doubtful whether it is still legitimate to regulate broadcasting operations. As a result, there are tendencies to forgo the requirement of a license for broadcasting and to propose in its place a different set of rules, particularly rules dealing with the modalities of using transmission facilities. A prototype of these is the body of U.S. cable laws.

Such rules can cover cases in which the medium of transmission is government owned (e.g., the PTT) or those in which transmission networks are operated by independent private companies. How decisions on access to the medium of transmission are made can always be legally controlled, particularly whether the owner (e.g., the operator of a cable network or a satellite) may itself decide who is entitled to use the facility, which programs may be transmitted, and whether the owner may in addition don the guise of broadcaster. In order to avoid abuses in the relationship with broadcasters or consumers, the power to control the medium of transmission is usually limited or even channeled by legal rules.

Thus, obligations on the part of cable operators to set up open-access channels are based either on legal requirements or governmentally set conditions in the cable franchise. Similarly, the cable operator can be obligated to act as broadcaster as well, in order, for example, to ensure local programming. Governmental regulations also form the basis of must-carry obligations.

Rules regarding cable operators' programming duties must be distinguished from another type of regulations, which have to do with preventing possible abuse of the (at least, de facto) monopolies held by cable operators. This mainly involves the guarantee of minimum conditions for equal access to transmission facilities. If there are transmission monopolies for a certain technology, as is usually the case with cable, it is possible to prevent monopoly abuse by setting rates or establishing the scope of conditions of use (e.g., vis-à-vis subscribers). Since cable operators gain an economic advantage by transmitting programming, the relationship to broadcasters (now sometimes called video programmers) is also regulated, as at the copyright level and with an eye toward the rates. This no longer has to do with the regulation of transmitted contents but, rather, with the basic conditions of transmission itself.

The EC is also orienting its unifying efforts toward the distribution of programming. However, this involves a totally different approach, namely, the free transmission of programming throughout all of Europe. The EC television directive omits issues of broadcasting programming organization and financing, leaving the regulation of these to the member states. It aims at eliminating hurdles to transmission that member states could erect: All programming is transmitted and received within the EC territory without border restrictions. In so doing, the EC accepts with approval the fact that the freedom to allocate across borders creates indirect pressure to abolish governmental regulations in the area of production. As a sacrifice for the purpose of easier political implementation of market freedom, it has, however, permitted states to enact a limited number of rules that do not comport with open markets, such as with advertising and juvenile protection. Harmonization of the laws of member states would otherwise have been politically unattainable.

Addressing only the distribution of programs is in yet another respect typical in broadcasting regulation. For instance, most measures of negative juvenile protection only still apply at the level of program dissemination: By way of limitations on time of broadcast or concentrating inappropriate programming in certain channels, the manner of transmission of programming not suited to juveniles is restricted, but not its production or basic accessibility.

2.8. Motives for Regulation: From Broadcasting Policy to Structural and Fiscal Policy

For at least the near future, there will continue to be some situations of scarcity, particularly with the use of terrestrial frequencies. In addition, monopoly-like structures will arise in some market segments or, especially in the cable sector, will be tolerated or even protected by the government. Most players in the broadcasting field accordingly make use of a privilege, which usually takes the form of a license rendered by virtue of governmental regulation. As compensation, the holder of such a privilege is often subjected to burdens. The traditional public service obligations in broadcasting aimed at such a noneconomic exchange for the use of an economically valuable good.

The elimination of public service obligations relieves broadcasters of this type of burdensome exchange, making it all the more equitable to expect a different type of compensation for the privilege. For instance, the German state media authorities use their licensing power to obtain pledges of broadcasters to set up production facilities in the relevant state. The imposition of production quotas (e.g., to ensure national or European productions) is another example of skimming off some of the license's commercial value in the interest of developing independent production infrastructures. A further example of reaping the benefits occurs when a satellite operator is required to disseminate (in addition to commercial programming) especially desirable programming (e.g., public service programs) at no cost or at low-cost rates (carriage obligations).[8]

It is most typical to demand an exchange in financial terms. The imposition of licensing fees or even the auctioning of licenses is an obvious step to take in replacing public service obligations with market-oriented regulations. It permits the governmental side to consider and use broadcasting as a source of fiscal revenues. The greater the government's need for funds, the greater the temptation to uncouple broadcasting from its traditional structures and commitments, thereby increasing profitability and ultimately using the government's regulatory power as a means of generating profits, or, for fiscal policy. Traditional broadcasting policy thus becomes fiscal policy.

As long as there are hurdles to access—particularly situations of scarcity or even monopoly—in the dissemination of broadcasting, it makes it difficult to justify that those who benefit from the privilege are provided cost-free with access to the scarce good or the right to establish an exclusive niche. It is more appropriate to derive some countervalue with the goal of protecting the viability of the broadcasting order in the interest of all citizens. At a time in which specific legal commitments of broadcasting are being abolished, it squares with the political logic of development to use the growth market of information and communication also as a financial source to realize goals the market is unable to attain. Part of this is use of the profits so derived to support programming that does not adequately stimulate the market mechanism but that is considered important for supervening (e.g., social policy) considerations. For example, support can be provided to programming for low-income minorities, broadcasters in local markets, or especially costly (e.g., cultural) programming. It is debatable whether such considerations in broadcasting policy will have the chance to be adopted at times when the government urgently needs money for other purposes.

2.9. Reacting to Market Failure: From Instruments and Regulations to Ethics and Appeals

The above-described trends in broadcasting regulation can be interpreted not only as a realignment of objectives but also, in a similar vein, as a change in the ways and instruments of regulation. Broadcasting regulation was a response to market deficits or even structural market failure. After initial efforts to exercise a regulatory effect on broadcasters in a commercial broadcasting system, it has become ever more apparent to legislatures and especially supervisory bodies that regulation does not lead to the desired results. The deregulatory strategy since followed is not simply the expression of changed sociopolitical appraisals of the relationship between government and market but, moreover, a means for finding a way out of the dead-end into which practical regulation has led. It is hoped that the failure of regulation, which can be observed everywhere, will seem to be less dramatic, if those expectations in terms of regulation are withdrawn. However, there will then be nothing to counteract any future market deficits or failures.

Therefore, it comes as no surprise that there continues to be criticism of the media, particularly regarding the concentration of power in the hands of a select few companies, the narrowing of the political spectrum represented in broadcasting, the ignoring of cultural goals such as those of education, the orientation on segments of society, and the substantive quality of programming. The development was foreseeable and was in fact foreseen. In countries that licensed private broadcasting only after vigorous discussions of media policy, as in Germany, the risks were reviewed in their diverse aspects. The establishment of media supervision and the equipping of supervisory bodies with regulatory instruments were responses to this criticism.

It has now become apparent that this response is of little or no use. Two particular solutions have been recommended as alternatives. The first refers to the responsibility of recipients to learn to deal with media offerings in a "mature" fashion and to protect themselves by selecting the programming they find appropriate. The government's duty is seen as enabling citizens through media education (e.g., in schools) and through instructive media counseling to act responsibly as media consumers. "Media pedagogy" is hailed as a means of dealing with the risks of media use. At the same time, appeals are made to parents to discharge their duty to educate their children and to teach them to consume media wisely.

Second, these educational appeals to recipients are supplemented by ethical appeals to media makers to be aware of their responsibility to the general public. They are to develop professional standards of ethics regarding the consideration of society's plural interests, truthful use of news, respect for privacy, restraint in using depictions of violence and morally offensive content, and so forth. Some calls have been made to support the effectiveness of these standards by institutions of self-regulation, as with the press councils for journalism. Examples include the press council and the ombudsman in Sweden (see Weibull & Börjesson, 1992), and the Consumer Council for Broadcasting being considered in Great Britain. Civic organizations, such as consumer associations, are called upon to act as advocates of recipients and to monitor broadcasting's obligation to heed the public interest. Examples include the Voice of the Listeners and Viewers in Great Britain and the Association Nationale des Téléspectateurs in France.

In attempting to make media publicly accountable, such ethical appeals are raised even though it is apparent that governmental regulation has been of little success for the very reason that it foundered in the face of the (mainly economic) interests of those being regulated. It is thus reasonable to ask why this should not also apply to ethical appeals, that is, why they should be any more successful than governmental supervision. There are no indications that this is the case. In the discussion on expanded public accountability for broadcasting, hardly any mention is made of such conditions. In particular, it is rarely asked which rules have to be in place for journalistic work in order for media companies to be able to afford to function according to professional standards and ethics and not according to commercial interests, that is, to be able to act contrary to market principles. When reliance is placed upon the public as a kind of guardian of broadcasters, it is rarely questioned whether it is likely that in highly integrated, multimedia companies the media itself can act as its own overseer.

The ethics discussion targets objectives that were once the focus of traditional broadcasting regulation; but in a market-oriented broadcasting order, these cannot be ensured either as a consequence of simple market forces or by regulations working against market forces. With the ethics discussion and the search for media accountability becoming ever more fashionable, it seems as if the new accountability mechanisms are not subject to the same test of effectiveness as were the traditional instruments criticized publicly as ineffective and since abolished in the course of deregulation.

In addition, the pedagogical appeals directed at recipients, particularly parents, are not subjected to the same test of effectiveness as the often-criticized instruments of media supervision. For instance, it has yet to be plausibly demonstrated that preschool and primary school education in the media is suitable for ensuring critical, self-responsible use of media and that will be effective for life. Moreover, it is not clear how parents, who are already faced with well-known problems in rearing children, can have a positive effect on their children's use of media. Nor is there an obligation to use the profits of media companies to finance supportive pedagogical measures that would counteract the long-term effects of broadcasting.

In short, deregulation has been accompanied by a shift in responsibility. While previous attempts had sought to ensure with governmental regulation that media companies shoulder a responsibility to the general public and to view the government as guaranteeing enforcement of such duties, the responsibility now has been shifted to individual actors: on one hand, programmers; and, on the other, recipients. To this extent, appeals to those concerned have been made, and organizations have been created by the industry itself to supervise ethical standards without being entrusted with the power of sanction. The possibility of implementation apparently seems to be of secondary concern. The basic structural conditions necessary for effecting ethical standards have yet to be detailed.

3. APPRAISAL

The trends described in this chapter reflect changes that have taken place in various countries at different times and with varying intensity. This depiction necessarily has been undertaken in a generalized form; differences among countries have been largely left aside in order to convey typical trends. The object of this concluding chapter is not to undertake a description of the events in each country studied. Rather, circumstances will be denoted that can be considered typical, even though these may show differences in details among the various countries.

In view of the differing stages of development in the countries examined, as well as of their basic political, legal, cultural, and economic conditions, this study does not conclude with a detailed assessment of broadcasting supervision, particularly its successes and failures. Such evaluations could only responsibly be undertaken by authorities familiar with all these dimensions—

usually writers from the respective countries. The author of this comparative study cannot claim to be competent to adjudge all the broadcasting orders analyzed. He can only attempt to describe them from his perspective—with an effort to be fair to other perspectives—and at most to formulate cautious appraisals of a comparative nature. Only such cautious evaluations are therefore presented.

As repeatedly mentioned, a process of redefining the objectives of broadcasting supervision can be noted in all the countries studied here, a process in which those supervised are themselves taking part. This surely is related to general trends, such as the worldwide phenomenon of de- and reregulation in the social and economic spheres. It is important that broadcasting policy dovetail with general regulatory policy. Such conformity among supervisory authorities even takes place where these are independent of the government in contrast to Australia. It occurs "voluntarily" or is at least seen as the alternative to forced compliance with political objectives by statutory, financial, or personal intervention.

In the broadcasting sector, the process of redefining objectives seems to be marked by the realization that the goals originally sought in the broadcasting order cannot be achieved through regulation, at least not without dysfunctional side effects. Accordingly, broadcasting supervision is (and has been) in a state of retreat. The process has been made somewhat easier by the fact that regulatory requirements are usually formulated quite loosely. The resulting built-in flexibility was consistently used by supervisory authorities to relax the requirements supposedly designed for the status quo. The express granting of exemptions and tolerance of rules violations or at least the refusal to enforce sanctions were an expression of this sort of strategy of retreat. Amendments to the rules often followed in due course. Most often, the rules were adjusted to conform to the changed reality, instead of the rules being used to change reality. Amended rules then usually provided for a grandfathering of what would then be unacceptable circumstances.

In some cases, it was decided to do away altogether with binding rules, formulating instead appeals or rules of thumb. The now stronger emphasis on self-regulation of the broadcasting industry and the use of law "privately" enacted by it as the standard for governmental supervision signaled the delegation of government duties to private parties. This took place with the clear intent of harmonizing these new rules with the interests of those so regulated. Moreover, the consultative participation of supervisory authorities in enacting rules—or the threat of subsidiary governmental rules—did little to change the respect paid to the interests of those regulated. However, the latter represent only a section of the interests thereby affected. Citizens, as listeners and viewers, are usually excluded from taking part in such rule making. And when there are legal possibilities for participation, as in the United States and Australia, these are considerably weaker than the industry's opportunity for input.

The diverse forms of informal arrangements, cooperative networks, and flexible sanctions of rules' violations create ways to take interests of those being regulated into account. By using these informal and cooperative avenues, the supervisory authorities have seen to it that supervision operates with the requisite caution and moderation. In this manner, they have come to realize that the organization, let alone the system of supervision is only able to survive when it pays a certain amount of consideration to the broadcasting industry's structure and basic concerns. As a result, the protection of societal values can only be weak: effective protection depends on the extent to which these values are accepted by the broadcasting industry itself.

There are many justifications for such consideration. The protection of freedom of communication is an especially crucial one. However, it can justify a relinquishment of the claim to regulate only when freedom of communication is primarily understood as a right of the communicator not the right of the recipient. As stressed above, this restriction is by no means unassailable from the standpoint of the theory and history of basic rights. If the interest of the communicator is made equivalent to his or her interest in making economic use of broadcasting's dissemination possibilities and thus in making profits, then this conception of basic rights promotes a commercialization of a right of liberty that historically was fought for and won in a different setting, namely, as a basic political right. Under such a reading of basic rights, recipients are only indirectly protected as objects of developing communicators. That there are other concepts of basic rights is shown in particular by the case law of the German Federal Constitutional Court. In view of the worldwide commercialization of basic rights, which is also being driven forward in the EC, it will, however, be difficult to cling to alternative interpretations and enforce these in structuring a broadcasting order. The importance of freedom of communication as a political civil right and as an element of democracy is on the decline in the field of broadcasting.

A further justification for waning supervision is the consideration paid to the way in which market forces work. Such consideration necessarily requires a redefinition of formerly recognized objectives and the restriction of available instruments. To this extent, it can also be interpreted as a part of the strategy of retreat. Proponents of the market model do, however, raise the point that a redefinition of objectives is a necessary first step to making successful regulation possible (though only with regard to limited objectives). This regulation is market regulation, sometimes supplemented with specific regulatory requirements to protect especially vulnerable values.

If broadcasting regulation is not measured against these new conditions but, rather, against the traditional requirements of public service broadcasting, then the overall assessment of broadcasting regulation must be skeptical. It seems to all appearances that licensing and supervision of broadcasting have been unable—and often have not even attempted—to raise the palette of programming made available by commercial broadcasting to a standard obtained by public broadcasting in some countries, such as the BBC in Great Britain, and have not been oriented toward embedding freedom of communication in the functioning of democracy and the self-realization of all citizens. In this regard, supervision of commercial broadcasting has its hands tied. In particular, it is unable to establish corresponding programming duties or to enforce these through intensive control, since supervisors would otherwise run the risk of acting somewhat like censors and, through substantive intervention, threatening the self-regulating formation of public opinion. Nor have the accomplishments of public broadcasting been attained through external supervision: These were possible because public broadcasting created room for the public service idea to grow on the basis of its financial and organizational structure and a professional orientation tailored to this situation.

Even more modest objectives of public service broadcasting have only been realized to a limited extent by regulation and supervision, whether in the area of advertising restrictions, reducing violent programming, or promoting children's programming. Existing regulations have been implemented only reluctantly. Broadcasting supervision was, however, able to contribute to preventing obvious violations of some of the rules. At the same time, though, this was only possible in connection with a relinquishment of the earlier regulatory claim. Easiest to enforce were the observance of those rules that the broadcasting industry essentially accepted. Sanctions against "black sheep" could also count on support in the broadcasting industry as well. The enforcement of rules particularly found support in the established broadcasting industry when this helped to ensure its overriding interest in maintaining the status quo. To this extent, supervision of broadcasting in all the countries studied here has contributed to a relative ordering of the broadcasting system.

There are, however, also examples of supervisory activities that were maintained despite considerable resistance by the broadcasting industry. These include advertising restrictions and the establishment of production quotas. With these measures, broadcasting supervision could expect strong political support, as from the government or sections of the general public. Supervision saw itself under pressure to enact rules and to enforce these against opposition. This is shown by the significance of the political climate as well as the quality of public participation in the chances for successful broadcasting supervision. An active public discussion can lend additional support. Nevertheless, the broadcasting industry has consistently learned how to sidestep rules in such fields or to follow only their letter. The success of broadcasting supervision has therefore remained very limited in this area.

Since supervision of broadcasting is often in the interest of those regulated (see Chapter 11) and since there continue to be political expectations that some market developments be counteracted, it can be expected that certain forms of broadcasting supervision will exist in the future. Despite an extension of market-oriented avenues of control, there will continue to be a need for regulation for the time being. The broadcasting industry is most likely to tolerate the role of supervisory authority as traffic police. Moreover, it cannot be ruled out that influential sections of the broadcasting industry are also interested in the legitimate function of supervision in easing burdens, that is, that broadcasting supervision can help to buffer criticism of the performance of the broadcasting industry. In other words, the existence of an institution that is supposed to protect the public interest through supervision can also serve to ease the situation of those to whom the public interest expectations are addressed, namely, companies in the broadcasting industry. Should the supervisory authority, in keeping with the current trend, define public interest narrowly—for example, solely in the sense of quantitative advertising restrictions and a prohibition of advertising for alcohol, tobacco, and pharmaceutical products or punishing indecent speech—this indirectly means that other public service expectations remain ignored.

If the supervisory authority enacts rules and these fail to address the desired objectives, then at least this can also be interpreted as a failure of the supervisory body and as a reinforcement of the populist rejection of the role of the state: It would be proof that once again a governmental bureaucracy failed. Its existence therefore makes it possible to divert attention from who the beneficiary of certain objectives is and who fails to reach them.

The reliance on regulation, coupled with a retreat of the regulatory claim, therefore seems most likely to enable the continued existence of supervision and supervisory authorities. Most supervisory entities have already accepted this situation. However, they can justify their existence in the long term only when they can show minimal success with specific supervisory measures. They cannot long afford to give the impression that they are only paper pushers.

There remains, however, the finding that in most traditional regulatory fields, broadcasting supervision can be successful only to a very limited extent. Nevertheless, despite deregulation, some traditional regulatory objectives continue to be praised, including in part the public service philosophy. The justification for broadcasting supervision seems also to depend on stipulating objectives that cannot be attained with the available instruments. The rhetoric behind the supervision of broadcasting extends further than its instruments are able to. As a result, acts of a symbolic nature are included among the arsenal of actions available to broadcasting supervision.

4. PROSPECTS

As mentioned repeatedly, the traditional public service requirements and the social responsibility derived from them can only be enforced in a commercial broadcasting order to the extent that this conforms to economic rationality. This study has not shown any approaches for avoiding such a dilemma. This appraisal will come as a sobering thought, or even a disappointment, for all those who are involved in the maintenance and development of commercial broadcasting yet who also wish to see it adhere to its traditional public responsibility. In many countries that still face or are just commencing changes in the broadcasting landscape, paths are being sought to combine commercial financing with successful regulation. In the 1990s, this particularly applies to countries in Eastern Europe and to many African, Asian, and South and Central American nations that seek to introduce democracy and free market economics. For these countries, there are no patent recipes.

However, it would be somewhat rash to take the skeptical synopsis offered by this study as reason for insisting on the abolishment of any sort of regulation and its replacement with pure market-based broadcasting. In this respect, it must once again be emphasized that the efforts behind governmental regulation are usually justified by the deficits and failures accompanying pure market self-regulation, and they are related to some special features of broadcasting. This problem of market failure and externalities is not overcome simply by the fact that the paths taken thus far have led only to limited successes. Rather, the risk of market failure and externalities becomes even more apparent: The pledge that, with regulation, broadcasting would be able to avoid developments deemed negative by society has proved to be fulfilled to only a limited extent. But this does not mean that developments have become any less problematic. There is no reason not to attempt to realize goals that the democratic decision-making process defines as important for the broadcasting system. However only a few of these goals can be achieved through supervision. Nonetheless, there is no reason not to try to attain at least the few supervisory objectives that can be ensured via supervision. Furthermore, the finding that regulation has failed also leads to the basic question as to the structures of an appropriate media order, which will most likely become a burning issue in the future. Indeed this question is more topical than ever. The discussion about the future of broadcasting does not become an irrelevance simply because the broadcasting market is on the march worldwide.

The broadcasting orders studied here do not simply accept the finding of limited regulatory success in a commercial broadcasting system with shrugged shoulders. On the contrary, a number of countries are attempting to maintain public broadcasting alongside commercial broadcasting, and this is financed, only to a limited extent, if at all, with advertising dollars and remains committed to the public service concept. In Germany, this duality in the broadcasting order has even been deemed by the Federal Constitutional Court to be constitutionally mandated: As long as it cannot be guaranteed that private broadcasting will fulfill public service obligations—the court doubts that this will be the case in the foreseeable future—there must be a viable public broadcasting system whose organization and financing make it easier than for commercial broadcasters to orient programming toward the area of public accountability.

The Western European and Australian experiences with dual broadcasting orders and thus with a public service pillar do, however, show that public service programming does not come about simply as a matter of course. It needs special organizational and financial assurances. It must also be taken into account that the changes described here in regulatory philosophy and supervision of broadcasting in the private arena will not be without consequence for public broadcasting. If public broadcasting is to remain strong— whether this is desirable will have to be decided politically—then provision must be made for adequate financing and for autonomy with regard to content. In addition, public service broadcasting must itself be sensitive to the need for its own reform. Its system stems from a time in which public service broadcasting in Western Europe was the norm. A glance at public broadcasting in the United States and Australia shows that it plays only a marginal role in broadcasting orders in which the commercial emphasis is the norm. Recent trends in broadcasting, which will undoubtedly become even stronger through new transmission technologies and the international networking of broadcasting markets, indicate that commercial broadcasting will also become the standard in Western Europe.

Commercial broadcasting has already reshaped the broadcasting landscape to a considerable extent. There has been a substantial increase in the number of broadcasters and programs. The programs provided by commercial broadcasters differ appreciably from those of the traditional public service broadcasters, although the latter have, to a certain degree, adapted to their competitors. Any assessment of the associated changes is bound to be controversial. The broadening of the range of programs on offer and, consequently, the increased scope of choice for recipients may be welcomed by many, yet there are still points of criticism. Such criticism is primarily leveled against the considerable change in the quality of information programs and the extensive subordination of information to entertainment priorities. Critics claim, for example, that education and information, in mass-oriented programs at least, are neglected and that the interests of numerous minority groups are ignored altogether. There is concern about repercussions of this on the political culture and its implications for the democratic process. Compounded by a fragmentation and segmentation of society, this trend in broadcasting could have adverse effects on the stability of political structures.

This is not the place for an in-depth discussion and analysis of such criticism. However, it is fair to contend that the liberalization of the broadcasting market has not and will probably never lead to a situation in which individual commercial broadcasters will feel similarly committed to the public service idea as public broadcasters in the past. The shortcomings it evidences are a challenge to all those who wish to adhere to the traditional objectives are who rely, therefore, on effectively supplementing private broadcasting with a viable public broadcasting system. In this sense, acceptance of the fact that the regulation of commercial broadcasting is problematic can be a stimulus for maintaining and further developing the alternative of public broadcasting (see also Blumler & Hoffmann-Riem, 1992). However, it also presupposes some thought as to how it should be structured in order to stand a chance of survival in the future.

This consideration leads to a more fundamental one. The extension and reorganization of the communications system are in full swing. Visions of an (American) National Information Infrastructure (NII) or a (European) Global Information Society[9] may, in some respects, seem unrealistic. However, it is plausible that the transformation of the media order and a certain convergence of mass and individual communications—so far generally treated separately— will take place in the near future. To some extent this has already occurred. It is still not clear, however, which regulation requirements will exist in future. There has been ample conjecture regarding the character of the future media landscape as well as the grounds and possibilities for regulation.

In view of the considerable increase in transmission opportunities, media regulation can no longer be justified by reference to the scarcity of transmission possibilities. Of course, with an abundance of frequencies, a need to ensure interference-free communications traffic will still remain. There are also grounds for safeguarding access to transmission channels for communicators as well as recipients. Guaranteeing a minimum of interconnectivity for the various paths of communication and for the software used is also particularly important in this context. However, such grounds for regulation cannot serve to justify content-related demands on programs or on program providers. As previously mentioned (Chapter 7), there is also justification for broadcasting regulation that does not depend on the argument of scarcity.

Apart from these changes in justification, there will also be changes in the concept of regulation. For example, in view of the interconnection of previously separate paths of transmission, the regulation of broadcasting can no longer be oriented to the respectively used technology. The distinctions customarily drawn in many countries, such as the United States, between over-the-air broadcasting, cable casting, and satellite broadcasting can now only be a limited reference point for a differing density of regulation.

If broadcasting regulation is accurately to identify all risk potentials, it can no longer concentrate solely on the programming decisions of broadcasters. Program production, acquisition, and distribution are just as important for the provision of communication as the decision of program arrangement. The risks of manipulation, violation of legal rights, or neglect of certain communication interests can be rooted at all these levels. This becomes even more pertinent insofar as new technologies and market developments eliminate the dividing lines between the various fields. The prospect of increased interactivity and, consequently, the exchange between the roles of communicator and recipient further highlight the fact that traditional demarcations are now obsolete. However, it is as yet unclear which areas of the media order can at all be regulated with any prospect of success. The diversification and segmentation of the communications order together with the associated decentralization of decision-making processes have made the task of statutory intervention much more difficult.

In the future, too, particular risks may emanate from the extension of market power. It is probable that the diversification of media markets will lead to the discovery of more and more niche markets and that newcomers too will have the opportunity to exploit them. However, the size (i.e., the globalization) of the markets, the economies of scale, in particular the synergy effects resulting from the integration of various media activities, and the vast concentration potential in the field of communications in general indicate that substantial market power will probably be created by individual companies. It is debatable whether statutory regulation can prevent this. However, the absence of counteracting statutory or international forms of regulation from the outset would eliminate any possibility of safeguarding the workability of competition. It is doubtful whether general competition law would suffice; presumably, special media-specific regulations are required to combat media-specific concentration potentials.

In Chapter 10, a number of past broadcasting goals were outlined. Political decisions are required to determine which of these goals should remain prominent or whether their number should be extended. Many of these—albeit not all—are unlikely to be controversial. What is already controversial, however, is whether achieving these goals requires statutory regulation or whether reliance on market forces will suffice and is, therefore, preferable. National statutes and, if required, international regulation are justified insofar as the pursuit of these goals complies with the law and contributes to the attainment of these goals without excessively jeopardizing yet other goals. This sentence is deliberately vague: The individual legal systems differ to a certain extent in terms of which goals are legitimate in respective political orders, which demands are to be made on the suitability for goal attainment, and which side effects can be taken into account. There are also certain differences with regard to the probability that certain goals are jeopardized or that certain measures contribute to achieving goals. Not only the overall institutional setting, therefore, is important—for example, the nature of the division of decision-making powers among the legislature, executive, and judiciary—but also the basic theoretical provisions, are important, especially in interpreting the constitutional rights of freedom of the media when statutory and social power are allocated.

Whether such differences can remain effective in view of the internationalization and globalization of the media remains to be seen. Growing international links among markets, however, mean a greater need for coordination. The media enterprises operating on the international stage will themselves try to ensure coordination. However, they will very probably then coordinate with state decision-makers if this improves their own position. The negotiations surrounding the General Agreement on Tariffs and Trade (GATT) and the controversies on special regulations for audiovisual services constitute paradigm examples.

The structures of the international media markets and the conduct of the media companies will probably become increasingly inaccessible to statutory intervention. Still, it is unlikely that an institution will be created at an international level to replace previous state regulatory bodies. Even if it were created, it is doubtful whether it would possess the necessary powers successful regulations demand. The example of Internet serves as a good case study of the limits of possible external regulation of complex and dynamically structured communications networks.

There are many indications that traditional broadcasting regulation will have to be substantially changed in the new media order. The traditional type of regulation is most likely to be possible—and necessary—where there is a need to protect against interference in private interests: Protection of personal integrity (especially of privacy), safeguarding of copyrights, as well as consumer protection are examples. In such cases, recourse to the protective tools of the general legal system is a possibility, although media-specific modifications will have to be made; in the age of cyberspace and virtual reality, for example, questions of copyright need to be completely redefined and new answers found.

Even if efforts to protect individual interests succeed (at least to a limited degree) in the future, it hardly seems possible that there will be any desire to regulate the communications order's ability to function in terms of certain content-related goals, for example, the quality of programming or even qualitative diversity. An attempt can thus only be made to structure the communications order so as to allow the chance of realizing such goals.

Insofar as the market proves to be an insufficient regulator to bring about the necessary range of programs, other stimuli could, if required, be arranged. With regard to the public service idea, the previous chapters have pointed out the possibility of providing a public broadcasting service that complements commercial broadcasting but is financed on a different basis. There is no need to limit such a concept to the field of traditional broadcasting. The imminent diversification of services and the vague transitions between individual and mass communication will presumably also result in deficits in those areas that are not, strictly speaking, included in the field of broadcasting.

A media order that promises its consumers that they will be able to define their own needs and satisfy them through access to a wide range of programs requires a correspondingly wide range. In view of differing levels of power exerted by different social groups it is not enough to rely on the provision of programs simply in response to demand alone. If the new communications order is to help protect the social opportunities of different members of society and not lead to an increasing inequality of opportunities—which would require political decisions—it is crucial to take steps to guarantee a broad range of programs and appropriate means of access. In a media order characterized by freedom from state control, the government, of course, should not act as a program provider. It can, however, help create a financial and, if necessary, organizational framework that offers consumers a range of programs, on appropriate terms, that are important for the satisfaction of needs but that are not offered on the market without additional support. These could be financed by siphoning off profits from those programs that retain a large profit margin by limiting themselves to profitable segments of the market.

Attention should not only be paid to financial possibilities. New forms of technology demand special skills from their users in order to be able to access available services. Interactive services that are currently in the planning stage are not just based, for example, on the willingness to interact and make active use of the media but, also, on the consumers' technological and intellectual abilities. This raises new questions of media competence. One question already asked is whether the new possibilities might not lead to a further widening of the social gap between the haves and the have-nots. Indeed, the scientific and information society of the future could considerably shift this division and not necessarily in favor of those already classed as underprivileged. A society that commits itself to the goal of equal opportunities in the field of communications (and this can only be the product of a political decision) must, if necessary, ensure that what is offered in the field of communications can also be actually used in such a way that new inequalities are prevented. The corresponding media competence will certainly not simply appear from nowhere. However, a system could be designed that minimizes obstacles. Which state institutions can play a role here requires further clarification.

These few comments alone show that a new communications order raises questions that touch upon the fundamental aspects of a modern society. Broadcasting regulation will depend to a large degree on the learning powers of all those involved, in other words, on the regulating government too. Effective and critical public opinion will also be important. In view of the concentration of power in the media sector, it remains to be seen whether the public's role will tend to be fostered or jeopardized by the possibilities of communication.

If broadcasting is to continue to be regulated in the future, such regulation will require a new quality. The state will probably only be obliged to exert less influence on the behavior of actors; it could, however, retain its importance as a mediator in conflicts of interest between various actors. It will also have to provide protection for those who need it. In particular, it must ensure that structures are established that avoid concentration of power and the creation of new forms of inequality. This at least applies if freedom of communication is primarily to serve the development of all citizens, enabling them to play a responsible part in the process of shaping state and societal structures. Governments may address some of the problems of the communications orders of the future reactively; others have to be addressed proactively. Even some of the traditional ideas of public service broadcasting can be upheld, provided the focus is on traditional goals rather than on traditional tools.

NOTES

1. See Report and Order, In the Matter of Deregulation of Radio, 84 F.C.C.2d 968, 1022f. (1981), Appendix D.

2. Report and Order, In the Matter of Deregulation of Radio, 84 F.C.C.2d 968, 1022 (1981).

3. Report and Order, In the matter of Deregulation of Radio, 84 F.C.C.2d 968 (1981); Report and Order, In the matter of Amendment of Section 73.3555 (formerly Sections 73.35, 73.240, and 73.636), of the Commission's Rules Relating to Multiple Ownership of AM, FM, and Televison Broadcast Stations, 100 F.C.C.2d 17 (1984).

4. Report and Order, In re Revision of Radio Rules and Policies, 7 F.C.C.Rcd. 2755 (1992)

5. Commercial Television Stations, 98 F.C.C.2d 1076 (1984).

6. See, for example, the Canadian rules to counter gender stereotypes.

7. *Red Lion Broadcasting Co. v. FCC,* 395 U.S. 367, 89 S.Ct. 179 (1969).

8. See the U.S. considerations in this direction. Notice of Proposed Rule Making, In the Matter of Implementation of Section 25 of the Cable Television Consumer Protection and Competition Act of 1992, 8 F.C.C.Rcd. 93 (1993).

9. See Bangemann, 1994.

Cases

UNITED STATES

National Broadcasting Co. v. United States, 319 U.S. 190, 63 S.Ct. 997 (1943).

United States v. RCA, 358 U.S. 334, 79 S.Ct. 457 (1959).

Office of Communications, United Church of Christ v. FCC, 359 F.2d 994 (D.C.Cir. 1966).

United States v. Southernwestern Cable Co., 392 U.S. 157, 88 S.Ct. 1994 (1968).

Red Lion Broadcasting Co. v. FCC, 395 U.S. 367, 89 S.Ct. 179 (1969).

Brandywine-Main Line Radio, Inc. v. FCC, 473 F.2d 79f. (D.C.Cir. 1972).

Miller v. California, 413 U.S. 15, 93 S.Ct. 2607 (1973).

Writers Guild of America West, Inc. v. FCC, 423 F.Supp. 1064 (C.D. Cal 1976).

FCC v. Pacifica Foundation, 438 U.S. 726, 93 S.Ct. 3026 (1978).

FCC v. Midwest Video Corp., 440 U.S. 689, 99 S.Ct. 1435 (1979).

Writers Guild of America West, Inc. v. ABC, 609 F.2d 355 (9th Cir. 1979).

FCC v. WNCN Listeners Guild, 450 U.S. 582, 101 S.Ct. 1266 (1981).

CBS, Inc. v. FCC, 453 U.S. 367, 101 S.Ct. 2813 (1981).

Central Florida Enterprises, Inc. v. FCC, 683 F.2d 503 (D.C.Cir. 1982), *cert. denied,* 460 U.S. 1984 (1983).

Office of Communications, United Church of Christ v. FCC, 707 F.2d 1413 (D.C.Cir. 1983).

NAB v. FCC, 740 F.2d 1190 (D.C.Cir. 1984).

Quincy Cable TV, Inc., v. FCC, 768 F.2d 1434 (D.C.Cir. 1985).

Steele v. FCC, 770 F.2d 1192 (D.C.Cir. 1985).

Office of Communications, United Church of Christ v. FCC, 779 F.2d 702 (D.C.Cir. 1985).

Telecommunications Research and Action Center v. FCC, 801 F.2d 501 (D.C.Cir. 1986), *cert. denied* 107 S.Ct. 3196 (1987).

Pappas v. FCC, 807 F.2d 1019 (D.C.Cir. 1986).

City of Los Angeles v. Preferred Communications, Inc., 476 U.S. 488, 106 S.Ct. 3034 (1986).

Century Communications Corp. v. FCC, 835 F.2d 292 (D.C.Cir. 1987).

Action for Children's Television v. FCC, 821 F.2d 741 (D.C.Cir. 1987).

Telecommunications Research and Action Center v. FCC, 836 F.2d 1349 (D.C.Cir. 1988).

National Association for Better Broadcasting v. FCC, 849 F. 2d 665, 64 R.R.2d 1570 (D.C.Cir. 1988).

Action for Children's Television v. FCC, 852 F.2d 1332 (D.C.Cir. 1988).

Syracuse Peace Council v. FCC, 867 F.2d 654 (D.C.Cir. 1989).

Sable Communications v. FCC, 492 U.S. 115, 109 S.Ct. 2829 (1989).

Metro Broadcasting, Inc. v. FCC, 110 S.Ct. 2997 (1990).

Monroe Communications Corp. v. FCC, 900 F.2d 351 (D.C.Cir. 1990).

Action for Children's Television v. FCC, 932 F.2d 1504 (D.C.Cir. 1990), *cert. denied*, 112 S.Ct. 1281.

Lamprecht v. FCC, 958 F.2d 382 (D.C.Cir. 1992).

Schurz Communications, Inc. v. FCC, 982 F.2d 1043 (7th 1992).

Action for Children's Television v. FCC, 304 U.S. App. D.C. 126 (D.C.Cir. 1993), *vacated* 15 F.3d 186.

Turner Broadcasting System, Inc. v. FCC, 114 S.Ct. 2445 (1994).

GREAT BRITAIN

R. v. Secretary of State for the Home Office, ex parte Brind (1991) 1 A.C. 696, H.L.

GERMANY

BVerfGE 12, 205; NJW 1961, 547—Deutschland-Fernsehen; Urteil vom 28.2.1961.

BVerfGE 31, 314; NJW 1971, 1793—Mehrwertsteuer; Urteil vom 27.7.1971.

BVerfGE 35, 202; NJW 1973, 1226—Lebach; Urteil vom 5.6.1973.

BayVerfGE 39, 69; ZUM 1986; Entscheidung vom 21.11.1986.

BVerfGE 57, 295; NJW 1981, 1774—FRAG; Urteil vom 16.6.1981.

BVerfGE 73, 118; NJW 1987, 239—Niedersächsisches Landes-Rundfunkgesetz; Urteil vom 4.11.1986.

BVerfGE 74, 297; NJW 1987, 1251—Landesmediengesetz Baden-Württemberg; Urteil vom 24.3.1987.

BVerfGE 83, 238; NJW 1991, 899—Nordrhein-Westfälische Rundfunkgesetzgebung; Urteil vom 5.2.1991.

BVerfGE 90, 60; NJW 1994, 1942—Gebühren; Urteil vom 22.2.1994.

BVerfGE, ZUM 1994, 639—DSF-Zulassung; Beschluss vom 9.7.1993.

OVG Lüneburg, DVBl. 1986, 1112; Beschluss vom 16.9.1986.

BayVerwGH, NVwZ 1987, 435; Beschluss vom 27.8.1987.

VG München, NVwZ 1987, 438; Urteil vom 4.12.1987.

VGH Mannheim, NJW 1990, 340, Beschluss vom 14.12.1988.

BayVerfGH, ZUM 1992, 378; Beschluss vom 30.3.1992.

VG München, ZUM 1993, 294; Beschluss vom 16.2.1993.

BayVerwGH, ZUM 1993, 296; Beschluss vom 24.3.1993.

FRANCE

Décision No. 86-217 DC, 18 September 1986, J.O. 19 September 1986, 11294 ff. Liberté de Communication.

Décision No. 88-248 DC, 17 January 1989, J.O. 18 January 1989, 755 ff.

Décision No. 90-182 DC, 19 June 1990, J.O. 24 June 1990, 7351 ff.

CANADA

Judicial Committee of the Privy Council re Regulation and Control of Radio Communication, A.G. Que. v. A.G. Can. et al., February 9, 1932, in: (1932) 2 D.L.R., 81.

Federal Court of Appeal re Capital Cities Communications Inc. et al. v. Canadian Radio Television Commission, January 17, 1975, in: (1975) 52 D.L.R. 3d, 415.

Ontario Court of Appeal re C.F.R.B. v. Attorney General of Canada et al., June 21, 1977, in: (1973) 38 D.L.R. 3d, 335.

Supreme Court of Canada re Capital Cities Communications Inc. et al. v. Canadian Radio-Television Commission et al., November 30, 1977, in: (1977) 81 D.L.R. 3d, 609.

Supreme Court of Canada re Public Service Board et al., Dionne et al. v. Attorney General of Canada et al., November 30, 1977, in: (1977) 83 D.L.R. 3d, 178.

Federal Court of Appeal re CTV Television Network Ltd. v. Canadian Radio–Television and Telecommunications Commission et al., November 7, 1980, in: (1980) 116 D.L.R. 3d, 741.

Supreme Court of Canada re CTV Television Network Ltd. v. Canadian Radio–Television and Telecommunications Commission et al., April 5, 1982, in: (1982) 134 D.L.R. 3d, 193.

Supreme Court of Canada, re Irwin Toy Ltd. v. Quebec A.G. (1989) 58 D.L.R. 4th, 577.

AUSTRALIA

R. v. Australian Broadcasting Tribunal, ex parte 2HD Pty Ltd. (1979) 27 Austl.L.R. 321.

Australian Broadcasting Tribunal v. Alan Bond (1990) 94 Austl.L.R. 11.

Australian Capitol Television Pty Ltd. and Ors v. The Commonwealth (1992) 66 Austl.L.Jur.R. 695 (HC).

Austereo v. Trade Practices Commission (1993) 115 Austl.L.R. 14.

The Herald and Weekly Times Ltd. v. Commonwealth 17 Austl.L.R. 281.

EUROPEAN COURTS

European Court of Human Rights, Judgment of 28 March 1990, Nr. 14/1988/158/214, Ser. A 173; EuGRZ 1990, 255—*Groppera Radio AG et al. v. Suisse.*

Authorities

UNITED STATES (FCC)

Public Notice, Policy Statement on Comparative Broadcast Hearings, 1 F.C.C.2d 393 (1965).

Second Report and Order, In the Matter of Community Antenna Television; Amendment of parts 21, 74, and 91 of the Commission's rules, 2 F.C.C. 725 (1966).

Children's Television Report and Policy Statement, In the Matter of Petition of Action for Children's Television (ACT) for Rulemaking Lookind toward the Elimination of Sponsorship and Commercial Content in the Children's Programming and the Establishment of a Weekly 14-Hour Quota of Children's Television Programs, 50 F.C.C.2d 1 (1974).

Memorandum and Opinion Order, In the Matter of Citizen's Complaint against Pacifica Foundation Station WBAI (FM), New York, N.Y., 56 F.C.C.2d 946 (1975).

Report, In the Matter of Inquiry into the Economic Relationships between Television Broadcasting and Cable Television, 71 F.C.C.2d 632 (1979).

Notice of Proposed Rulemaking, In the Matter of Children's Television Programming and Advertising Practices, 75 F.C.C.2d 138 (1979).

Report and Order, In the Matter of Deregulation of Radio, 84 F.C.C.2d 968 (1981).

Decision, In re Applications of Cowles Broadcasting Inc. (WESH-TV), for Renewal of License Central Florida Enterprises, Inc., 86 F.C.C.2d 993 (1981).

Report and Order, In the Matter of Inquiry into the Development of Regulatory Policy in Regard to Direct Broadcast Satellites for the Period Following the 1983 Regional Administrative Radio Conference, 90 F.C.C.2d 676 (1983).

Policy Statement and Notice of Proposed Rule Making, In the Matter of Commission

Policy Regarding the Advancement of Minority Ownership in Broadcasting, 92 F.C.C.2d 849 (1982).

Second Report and Order, In the Matter of Amendment of the Commission's Rules to Allow the Selection from Among Certain Competing Applications Using Random Selection or Lotteries Instead of Comparative Hearings, 93 F.C.C.2d 952 (1983).

Notice of Proposed Rule Making, In the Matter of Amendments of Sections 73.35, 73.240, and 73.636 of the Commission's Rules Relating to Multiple Ownership of AM, FM, and Television Broadcast Stations, 95 F.C.C.2d 360 (1983).

Report and Order, In the Matter of Children's Television Programming and Advertising Practices, 96 F.C.C.2d 634 (1984).

Report and Order, In the Matter of the Revision of Programming and Commercialization Policies, Ascertainment Requirements, and Program Log Requirements for Commercial Television Stations, 98 F.C.C.2d 1076 (1984).

Report and Order, In the Matter of Amendment of Section 73.3555, (formerly Sections 73.35, 73.240, and 73.636) of the Commission's Rules Relating to Multiple Ownership of AM, FM, and Television Broadcast Stations, 100 F.C.C.2d 17 (1984).

Report and Order, In the Matter of Amendment of Part 73 of the Commission's Rules Concerning the Filing of Network Affiliation and Transcription Contracts, 101 F.C.C.2d 516 (1985).

Report, In the Matter of Inquiry into Section 73.1910 of the Commission's Rules and Regulations Concerning the General Fairness Doctrine Obligations of Broadcast Licensees, 102 F.C.C.2d 145 (1985).

Memorandum Opinion and Order, In the Matter of Revision of Programming and Commercialization Policies, Ascertainment Requirements, and Program Log Requirements for Commercial Television Stations, 104 F.C.C.2d 358 (1986).

Memorandum Opinion and Order, In the Matter of Deregulation of Radio, 104 F.C.C.2d 505 (1986).

Notice of Inquiry, In the Matter of Reexamination of the Commission's Comparative Licensing, Distress Sales and Tax Certificate Politics Premised on Racial, Ethnic or Gender Classifications, 1 F.C.C.Rcd. 1315 (1986).

New Indecency Enforcement Standards to be Applied to All Broadcast and Amateur Radio Licenses, 62 R.R.2d 1218 (1987).

Memorandum Opinion and Order, In the Matter of Infinity Broadcasting Corporation of Pennsylvania; Pacifica Faundation, Inc.; The Regents of the University of California, 3 F.C.C.Rcd. 930 (1987).

Memorandum Opinion and Order, In re Complaint Syracuse Peace Council against WTVH Syracuse, 63 R.R.2d 542 (1987).

Order, In the Matter of Enforcement of Prohibitions against Broadcast Obscenity and Indecency in 18 U.S.C. 1464, 4 F.C.C.Rcd. 457 (1988).

First Report and Order, In the Matter of Formulation of Policies and Rules Relating to

Broadcast Renewal Applicants, Competing Applicants, and Other Participants to the Comparative Renewal Process and to the Prevention of Abuses of the Renewal Process, 4 F.C.C.Rcd. 4780 (1989).

Report and Order, In the Matter of Review of Rules and Policies Concerning Network Broadcasting by Television Stations: Elimination or Modification of sec. 73.658(c) of Commission's Rules, 4 F.C.C.Rcd. 2755 (1989).

Report and Order, In the Matter of Amendment of Sections 1.420 and 73.3584 of the Commission's Rules Concerning Abuses of the Commission's Processes, 5 F.C.C.Rcd. 3911 (1990).

Report of the Commission, In the Matter of Enforcement of Prohibition against Broadcast Indecency in 18 U.S.C. 1464, 5 F.C.C.Rcd. 5297 (1990).

Report and Order, In the Matter of Amendment of Section 73.3525 of the Commission's Rules Regarding Settlement Agreements among Applicants for Construction Permits, 6 F.C.C.Rcd. 85 (1991).

Report and Order, In the Matter of Policies and Rules Concerning Children's Television and Commercialization Policies, Ascertainment Requirements and Program Log Requirements for Commercial Television Stations, 6 F.C.C.Rcd. 2111 (1991).

Report and Order, In the Matter of Evaluation of the Syndication and Financial Interest Rules, 6 F.C.C.Rcd. 3094 (1991).

Office of Plans and Policy Working Paper No. 26, Broadcast Television in a Multichannel Marketplace, 6 F.C.C.Rcd. 3996 (1991).

Memorandum Opinion and Order, In the Matter of Policies and Rules Concerning Children's Television Programming; Revision of Programming and Commercialization Policies, Ascertainment Requirements, and Program Log Requirements for Commercial Television Stations, 6 F.C.C.Rcd. 5093 (1991).

Memorandum Opinion and Order, In the Matter of Evaluation of the Syndication and Financial Interest Rules, 7 F.C.C. Rcd. 345 (1991).

Notice of Proposed Rule Making and Notice of Inquiry, In the Matter of Review of Commission's Regulations and Policies Affecting Investment in the Broadcast Industry, 7 F.C.C.Rcd. 2654 (1992).

Report and Order, In re Revision of Radio Rules and Policies, 7 F.C.C.Rcd. 2755 (1992).

Video Dialtone Order, In the Matter of Telephone Company Cable Television Cross-Ownership Rules Sectors 63.54–63.58, 7 F.C.C. Rcd. 5781 (1992).

Report and Order, In the Matter of Amendment of Part 76, Subpart J, Section 76.501 of the Commission's Rules and Regulations to Eliminate the Prohibition on Common Ownership of Cable Television Systems and National Television Networks, 7 F.C.C. Rcd. 6156 (1992).

Further Proceedings on Cable-Cross-Ownership Rules, 69 R.R.2d 1613 (1993).

Second Report and Order, In the Matter of Evaluation of the Syndication and Financial Interest Rules, 8 F.C.C. Rcd. 3282 (1993).

Notice of Proposed Rule Making, In the Matter of Implementation of Section 25 of the Cable Television Consumer Protection and Competition Act of 1992, 8 F.C.C. 93 (1993).

Notice of Proposed Rule Making, In the Matter of Implementation of Section 4 (g) of the Cable Television Consumer Protection and Competition Act of 1992, Home Shopping Station Issues, 8 F.C.C. Rcd. 5321 (1993).

Memorandum Opinion and Order, In the Matter of Evaluation of the Syndication and Financial Interest Rules, 8 F.C.C. Rcd. 8270 (1993).

CANADA (CRTC)

CRTC, Public Announcement, Ottawa, March 26, 1979, A Review of Certain CableTelevision Programming Issues.

CRTC, Public Notice 1983-18, Ottawa, January 31, 1983, Policy Statement on Canadian Content in Television.

CRTC, Public Notice 1984-94, Ottawa, April 15, 1984, Recognition for Canadian Programs.

CRTC, Decision 1986-367, Ottawa, April 18, 1986, Applications for Authority to Transfer Effective Control of Télé-Métropole Inc. to Power Corporation of Canada.

CRTC, Decision 1986-586, Ottawa, June 19, 1986, Multilingual Television (Toronto) Limited.

CRTC, Decision 1986-789, Ottawa, August 26, 1986, QCTV Ltd.

CRTC, Public Notice 1986-247, Ottawa, September 19, 1986, Broadcast Advertising of Beverages and Food and Drugs.

CRTC, Public Notice 1986-248, Ottawa, September 19, 1986, Regulations Respecting Radio Broadcasting.

CRTC, Decision 1986-1086, Ottawa, November 14, 1986, Global Communications Limited.

CRTC, Public Notice 1986-351, Ottawa, December 22, 1986, Policy on Sex-Role Stereotyping in the Broadcast Media.

CRTC, Decision 1987-62, Ottawa, January 27, 1987, Applications for Authority to Transfer Effective Control of Télé-Métropole Inc. to Le Groupe Vidéotron Ltée.

CRTC, Decision 1987-200, Ottawa, March 24, 1987, CTV Television Network Limited.

CRTC, Public Notice 1989-109, Ottawa, September 28, 1989, Elements Assessed by the Commission in Considering Applications for the Transfer of Ownership of Control of Broadcasting Undertakings.

CRTC, Public Notice 1990-111, Ottawa, December 17, 1990, An FM Policy for the Nineties.

CRTC, Public Notice 1990-81, Ottawa, August 15, 1990, Call for Comments on Ownership Transfer Policy.

CRTC, Public Notice 1990-99, Ottawa, October 26, 1990, Industry Guidelines for Sex-Role Portrayal.

CRTC, Public Notice 1991-11, Ottawa, January 23, 1991, Proposed Amendments to Broadcast Code for Advertising to Children.

CRTC, Public Notice 1991-90, Ottawa, August 30, 1991, Canadian Broadcast Council.

CRTC, Public Notice 1992-28, Ottawa, February 26, 1992, New Flexibility with regard to Canadian Program Expenditures to Canadian Television Nations.

CRTC, Public Notice 1993-74, Ottawa, June 3, 1993, Structural Public Hearing.

CRTC, Telecom Decision 1994-19, Ottawa, September 16, 1994, Regulatory Framework.

AUSTRALIA (ABT)

ABT-Report No. 41/90 (1980), Ansett Transport Industries Ltd. / Control Investments Pty Ltd.

References

Acheson, Keith, and Christopher Maule. (1990). Canadian content rules: A time for reconsideration, *Canadian Public Policy, 16*(3): 284–297.

Alston, Richard. (1989). Fit and proper person, *Communications Law Bulletin, 9*(3): 1–2.

Alstyen, William W. Van. (1984). *Interpretations of the First Amendment.* Durham, NC: Duke University Press.

Alvarado, Manuel, and Edward Buscombe. (1978). *Hazel: The Making of a TV Series.* London: British Film Institute.

Alvarado, Manuel, Gareth Locksley, and Silvia Paskin. (1992). Great Britain, pp. 295–333 in Alessandro Silj (Ed.), *The New Television in Europe.* London: John Libbey.

Anderson, David A. (1983). Origin of the press clause, *U.C.L.A. Law Review, 30*: 455–481.

Annan Report. (1977). *Report of the Committee on the Future of Broadcasting.* London: Her Majesty's Stationery Office.

Anstalt für Kabelkommunikation Berlin. (1989). *Jahresbericht 1988.* Berlin: Author.

Appleton, Gillian. (1988). How Australia sees itself: The role of commercial television, pp. 185–246 in Australian Broadcasting Tribunal (Ed.), *The Price of Being Australian.* Sydney: ABT.

Appleton, Gillian. (1991). Broadcasting ownership: Same Titanic, different deckchairs, *Communications Update, 1991*(63): 2–4.

Armstrong, Mark. (1979). The Broadcasting and Television Act, 1948–1976: A case study of the Australian Broadcasting Control Board, pp. 124–145 in R. Tomasic (Ed.), *Legislation and Society in Australia.* Sydney: Allen & Unwin.

Armstrong, Mark. (1987–1991). *Communications Law and Policy in Australia (Loose-Leaf Service 1987–1991).* Sydney: Butterworth.

Armstrong, Mark. (1987/1993). *Communications Law and Policy in Australia (Loose-Leaf Service 1987–1993).* Sydney: Butterworth.

Armstrong, Mark. (1991a). Das Rundfunksystem Australiens, pp. 42–48 in Hans-Bredow-Institut (Ed.), *Internationales Handbuch für Rundfunk und Fernsehen 1992/93*. Baden-Baden: Nomos.

Armstrong, Mark. (1991b). Access to decision-making about communications: Form and substance in the Australian experience, pp. 114–121 in Pacific Telecommunications Council (Ed.), *Proceedings of the Thirteenth Annual Pacific Telecommunications Conference*. Honolulu: Author.

Armstrong, Mark, M. Blakeney, and R. Watterson. (1988). *Media Law in Australia*, 2nd ed. Melbourne: Oxford University Press. (Originally published 1983)

Aufderheide, Patricia. (1990). After the fairness doctrine: Controversial broadcast programming and the public interest, *Journal of Communication, 40*: 47–73.

Auletta, Ken. (1992). *Three Blind Mice: How the TV Networks Lost Their Way*. New York: Vintage Books.

Australian Broadcasting Corporation (ABC). (1990). *Annual Report 1989–90*. Sydney: Author.

Australian Broadcasting Tribunal (ABT). (1977). *Self-Regulation for Broadcasters? A Report on the Public Inquiry into the Concept of Self-Regulation for Australian Broadcasters*. Canberra: Australian Government Publishing Service; Sydney: Author.

Australian Broadcasting Tribunal (ABT). (1984). *Satellite Program Services. Summary of Main Findings and Conclusions of the Report to the Minister for Communications*. Sydney: Author.

Australian Broadcasting Tribunal (ABT). (1987). *Advertising Time on Television. A Review of the Advertising Time Standards*. Canberra: Australian Government Publishing Service.

Australian Broadcasting Tribunal (ABT). (1988a). *Australian Content Inquiry. Discussion Paper. Ratings of Australian Drama Series, Mini-Series, Films and Tele Movies*. Sydney: Author.

Australian Broadcasting Tribunal (ABT). (1988c). *Australian Broadcasting Tribunal Manual*. Sydney: Author.

Australian Broadcasting Tribunal (ABT). (1989). *Annual Report 1988–89*. Sydney: Author.

Australian Broadcasting Tribunal (ABT). (1990a). *Australian Broadcasting Tribunal Manual*. Sydney: Author.

Australian Broadcasting Tribunal (ABT). (1990b). *TV Violence in Australia. Report to the Minister for Transport and Communications, Vols. 1–6*. Sydney: Author.

Australian Broadcasting Tribunal (ABT). (1990c). *The Bond Inquiry. Final Report*. Sydney: Author.

Australian Broadcasting Tribunal (ABT). (1990d). *Inquiry into Accuracy, Fairness and Impartiality in Current Affairs Programs on Television and Radio. Information Paper Update*. Sydney: Author.

Australian Broadcasting Tribunal (ABT). (1990e). *Review of Advertising Time on Television. Proposal*. Sydney: Author.

Australian Broadcasting Tribunal (ABT). (1990f). *Annual Report 1989–90*. Sydney: Author.

Azibert, Michel. (1987). La communication audiovisuelle: Éléments de conceptualisation d'un système, *Revue Française d'Administration Publique, 44*: 27–43.

Babe, Robert E. (1976). Regulation of private television broadcasting by the Canadian

Radio-Television Commission: A critique of ends and means, *Canadian Public Administration, 19*(4): 552–586.

Babe, Robert E. (1979). *Canadian Television Broadcasting. Structure, Performance and Regulation.* Ottawa: Minister of Supply and Services Canada.

Bagdikian, Ben H. (1990). *The Media Monopoly,* 3rd ed. Boston: Beacon Press. (Originally published 1983)

Bailey, S. H., D. J. Harvi, and B. L. Jones. (1991). *Civil Liberties: Cases and Materials,* 3rd ed. London: Butterworths.

Baldwin, Thomas F., Michael O. Wirth, and Jayne W. Zenaty. (1978). The Economics of per-program pay cable television, *Journal of Broadcasting, 22*(2): 143–154.

Bangemann, Martin. (1994). *Europa und die globale Informationsgesellschaft.* Brussels: Author.

Barile, Paolo, and Guiseppe Rao. (1992). Trends in the Italian mass media and media law, *European Journal of Communication, 7*(2): 261–283.

Barnett, Stephen R. (1988). Regulation of mass media, pp. 81–246 in Stephen R. Barnett, Michael Botein, and Eli M. Noam (Eds.), *Law of International Telecommunications in the United States.* Baden-Baden: Nomos.

Barrow, Roscoe L. (1975). The fairness doctrine: A double standard for electronic and print media, *Hastings Law Journal, 26*: 659–707.

Bartlett, Peter. (1990). Right of reply, *Communications Law Bulletin, 10*(3): 30.

Bates, Benjamin J. (1987). The role of theory in broadcasting economics: A review and development, pp. 146–171 in Margaret L. McLaughlin (Ed.), *Communication Yearbook 10.* Newbury Park, CA: Sage.

Bauer, Hartmut. (1987). Informelles Verwaltungshandeln im öffentlichen Wirtschaftsrecht, *Verwaltungs-Archiv, 78*: 241–248.

Bauman, Serge. (1981). *L'information manipulée.* Paris: Edition de la Revue Politique et Parlementaire.

Bausch, Hans. (1980). *Rundfunk in Deutschland. Band 3: Rundfunkpolitik nach 1945.* München: DTV.

Bayerische Landeszentrale für Neue Medien (BLM). (1989a). *Geschäftsbericht der Bayerische Landeszentrale für Neue Medien (BLM).* München: Author.

Bayerische Landeszentrale für neue Medien (BLM). (Ed.). (1989b). *BLM Rundfunkkongress am 10./11.10.1989: Rundfunk in den 90er Jahren—zwischen Kultur, Kommerz und Internationalisierung, Dokumentation.* München: Fischer.

Bayerische Landeszentrale für Neue Medien (BLM). (1990). *Geschäftsbericht der Bayerische Landeszentrale für Neue Medien (BLM).* München: Author.

Bayerische Landeszentrale für Neue Medien (BLM). (1991). *Geschäftsbericht der Bayerische Landeszentrale für Neue Medien (BLM).* München: Author.

Beazley, Kim. (1990). The overhaul of the Broadcasting Act, *Communications Law Bulletin, 10*(2): 15.

Beke, John A.. (1972). Government regulation of broadcasting in Canada, *Canadian Communications Law Review, 2*: 104–110.

Belot, Jean. (1988). Zwischen Staat und Kommerz. Die Fernsehentwicklung in Frankreich (1), *Evangelischer Pressedienst/Kirche und Rundfunk, 1988*(84): 3–6.

Benda, Ernst. (1989). Den Länderegoismus überwinden, *Evangelischer Pressedienst/Kirche und Rundfunk, 1989*(97): 5–8.

Berendes, Konrad. (1973). *Die Staatsaufsicht über den Rundfunk.* Berlin: Duncker & Humblot.

Bernstein, Marver H. (1955). *Regulating Business by Independent Commissions.* Princeton, NJ: Princeton University Press.

Berwanger, Dietrich. (1988). *Televison in the Third World,* 2nd ed. Bonn: Friedrich-Ebert-Stiftung. (Originally published 1987)

Bird, Roger. (Ed.). (1989). *Documents of Canadian Broadcasting.* Ottawa: Carleton University Press.

Bittner, John R. (1994). *Law and Regulation of Electronic Media,* 2nd ed. Englewood Cliffs, NJ: Prentice-Hall.

Blaise, Jean-Bernard, and Michel Formont. (1992). *Das Wirtschaftsrecht der Telekommunikation in Frankreich.* Baden-Baden: Nomos.

Blasi, Vincent. (1977). *The Checking Value in the First Amendment Theory.* Washington, DC: Weaver (American Bar Foundation).

Blumler, Jay G. (Ed.). (1992). *Television and the Public Interest: Valuable Values in West European Broadcasting.* London: Sage.

Blumler, Jay G., and Wolfgang Hoffmann-Riem. (1992). New roles for public television in Western Europe: Challenges and prospects, *Journal of Communication, 42*(1): 20–35.

Bock, Gabriele, and Siegfried Zielinski. (1987). Der britische Channel 4, *Media Perspektiven, 1987*(1): 38–54.

Bohne, Eberhard. (1981). *Der informale Rechtsstaat.* Berlin: Duncker & Humblot.

Bohne, Eberhard. (1984). Informales Verwaltungs-und Regierungshandeln als Instrumente des Umweltschutzes, *Verwaltungsarchiv, 75*(4): 343–374.

Bombardier, Denise. (1975). *La voix de la France.* Paris: Laffont.

Bonney, Bill. (1987). Commercial television: Regulation, technology and market forces, pp. 51–78 in T. Weelwright and K. Buckley (Eds.), *Communications and the Media in Australia.* Sydney: Allen & Unwin.

Boudet, Daniel. (1987). L'enjeu des concentrations: Le difficile avènement d'une loi, *Revue Française d'Administration Publique, 44*: 63–72.

Branscomb, Anne Wells. (1994). *Who Owns Information? From Privacy to Public Access.* New York: Basic Books.

Brants, Kees, and Denis McQuail. (1992). The Netherlands, pp. 152–166 in Euromedia Research Group (Ed.), *The Media in Western Europe: The Euromedia Handbook.* London: Sage.

Bredin Rapport. (1985). *Les nouvelles télévisones hertziennes.* Paris: La Documentation Française.

Briggs, Asa. (1985). *The BBC: The First Fifty Years.* Oxford: Oxford University Press.

Briggs, Asa, and Joanna Spicer. (1986). *The Franchise Affair, Creating Fortunes and Failures in Independent Television.* London: Century Hutchinson.

Brinkmann, Thomas. (1983). Der Zugang der Presse zum Rundfunk und die Chancen für Vielfalt durch Wettbewerb, *Media Perspektiven, 1983*(10): 677–689.

British Broadcasting Corporation (BBC). (1992). *Extending Choice: The BBC's Role in the New Broadcasting Age.* London: BBC-Edition.

Broadband Services Expert Group. (1994). *Networking Australia's Future—Interim Report.* Sydney: Author.

Broadcasting Standards Council (BSC). (1989). *A Code of Practice.* London: BSC-Edition.

Brotman, Stuart N. (1988). *Communications Policy Making at the FCC.* St Paul, MN: West Publishing.

Brown, Allan. (1987). Television equalisation: A case study of industry restructure, pp. 174–184 in Australian Broadcasting Tribunal (Ed.), *The Price of Being Australian.* Sydney: Australian Broadcasting Tribunal.

Brown, Paul. (1988). Action stations, *Airwaves, 1988*(17): 19.

Bullinger, Martin. (1987). Freiheit und Gleichheit in den Medien, *Juristenzeitung, 42*(6): 257–265.

Bullinger, Martin. (1988). Europäische Rundfunkordnungen im Übergang, pp. 45–88 in Ernst-Joachim Mestmäcker (Ed.), *Offene Rundfunkordnung.* Gütersloh: Verlag Bertelsmann Stiftung.

Bultmann, Fritz A., and Gerd-Jürgen Rahn. (1988). Rechtliche Fragen des Tele-shopping, *Neue Juristische Wochenschrift, 41*(39): 2432–2438.

Bureau, André. (1988). *Notes for an Address to the Law Society of Upper Canada's Conference on Communications Law and Policy, March 25, 1988.* Ottawa: CRTC.

Bureau of National Affairs, Inc. (1991). *Antitrust and Trade Regulation Report.* Washington, DC: Author.

Bureau of Transport and Communications Economics. (1990). *Communications Services in Australia.* Canberra: Australian Government Publishing Service.

Burnett, Claire. (1989). Second Base, *Cable and Satellite Europe, 1989*(5): 22–24.

Cable Authority. (1988a). *Advertising & Sponsorship Codes and Guidelines.* London: Author.

Cable Authority. (1988b). *Programme Guidelines and Codes.* London: Author.

Cable Authority. (1989). *Annual Report and Accounts 1988–89.* London: Author.

Cable Authority. (1990). *Annual Report and Accounts 1989–90.* London: Author.

Canadian Association of Broadcasters. (1987). *The Broadcast Code for Advertising to Children.* Ottawa: Author.

Canadian Broadcasting Corporation. (1988). *Annual Report 1987–1988,* Ottawa: Author.

Canadian Broadcasting Corporation. (1990). *Annual Report 1989–1990,* Ottawa: Author.

Canadian Radio and Television Commission (CRTC). (1971). *La Radiodiffusion Canadienne: "Un Système Unique." Enonce de Politique sur la Télévision par Cable, 16 juillet 1971.* Ottawa: Minister of Supply and Services Canada.

Canadian Radio and Television Commission (CRTC). (1975). *Policies Respecting Broadcasting Receiving Undertakings (CableTelevision), December 16, 1975.* Ottawa: Minister of Supply and Services Canada.

Canadian Radio-Television and Telecommunications Commission (CRTC). (1979). *Special Report on Broadcasting in Canada 1968–1978, Vol. 1.* Ottawa: Minister of Supply and Services Canada.

Canadian Radio-Television and Telecommunications Commission (CRTC). (1980). *Canadian Broadcasting and Telecommunications: Past Experience, Future Options.* Ottawa: Minister of Supply and Services Canada.

Canadian Radio-Television and Telecommunications Commission (CRTC). (1988). *Annual Report 1987–1988.* Ottawa: Minister of Supply and Services Canada.

Canadian Radio-Television and Telecommunications Commission (CRTC). (1989). *Annual Report 1988–1989.* Ottawa: Minister of Supply and Services Canada.

Canadian Radio-Television and Telecommunications Commission (CRTC). (1990). *Annual Report 1989–1990.* Ottawa: Minister of Supply and Services Canada.

Carter, Barton T., Juliet Lushbough Dee, Martin J. Gaynes, and Harvey L. Zuckman. (1994). *Mass Communication Law,* 4th ed. St. Paul, MN: West Publishing.

Carter, Barton T., Marc A. Franklin, and Jay B. Wright. (1994). *The First Amendment and the Fourth Estate: The Law of Mass Media,* 6th ed. Westbury, NY: Foundation Press. (Originally published 1977)

Cazenave, François. (1984). *Les radio libres. Des radio pirates aux radios locales privées,* 2nd ed. Paris: Presse Universitaires de France. (Originally published 1980)

Chadwick, Paul. (1989). *Media Mates: Carving up Australia's Media.* Melbourne: Macmillan.

Chamberlin, Bill F. (1982). Lessons in regulating information flow: The FCC's weak track record in interpreting the public interest standard, *North Carolina Law Review, 60*: 1057–1113.

Channel Four. (1990). *Report and Accounts for the Year Ended 31st March 1990.* London: Author.

Chaumont, Jean-Pierre. (1989). La création du Conseil Supérieur de l'Audiovisuel, *Les Petit Affiches, 27*: 7–15.

Checkland, Michael. (1989). The BBC—licence to keep viewers, *Combroad, 85*: 19–22.

Chevallier, Jacques. (1982). Le statut de la communication audiovisuelle, *L'Actualité Juridique—Droit Administratif, 1982*(10): 555–576.

Chevallier, Jaques. (1989). De la C.N.C.L. supérieur des l'audiovisuel, premiers commentaires de la loi du 17 janvier 1989, *L'Actualité Juridique—Droit Administratif, 1989*(2): 59–76.

Chomsky, Noam. (1989). *Necessary Illusion: Thought Control in Democratic Societies.* Boston: South End Press.

Chomsky, Noam, and Eduard V. Herman. (1988). *Manufacturing Consent: The Political Economy of the Mass Media.* New York: Pantheon.

Cojean, Annick, and Frank Eskenazi. (1986). FM—*La folle histoire de radios libres.* Paris: Grasset.

Cole, Barry, and Mal Oettinger. (1978). *Reluctant Regulators: The FCC and the Broadcast Audience.* Reading, MA: Addison-Wesley.

Collins, Richard. (1990). *Culture, Communication and National Identity: The Case of Canadian Television.* Toronto: University of Toronto Press.

Collins, Richard. (1993). *Broadcasting and Audiovisual Policy in the European Single Market.* London: Libbey.

Commission de Réflection et d'Orientation Presidée par Pierre Moinot. (1981). *Pour une reforme de l'audiovisuel.* Paris: La documentation Française.

Commission Nationale de la Communication et des Libertés (CNCL). (1987). *Rapport Annuel Novembre 1986–Novembre 1987.* Paris: Author.

Commission Nationale de la Communication et des Libertés (CNCL). (1988). *2e Rapport Annuel Novembre 1987–Novembre 1988.* Paris: Author.

Commission of the European Community. (1984). *Television without Frontiers—Green*

Paper of the Establishment of the Common Market, Especially by Satellite and Cable. Brüssels: Author.

Committee of Review of the Australian Broadcasting Commission. (1981). *Report by the Committee of Review of the Australian Broadcasting Commission: The ABC in Review, National Broadcasting in the 1980s (Dix Report).* Canberra: Australian Government Publishing Service.

Committee on Broadcasting. (1965). *Report of the Committee on Broadcasting.* Ottawa: Roger Duhamel, F.R.S.C.

Communications Law Centre. (1988). *Review of the Australian Broadcasting Tribunal.* Submission to the House of Representatives Standing Committee on Transport, Communications and Infrastructure—Inquiry into the Role and Functions of the Australian Broadcasting Tribunal. Canberra: Author.

Communications Law Centre. (1989). *Efficiency Reforms—Review of Broadcasting Regulation.* Submission to the Department of Transport and Communications. Canberra: Author.

Conklin, David D. (1992). The Broadcasting Act and the changing politicial pathology of cabinets appeals, *Media and Communication Law Review, 2*: 297–333.

Conseil Supérieur de l'Audiovisuel (CSA). (1990). *Conseil Supérieur de l'Audiovisuel, 2e Rapport Annuel.* Paris: La Documentation Française.

Conseil Supérieur de l'Audiovisuel (CSA). (1991a). *L'Ame des Chaines, Evolution des Programmes de TF1—Antenne 2—FR3 de 1977 à 1990.* Paris: La Documentation Française.

Conseil Supérieur de l'Audiovisuel (CSA). (1991b). *Conseil Supérieur de l'Audiovisuel, 3e Rapport Annuel.* Paris: La Documentation Française.

Conseil Supérieur de l'Audiovisuel (CSA). (1993). *Conseil Supérieur de l'Audiovisuel, 4e Rapport Annuel.* Paris: La Documentation Française.

Conseil Supérieur de l'Audiovisuel (CSA). (1994a). *Conseil Supérieur de l'Audiovisuel, 5e Rapport d'Activité.* Paris: La Documentation Française.

Conseil Supérieur de l'Audiovisuel (CSA). (1994b). *Cinq chaîne publiques européenes fâce à la concurrence privée.* Paris: La Documentation Française.

Coopers & Lybrand. (1988). *Licensing Approach to Broadcasting Regulation in the 1990's.* London: Author.

Coppey, Clémence, Bertand Delcors, Jean Pierre, Thierry-Pierre Jouaudet, Frédéric Rauchet, and Hervé Rony. (1989). *Cent question clés de la communication audiovisuel.* Paris: Dixit Mediapouvoirs.

Corker, John, and Giles Tanner. (1990). Australian Broadcasting Tribunal vs. Alan Bond, *Communications Law Bulletin, 10*(3): 17.

Cotta, Michèle. (1986). *Les miroirs de Jupiter.* Paris: Fayard.

Coulson, David. (1990). With hindsight, *Airwaves, 1990*(23): 6.

Cousin, Bertrand, and Bertrand Delcros. (1990). *Le Droit de la Communication—Tome 1.* Paris: Editions du Moniteur.

Cousin, Bertrand, Bertrand Delcros, and Thierry Jouandet. (1990). *Le Droit de la Communication—Tome 2.* Paris: Editions du Moniteur.

Coyne, David M. (1981). The future of content regulation in broadcasting, *California Law Review, 69*: 555–598.

Cromme, Franz. (1985). Die Programmüberwachung des Rundfunkrates, *Neue Juristische Wochenschrift, 38*(7): 351–360.

Culver, Robert. (1993). Casenotes, *Cumberland Law Review, 23*: 219–235.

Cunningham, S. (1992). *Framing Culture, Criticism and Policy in Australia.* Sydney: Allen & Unwin.

Davey, Jon. (1990). Laying the foundations, *Airwaves, 1990*(23): 7–8.

Debbasch, Charles. (1989). La liberté de la communication audiovisuelle en France, *Revue Internationale de Droit Comparé, 41*(2): 305–312.

Debbasch, Charles. (1991). *Droit de l'audiovisuel,* 2nd ed. Paris: Dalloz. (Originally published 1988)

Delcros, Bertrand. (1985). Le cadre juridique de la télédistribution en France, *L'Actualité Juridique—Droit Administratif, 1985*(5): 243–253.

Delcros, Bertrand, and B. Vodan. (1987). *La liberté de communication. La loi du 30 Sept. 1986, analysem et commentaire.* Paris: La Documentation Française.

Delivet, Jean-Pierre. (1989). Une nouvelle instance de régulation. Le Conseil Supérieur de l'Audiovisuel, *Mediapouvoirs, 14*: 1–8.

Delivet, Jean-Pierre, and Herve Rony. (1987). La Commission nationale de la communication et des libertés et la régulation de la communication audiovisuelle, *Revue Française d'Administration Publique, 44*: 127–139.

Delvolvé, Pierre. (1987). Les arrets et le système de la concession de service public, *Revue Française de Droit Administratif, 3*(1): 2–10.

Denton, T. M. (1984). Canadian responses to American DBS services: Protecting the infant identity, pp. 393–399 in Vincent Mosco (Ed.), *Policy Research in Telecommunications: Proceedings from the Eleventh Annual Telecommunications Policy Research Conference.* Norwood, NJ: Ablex.

Department of Communications. (1983). *Towards a New National Broadcasting Policy. New Policies and Initiatives to Provide Canadians with Greater Program Choice and Make the Canadian Broadcasting Industry More Competitive: A Response to New Technologies and a Changing Environment.* Ottawa: Author.

Department of Communications. (1986). *Satellite Broadcasting.* Canberra: Australian Government Publishing Service.

Department of Communications. (1988). *Des voix Canadiennes pour un choix véritable: Une nouvelle politique de la radiodiffusion pour le Canada.* Ottawa: Author.

Department of National Heritage. (1992). *The Future of the BBC: A Consulation Document.* London: Her Majesty's Stationery Office.

Department of Transport and Communications. (1989). *Future Directions for Pay Television in Australia, Vols. I–II.* Canberra: Australian Government Publishing Service.

Derkenne, Marine Corpet. (1986). L'affaire de la 5e chaine, *La Revue Administratif, 39*(232): 411–412.

Deutscher Juristentag. (Ed.). (1988). *Sitzungsbericht O zum 56. Deutschen Juristentag.* München: C. H. Beck.

Diller, Ansgar. (1980). *Rundfunk in Deutschland, Band 2: Rundfunkpolitik im Dritten Reich.* München: DTV.

Direktorenkonferenz der Landesmedienanstalten (DLM). (Ed.). (1988). *DLM Jahrbuch 1988.* München: Neue Mediengesellschaft.

Direktorenkonferenz der Landesmedienanstalten (DLM). (Ed.). (1993). *DLM Jahrbuch 92—Privater Rundfunk in Deutschland.* München: R. Fischer.

Dix, Alexander. (1980). Kommerzieller Rundfunk in Grossbritannien—sein rechtlicher Rahmen und seine Kontrolle in der Praxis, *Rundfunk und Fernsehen, 28*(1): 361–378.

Donner, Arthur, and Mel Kliman. (1984). The effect of Section 19.1 of the Income Tax Act on television advertising, *Canadian Tax Journal, 32*(6), 1084–1095.

Drouot, Guy. (1986). Les modifications récentes de la loi du juillet 1982 et le développement des télévision privées en France, *Actualité Legislative Dalloz, 1986*(7): 47–57.

Drouot, Guy. (1987). Le secteur public de l'audiovisuel dans la loi du 30 septembre 1986, *Revue Française de Droit Administratif, 3*(3): 399–408.

Dutton, Williams H., and Jay G. Blumler. (1988). The faltering development of cable television in Britain, *International Political Science Review, 9*(4): 279–303.

Dyson, Kenneth. (1988a). *Politics, Policy and New Media in Britain: A Reinterpretation* (unpublished manuscript).

Dyson, Kenneth. (1988b). Regulating new media: The implementation process, pp. 251–304 in Kenneth Dyson and Peter Humphreys (Eds.), *Broadcasting and New Media Policies in Western Europe.* London: Routledge.

Dyson, Kenneth, and Peter Humphreys. (1985). The new media in Britain and in France—two versions of heroic muddle? *Rundfunk und Fernsehen, 33*(3/4): 362–379.

Dyson, Kenneth, and Peter Humphreys. (1988a). Regulatory change in Western Europe, pp. 92–160 in Kenneth Dyson and Peter Humphreys (Eds.), *Broadcasting and New Media Policies in Western Europe,* London: Routledge.

Eastman, Susan Tyler. (1993). *Broadcast/Cable Programming: Strategies and Practices,* 4th ed. Belmont: Wadsworth.

Easton, Paul. (1989). Split-programming-targeting for success, *Music and Media, 2.9.1989.*

Eberle, Carl-Eugen. (1989). *Rundfunkübertragung: Rechtsfragen derNutzung terrestrischer Rundfunkfrequenzen.* Berlin: Duncker & Humblot.

Edelman, Murray. (1950). *The Licensing of Radio Services in the United States, 1927 to 1947: A Study in Administrative Formulation of Policy.* Urbana: University of Illinois Press.

Edelman, Murray. (1964). *The Symbolic Uses of Politics.* Urbana: University of Illinois Press.

Ellis, David. (1979). *Evolution of the Canadian Broadcasting System: Objectives and Realities, 1928–1968.* Ottawa: Minister of Supply and Services Canada.

Emerson, Thomas J. (1970). *The System of Freedom of Expression.* New York: Random House.

Enchin, Harvey, and Hugh Winsor. (1990). Loss of 1.100 posts the deepest cut, *Globe and Mail, December 6, 1990,* A10.5.

Engel, Christoph. (1993a). *Privater Rundfunk vor der Europäischen Menschenrechtskonvention.* Baden-Baden: Nomos.

Engel, Christoph. (1993b). Vorsorge gegen die Konzentration im privaten Rundfunk mit den Mitteln des Rundfunkrechts—Eine Analyse von 21 Rundfunkstaatsvertrag 1991, *Zeitschrift für Urheber und Medienrecht, 37*(Sonderheft): 557–585.

Engel, Christoph. (1994). Rundfunk in Freiheit, *Archiv für Presserecht, 25*(3): 185–191.

Engel, Christoph. (1995). Multimedia und das deutsche Verfassungsrecht, pp. 155–171 in Wolfgang Hoffmann-Riem and Thomas Vesting (Eds.), *Perspektiven der Informationsgesellschaft*. Baden-Baden: Nomos.

Engels, Wolfram. (1986). *Mehr Markt bei Rundfunk und Fernsehen*. Bonn: Arbeitsgemeinschaft selbständiger Unternehmer.

Euromedia Research Group. (Ed.). (1992). *The Media in Western Europe*. London: Sage.

European Community (EC). (1984). *Television without Frontiers: Green Paper*. Brüssels: Author.

Europäisches Medieninstititut. (1994). *Bericht über die Entwicklung der Meinungsvielfalt und der Konzentration im privaten Rundfunk gemäss 21 Abs. 6*. Düsseldorf: Author.

European Institute for the Media. (1990). *Before and After the Broadcasting Bill: Television in the United Kingdom, European Mediafacts 3*. Manchester: Author.

Faul, Erwin. (1989). Die Fernsehprogramme im dualen Rundfunksystem, *Rundfunk und Fernsehen, 37*(1): 25–46.

Faul, Erwin. (1991). Die neue Rundfunkordnung Deutschlands—Verkümmert eine nationale Verfassungsaufgabe im Dschungel partikularer Interessenstrategien? pp. 123–137 in Wienand Gellner (Ed.), *An der Schwelle zu einer neuen deutschen Rundfunkordnung—Grundlagen, Erfahrungen und Entwicklungsmöglichkeiten*. Berlin: Vistas.

Federal Communications Commission (FCC). (1979). *Television Programming for Children*. Washington, DC: Author.

Federal Communications Commission (FCC). (1984). *50th Annual Report/Fiscal Year 1984*. Washington, DC: Author.

Federal Cultural Policy Review Committee. (1989). Report, pp. 672–696 in Roger Bird (Ed.), *Documents of Canadian Broadcasting*. Ottawa: Carleton University Press.

Federation of Australian Commercial Television Stations. (1991). *A Code of Industry Practice: The Portrayal of Violence on Television*. Sydney: Author.

Ferrall, Victor E., Jr. (1989). The impact of television deregulation on private and public interests, *Journal of Communication, 39*(1): 8–38.

Fisher, David J. (1990). *Prior Consent to International Direct Satellite Broadcasting*. Dordrecht: Martinus Nijhoff.

Fiss, Owen N. (1987). Why the state? *Harvard Law Review, 100*: 781–794.

Follath, Erich. (1974). *Ein internationaler Vergleich von Rundfunksystemen. Die Interdependenz von Rundfunkpolitik und Gesamtpolitik in Grossbritannien, Frankreich, der Sowjetunion, der VR China und Indien* (dissertation, rer. pol.). Stuttgart-Hohenheim: Author.

Forbes, Jill. (1989). France: Modernisation across the spectrum, pp. 23–36 in G. Nowell-Smith (Ed.), *The European Experience*. London: British Film Institute.

Fornacciari, Marc. (1987). Le controle juridictionnel du coix du concessionaire, *Revue Française de Droit Administratif, 3*(1): 19–21.

Fowler, Robert M. (1965). *Report of the Committee on Broadcasting*. Ottawa: Roger Duchamel, F.R.S.C.

Fox, Francis. (1983). *Culture and Communications: Key Elements of Canada's Economic Future, Brief Submitted by The Honourable Francis Fox, Minister of Communications, to the Royal Commission on The Economic Union and Development Prospects for Canada*. Montreal, Quebec, November 3, 1983.

Franklin, Marc A., and David A. Anderson. (1990). *Cases and Materials on Mass Media Law,* 4th ed. Westbury, NY: Foundation Press. (Originally published 1977)

Fraser, Nicholas. (1989). Deregulation: Will it lead to destruction? *Broadcast, 8.12.1989:* 7.

Freiman, Mark. (1983). Canadian content in private television: An innisian analysis, *University of Toronto Faculty of Law Review, 41*(1): 19–33.

Frémont, Jacques. (1986). *Étude des objectifs et des principes proposés etadoptés relativement au système de la radiodiffusion Canadienne. Étude réalisée à l'intention du Goupe de travail sur la politique de la radiodiffusion.* Montréal: Université de Montréal Centre de Recherche en Droit Public Faculté de Droit.

Fritz, Roland. (1977). *Massenmedium Rundfunk—Die rechtliche Stellung der Rundfunkräte und ihre tatsächliche Einflussnahme auf die Programmgestaltung* (dissertation jur.). Frankfurt am Main: Author.

Garnham, Nicolas. (1980). *Structures of Television.* London: British Film Institute. (Originally published 1961)

Gavalda, Christian, and Martine Boizard. (1989). *Droit de l'audiovisuel,* 2nd ed. Paris: Lamy.

Gebel, Volkram. (1989). Probleme der Anwendung und des Vollzugs der Werberichtlinien—eine Quadratur des Kreises? pp. 354–358 in Bayerische Landeszentrale für neue Medien (BLM) (Ed.), *BLM Rundfunkkongress am 10./11.10.1989: Rundfunk in den 90er Jahren—zwischen Kultur, Kommerz und Internationalisierung, Dokumentation.* München: Fischer.

Gebel, Volkram. (1990). Veränderungen der Rundfunkwerbung—Herausforderung an Rundfunkveranstalter und Medienanstalten, pp. 52–56 in Die Landesmedienanstalten (Ed.), *DLM Jahrbuch 89/90—Privater Rundfunk in Deutschland.* München: Fischer.

Gebhardt, Hans-Peter. (1987). Telekommunikations- und Medienreform in Frankreich. Das Gesetz über die Freiheit der Kommunikation vom 30. September 1986, *Archiv für das Post- und Fernmeldewesen, 39*(2): 122–127.

Gellner, Winand. (1990). *Ordnungspolitik im Fernsehwesen: Bundesrepublik Deutschland und Grossbritannien.* Frankfurt am Main: Peter Lang.

Genevois, Bruno. (1989). Le conseil constitutionnel et la définition des pouvoirs du Conseil Supérieur de l'Audiovisuel, *Revue Française de Droit Administratif, 5*(2): 215–228.

Gerber, Volker. (1990). Das Rundfunksystem der Deutschen Demokratischen Republik, pp. A 92-107 in Hans-Bredow-Institut (Ed.), *Internationales Handbuch für Rundfunk und Fernsehen 1990/91.* Baden-Baden: Nomos.

Gerlach, Peter. (1990). *Rundfunksystemstrukturen und Rezipientengratifikationen in Kanada.* Frankfurt am Main: Peter Lang.

Gibbons, Thomas. (1991). *Regulating the Media.* London: Sweet & Maxwell.

Gillmor, Donald M., and James A. Baron. (1990). *Mass Communications Law,* 5th ed. St. Paul: West Publishing. (Originally published 1969)

Ginsburg, Douglas H., Michael H. Botein, and Mark D. Director. (1991). *Regulation of the Electronic Mass Media: Law and Policy for Radio, Television, Cable and the new Video Technologies.* St. Paul: West Publishing.

Glogauer, Werner. (1991). *Kriminalisierung von Kindern und Jugendlichen durch Medien.* Baden-Baden: Nomos.

Goodfriend, André. (1988). Satellite Broadcasting in the UK, pp. 144–174 in Ralph Negrine (Ed.), *Satellite Broadcasting*. London: Routledge.

Goodwin, Peter. (1990). Broadcasting complaints commission in the wars, *Broadcast, 24.8.1990*: 14.

Grafstein, Laurence. (1988). Out of focus: A thematic critique of the task force on broadcasting policy, *University of Toronto, Faculty of Law Review, 46*(1): 271–302.

Grams, Susanne, and Hans Hege. (1992). Jugendschutz in der Praxis des Rundfunks, pp. A100–108, in Hans-Bredow-Institut (Ed.), *Internationales Handbuch für Hörfunk und Fernsehen 1992/93*. Baden-Baden: Nomos.

Grant, Stewart Peter. (1968). The regulation of program content in Canadian television: An introduction, *Canadian Public Administration, 11*(3): 322–391.

Groebel, Jo. (1993). *Gewaltprofil des deutschen Fernsehprogramms: Eine Analyse des Angebots privater und öffentlich-rechtlicher Sender.* Opladen: Leske & Budrich.

Groebel, Jo, Wolfgang Hoffmann-Riem, Renate Köcher, Bernd-Peter Lange, Ernst Gottfried Mahrenholz, Ernst-Joachim Mestmäcker, Ingrid Scheithauer, and Norbert Schneider. (1994). *Bericht zur Lage des Fernsehens.* Bonn: Author.

Grossman, Lawrence K. (1994). Reflections on life along the electronic superhighway, *Media Studies Journal, 8*(Winter 1994): 27–39.

Grothe, Thorsten, and Wolfgang Schulz. (1993). Reflexives Recht—ein innovatives Steuerungskonzept für den Rundfunk? pp. 63–83 in Otfried Jarren, Frank Marzinkowski and Heribert Schatz (Eds.), *Landesmedienanstalten—Steuerung der Rundfunkentwicklung? Jahrbuch 1993 der Arbeitskreise "Politik und Kommunikation" der DVPW und der DGPuK.* Münster/Hamburg: LIT Verlag.

Grundfest, Joseph. (1977). Participation in FCC Licensing, *Journal of Communication, 27*(4): 85–88.

Guillou, Bernard, and Jean-Gustave Padioleau. (1988). *La régulation de la télévision.* Paris: La Documentation Française.

Gustedt, Volker. (1992). Deckmantel—Der Lokalfunk in Bayern am wirtschaftlichen Scheideweg, *Evangelischer Pressedienst/Kirche und Rundfunk, 1992*(56): 5–8.

Haeckel, Helmut. (1988). Der Übergang zum dualen Rundfunksystem. Geschichte und Perspektiven der rundfunkpolitischen Entwicklungen in der Bundesrepublik, pp. 17–26 in Direktorenkonferenz der Landesmedienanstalten (DLM). (Ed.), *DLM Jahrbuch 89/90—Privater Rundfunk in Deutschland.* München: R. Fischer.

Haeckel, Helmut. (1993). Konzentrationskontrolle im privaten Rundfunkmarkt, pp. 15–24 in Direktorenkonferenz der Landesmedienanstalten (DLM) (Ed.), *DLM Jahrbuch 92—Privater Rundfunk in Deutschland.* München: R. Fischer.

Hamburgische Anstalt für neue Medien. (HAM). (1992). *Erfahrungsbericht des Vorstandes der Hamburgischen Anstalt für neue Medien über die Amtsperiode 1986 bis 1992.* Hamburg: Author.

Hammond, Allen S. (1992). Regulating broadband communications networks, *Yale Journal on Regulation, 9*: 181—235.

Hardin, Herschel. (1985). *Closed Circuits. The Sellout of Canadian Television.* Vancouver: Douglas & McIntyre.

Hardt, Hanno. (1981). Öffentliches Interesse und Big Business: Lokales Radio in den USA, 1920–34, *Rundfunk und Fernsehen, 29*(4): 373–380.

Harrison, Kate. (1980). *The Points System for Australian Television Content—A Study in Symbolic Policy.* Brisbane: Royal Institute of Public Administration.

Harrison, Kate. (1987). The Changing Face of the Television Industry, *Media Information Australia, 1987*(44): 16–17.

Hase, Karl-Günther von. (1980). Das Rundfunksystem in Grossbritannien—Beispiel für die Bundesrepublik Deutschland? pp. 13–23 in Wolfgang Hoffmann-Riem and Will Teichert (Eds.), *Aktuelle Fragen der Rundfunkpolitik.* Hamburg: Hans-Bredow-Institut.

Haute Autorité de la Communication Audiovisuelle. (1983). *Rapport Septembre 1982—Septembre 1983.* Paris: Author.

Haute Autorité de la Communication Audiovisuelle. (1984). *Rapport Septembre 1983—September 1984.* Paris: Author.

Haute Autorité de la Communication Audiovisuelle. (1985). *Rapport Septembre 1984—Septembre 1985.* Paris: Author.

Head, Sydney W. (1985). *World Broadcasting Systems: A Comparative Analysis.* Belmont: Wadsworth.

Head, Sydney W., and Christopher H. Sterling. (1990). *Broadcasting in America: A Survey of Electronic Media.* Boston: Houghton Mifflin.

Hearst, Stephen. (1979). Diskussionsbeitrag, pp. 88–90 in Hans-Bredow-Institut (Ed.), *Hamburger Medientage 1979.* Hamburg: Hans-Bredow-Institut.

Hearst, Stephen. (1992). Broadcasting Regulation in Great Britain, pp. 61–78 in Jay G. Blumler (Ed.), *Television and the Public Interest: Vulnerable Values in West European Broadcasting.* London: Sage.

Heinrich, Jürgen. (1994). Keine Entwarnung bei Medienkonzentration, *Media Perspektiven, 1994*(6): 297–309.

Hellstern, Gerd-Michael. (1989a). Baden-Württemberg: Landesanstalt für Kommunikation, pp. 3–54 in Gerd-Michael Hellstern, Wolfgang Hoffmann-Riem, and Jürgen Reese (Eds.), *Rundfunkaufsicht—Vol. 16/I Begleitforschung des Landes Nordrhein-Westfalen zum Kabelpilotprojekt Dortmund.* Düsseldorf: Presse und Informationsamt der Landesregierung Nordrhein-Westfalen.

Hellstern, Gerd-Michael. (1989b). Berlin: Anstalt für Kabelkommunikation, pp. 115–214 in Gerd-Michael Hellstern, Wolfgang Hoffmann-Riem, and Jürgen Reese (Eds.), *Rundfunkaufsicht—Vol. 16/I Begleitforschung des Landes Nordrhein-Westfalen zum Kabelpilotprojekt Dortmund.* Düsseldorf: Presse und Informationsamt der Landesregierung Nordrhein-Westfalen.

Hellstern, Gerd-Michael, and Jürgen Reese. (1989). Ziele, Organisation und Leistung der Landesanstalten für Rundfunk, pp. 3–58 in Gerd-Michael Hellstern, Wolfgang Hoffmann-Riem, and Jürgen Reese (Eds.), *Rundfunkaufsicht—Vol. 16/III Begleitforschung des Landes Nordrhein-Westfalen zum Kabelpilotprojekt Dortmund.* Düsseldorf: Presse und Informationsamt der Landesregierung Nordrhein-Westfalen.

Hendriks, Birger. (1984). Landesrundfunkgesetze und Medienverflechtung: Anmerkungen zum 5. Hauptgutachten der Monopolkommission, *Media Perspektiven, 1984*(12): 922–930.

Herrmann, Günter. (1994). *Rundfunkrecht.* München: C.H. Beck.

Hesse, Albrecht. (1990). *Rundfunkrecht: die Organisation des Rundfunks in der Bundesrepublik Deutschland.* München: Vahlen.

Heyn, Jürgen, and Hans-Jürgen Weiss. (1980). Das Fernsehprogramm von ITV und BBC, *Media Perspektiven, 1980*(3): 145–155.

Hinkson, W. E., and W. Krasilovsky. (1982). Constructive Canadian cultural and commercial chauvinism, *Journal of Media Law and Practice, 3*(2): 205–214.

Hirsch, Nicola. (1991). *Lokaler Hörfunk in Nordrhein-Westfalen—Eine Analyse des Zwei-Säulen-Modells für privat-kommerziellen Rundfunk sowie seiner Realisierung aus politikwissenschaftlicher Sicht.* Bochum: Schallwig.

Hochstein, Reiner. (1991). Neue Werbeformen im Rundfunk—Ordnungspolitische und aufsichtsrechtliche Aspekte, *Archiv für Presserecht, 22*(3): 696–703.

Hoffmann-Riem, Wolfgang. (1972). *Redaktionsstatute im Rundfunk.* Baden-Baden: Nomos.

Hoffmann-Riem, Wolfgang. (1979). *Innere Pressefreiheit als politische Aufgabe.* Neuwied: Luchterhand.

Hoffmann-Riem, Wolfgang. (1981a). Fernsehkontrolle als Ritual? Überlegungen zur staatlichen Kontrolle im amerikanischen Fernsehen, *Juristenzeitung, 36*(3): 73–82.

Hoffmann-Riem, Wolfgang. (1981b). *Kommerzielles Fernsehen. Rundfunkfreiheit zwischen ökonomischer Nutzung und staatlicher Regelungsverantwortung: das Beispiel USA.* Baden-Baden: Nomos.

Hoffmann-Riem, Wolfgang. (1982). Stellungnahme zum Sondergutachten der Monopolkommission: Wettbewerbsprobleme bei der Einführung von privatem Hörfunk und Fernsehen, *Wirtschaft und Wettbewerb, 32*(3): 265–271.

Hoffmann-Riem, Wolfgang. (1985). Deregulierung als Konsequenz des Marktrundfunks. Vergleichende Analyse der Rundfunkrechtsentwicklung in den USA, *Archiv des öffentlichen Rechts, 110*: 528–576.

Hoffmann-Riem, Wolfgang. (1988). Teleshopping—Ende der Rundfunkregulierung? *Evangelischer Pressedienst/Kirche und Rundfunk, 1988*(45): 18–21.

Hoffmann-Riem, Wolfgang. (1989a). Kommentierung von Artikel 5 Abs. 1 und 2 (Meinungs- und Medienfreiheit), pp. 408–533 in Rudolf Wassermann (Ed.), *Alternativkommentar zum Grundgesetz,* 2nd ed. Neuwied: Luchterhand.

Hoffmann-Riem, Wolfgang. (1989b). *Konfliktmittler in Verwaltungsverhandlungen.* Heidelberg: C. F. Müller.

Hoffmann-Riem, Wolfgang. (1989c). Medienstädte im Wettbewerb. Am Beispiel Hamburgs, *Medien-Journal, 13*(2): 66–76.

Hoffmann-Riem, Wolfgang. (1990a). Freedom of information and new technological developments in the Federal Republic of Germany: A case law analysis, pp. 49–77 in A. Cassese and A. Clapham (Eds.), *Transfrontier Television in Europe: The Human Rights Dimension.* Baden-Baden: Nomos.

Hoffmann-Riem, Wolfgang. (1990b). *Erosionen des Rundfunkrechts. Tendenzen der Rundfunkrechtsentwicklung in Westeuropa.* München: C. H. Beck.

Hoffmann-Riem, Wolfgang. (1991a). Zwischen ökonomischer Deregulierung und politisch-moralischer Überregulierung—zur Neuordnung des Rundfunksystems in Grossbritannien, *Rundfunk und Fernsehen, 39*(1): 17–28.

Hoffmann-Riem, Wolfgang. (1991b). *Rundfunkrecht neben Wirtschaftsrecht. Überlegungen zur Anwendbarkeit des GWB und des EWG-V auf das Wettbewerbsverhalten des öffentlich-rechtlichen Rundfunks in der dualen Rundfunkordnung.* Baden-Baden: Nomos.

Hoffmann-Riem, Wolfgang. (1991d). The road to media unification: Press and broadcasting law reform in the GDR, *European Journal of Communication, 6*(4): 523–543.

Hoffmann-Riem, Wolfgang. (1991e). Rollenkonflikte und Transferprobleme zwischen Wissenschaft, Politik und Medienpraxis, pp. 51–69 in Dieter Ross and Jürgen Wilke (Eds.), *Umbruch in der Medienlandschaft.* München: Oelschläger.

Hoffmann-Riem, Wolfgang. (1992a). Protection of vulnerable values in the German broadcasting order, pp. 43–60 in Jay G. Blumler (Ed.), *Television and the Public Interest: Vulnerable Values in West European Broadcasting.* London: Sage.

Hoffmann-Riem, Wolfgang. (1992b). Defending vulnerable values. Regulatory measures and enforcement dilemmas, pp. 173–201 in Jay G. Blumler (Ed.), *Television and the Public Interest. Vulnerable Values in West European Broadcasting.* London: Sage.

Hoffmann-Riem, Wolfgang. (1994a). *Finanzierung und Finanzkontrolle der Landesmedienanstalten,* 2nd ed. Berlin: Vistas. (Originally published 1993)

Hoffmann-Riem, Wolfgang. (1994b). Stadien des Rundfunk-Richterrechts, pp. 17–33 in Otfried Jarren (Ed.), *Medienwandel—Gesellschaftswandel.* Berlin: Vistas.

Hoffmann-Riem, Wolfgang, and Stefan Engels. (1995). Einfachgesetzliche Regelungen der Fernsehwerbung mit Kindern und für Kinder, in Michael Charlton, Klaus Neumann-Braun, Stefan Aufenanger, and Wolfgang Hoffmann-Riem (Eds.), *Fernsehwerbung und Kinder,* Opladen: Leske & Budrich.

Hoffmann-Riem, Wolfgang, and Thomas Vesting. (Eds.). (1995). *Perspektiven der Informationsgesellschaft.* Baden-Baden: Nomos.

Hoffmann-Riem, Wolfgang, and Michael P. Ziethen. (1989). Hamburgische Anstalt für Neue Medien, pp. 215–308 in Gerd-Michael Hellstern, Wolfgang Hoffmann-Riem, and Jürgen Reese (Eds.), *Rundfunkaufsicht—Vol. 16/I Begleitforschung des Landes Nordrhein-Westfalen zum Kabelpilotprojekt Dortmund.* Düsseldorf: Presse und Informationsamt der Landesregierung Nordrhein-Westfalen.

Holgersson, Silke. (1993). Programmkontrolle der Landesmedienanstalten: Anspruch und Umsetzung, pp. 153–166 in Otfried Jarren, Frank Marzinkowski, and Heribert Schatz (Eds.), *Landesmedienanstalten—Steuerung der Rundfunkentwicklung? Jahrbuch 1993 der Arbeitskreise "Politik und Kommunikation" der DVPW und der DGPuK.* Münster/Hamburg: LIT Verlag.

Holleaux, André. (1987). Le droit de l'audiovisuel, entre la politique et la technique, *Revue Française d'Administration Publique, 44*: 9–25.

Holsinger, Ralph R. (1991). *Media Law,* 2nd ed. New York: McGraw-Hill. (Originally published 1987)

Holznagel, Bernd. (1991). Konzentrationsbekämpfung im privaten Rundfunk, *Zeitschrift für Urheber- und Medienrecht/Film und Recht, 35*(6): 263–271.

Home Affairs Committee. (1988). *The Future of Broadcasting, Third Report Session, 1987–88.* London: Her Majesty's Stationery Office.

Home Office. (1987). *Radio: Choices and Opportunities, Consultative Document.* London: Her Majesty's Stationery Office.

Home Office. (1988). *Broadcasting in the '90s: Competition Choices and Quality: The Government's Plans for Broadcasting Legislation.* London: Her Majesty's Stationery Office.

Hood, Stuart. (1980). *On Television.* London: Pluto Press.

Horwitz, Robert Britt. (1989). *The Irony of Regulatory Reform: The Deregulation of American Telecommunications.* New York: Oxford University Press.

Hoskins, Colin, and Stuart McFayden. (1982). Market structure and television programming performance in Canada and the U.K.: A comparative study, *Canadian Public Policy, 8*(3): 347–357.

Hoskins, Colin, and Stuart McFayden. (1986). The economic factors relating to Canadian television broadcasting policy: A nontechnical synthesis of the research literature, *Canadian Journal of Communication, 12*(1): 21–40.

Hoskins, Colin, and Stuart McFayden. (1989). Television in the new broadcasting environment: Public policy lessons from the Canadian experience, *European Journal of Communication, 4*(2):173–191.

Howse, Robert, Robert J. Prichard, and Michael J. Trebikock. (1990). Smaller or smarter government? *University of Toronto Law Journal, 40*(4): 498–541.

Huet, Pierre. (1989). La loi du 17 janvier 1989 sur la liberté de communication, *Recueil Dalloz Sirey, 1989*(27): 179–188.

Hughes, Chris. (1988). Doubling up, *Airwaves, 1988*(17): 8–19.

Humphreys, Peter. (1988). Das Rundfunksystem Grossbritanniens, pp. E 67–77 in Hans-Bredow-Institut (Ed.), *Internationales Handbuch für Rundfunk und Fernsehen 1988/89.* Baden-Baden: Nomos.

Humphreys, Peter. (1990). Das Rundfunksystem Grossbritanniens, pp. D 83–94 in Hans-Bredow-Institut (Ed.), *Internationales Handbuch für Rundfunk und Fernsehen 1990/91.* Baden-Baden: Nomos.

Independent Broadcasting Authority (IBA). (1985). *Television Programme Guidelines.* London: Author.

Independent Broadcasting Authority (IBA). (1986). *Annual Report and Accounts 1985–1986.* London: Author.

Independent Broadcasting Authority (IBA). (1987a). *The IBA Annual Report and Accounts 1986–87.* London: Author.

Independent Broadcasting Authority (IBA). (1987b). *The Future of UK Independent Radio.* London: Author.

Independent Broadcasting Authority (IBA). (1988). *Annual Report 1987–88.* London: Author.

Independent Broadcasting Authority (IBA). (1989). *Annual Report 1988–89.* London: Author.

Independent Broadcasting Authority (IBA). (1990). *Annual Report 1989–90.* London: Author.

Independent Television Commission (ITC). (1994). *Performance Review.* London: Author.

Independent Television Companies Association/Independent Television Association (ITCA/ITVA). (1981–1989). *Series of Notes of Guidance No. 1–10.* London: Author.

Ingber, Stanley. (1984). The marketplace of ideas: A legitimizing myth, *Duke Law Journal, 1984*(1): 1–91.

Institute National de l'Audiovisuel. (1988). *Dossiers de l'Audiovisuel–Télévision et Déontologie.* Paris: La Documentation Française.

Interdisziplinäre Berater- und Forschungsgruppe Basel AG (IBFG). (1989). *Studie zur wirtschaftlichen Tragfähigkeit von Lokalradios in Bayern (Schlussbericht vom 28. Juni 1988)*. München: Bayerische Landeszentrale für neue Medien (BLM).

Intven, Hank. (1994). Traffic rules on Canada's information highways: The regulatory framework for new cable and telephone services, *Media and Communications Law Review, 4*(2): 131–171.

Jakubowicz, Andrew. (1987). Days of our lives: Multiculturalism, mainstreaming and "special" broadcasting, *Media Information Australia, 1987*(45): 18–32.

Janisch, Hudson. (1979). Policy making in regulation: Towards a new definition of the status of independent regulatory agencies in Canada, *Osgood Law Journal,* 17(1): 46–101.

Janisch, Hudson. (1987). Culture or commerce? *Inter Media, International Institute of Communications,* 15(4/5): 42–43.

Janisch, Hudson. (1990). *Aid for Sisyphus: Incentives and Canadian Content Regulation in Broadcasting* (a revised version of a discussion paper prepared for the conference on "The Power of the Purse: Financial Incentives As Regulatory Instruments," University of Calgary, October 12–13, 1990). Toronto: Faculty of Law, University of Toronto.

Janisch, Hudson. (1991). *Communications Law, Vols. 1–3, 1991: Casebook.* Toronto: Faculty of Law, University of Toronto.

Jehoram, Herman. (1981). The unique Dutch broadcasting system on the eve of the revolution in teletechnics and freedom of information, *Journal of Media Law and Practice, 2*(3): 253–269.

Jens, Carsten. (1991). Die Entwicklung des Presseengagements beim privaten Hörfunk 1988–1991—Ergebnisse der zweiten Dokumentation über Verlagsbeteiligungen an Privatradios in der Bundesrepublik Deutschland, *Media Perspektiven, 1991*(9): 570–589.

Jezequel, Jean-Pierre. (1991). Canal Plus . . . de priviléges, *Angle Droit, 4*: 3–5.

Jezequel, Jean-Pierre, and Guy Pineau. (1992). French Television, pp. 429–522 in Alessandro Silj (Ed.), *The New Television in Europe.* London: John Libbey.

Johansen, Peter W. (1973). The Canadian Radio-Television Commission and the Canadianization of broadcasting, *Federal Communications Bar Journal, 1973*: 183–208.

Johnson, A. W. (1982). Canadian content in Canadian television, *Combroad, 54*: 1–6.

Johnson, Leland. (1994). *Toward Competition in Cable Television.* Cambridge, MA: MIT Press.

Juneau, Pierre. (1984). Audience fragmentation and cultural erosion: A Canadian perspective on the challenge for the public broadcaster, *EBU Review. Programmes, Administration, Law, 35*(2): 14–20.

Juneau, Pierre. (1985). Enduring principles; unanswered questions—problems for the CBC in 1985, *Combroad, 67*: 12–18.

Juneau, Pierre. (1989). Television without frontiers? *EBU Review, Programmes, Administration, Law, 40*(3): 25–29.

Kabbert, Rainer. (1987). *Rundfunkkontrolle als Instrument der Rundfunkpolitik: Einfluss im Prozess der öffentlichen Meinung.* Nürnberg: Verlag der Kommunikationswissenschaften.

Kaufman, Donna Soble. (1991). Broadcasting law in Canada, pp. 109–119 in Hudson

Janisch (Ed.), *Communications Law, Vol. 1: Casebook.* Toronto: Faculty of Law, University of Toronto.

Keidel, Hannelore. (1993). Politische Aspekte der öffentlichrechtlichen Rundfunkaufsicht über Privatfunk, pp. 97–112 in Otfried Jarren, Frank Marzinkowski, and Heribert Schatz (Eds.), *Landesmedienanstalten—Steuerung der Rundfunkentwicklung? Jahrbuch 1993 der Arbeitskreise "Politik und Kommunikation" der DVPW und der DGPuK.* Münster: LIT Verlag.

Keune, Reinhard. (1984). Asiavision—erstes eigenständiges Regionalsystem für Fernsehdrittaustausch in der Dritten Welt, *Rundfunk und Fernsehen, 32*(2): 214–220.

Kleinsteuber, Hans-J. (1988). *Nicht-kommerzielles Lokalradio, Vols. I–II,* Hamburg: University Press.

Kleinsteuber, Hans-J. (1988). Radiosystem und Radiopolitik in den USA, *Media Perspektiven, 1988*(3): 125–135.

Kleinsteuber, Hans J. (1989). Medien und Medienpolitik in Australien, *Media Perspektiven, 1989*(4): 207–216.

Kleinsteuber, Hans J. (1994). Das Rundfunksystem der USA, pp. C88–97 in Hans-Bredow-Institut (Ed.), *Internationales Handbuch für Rundfunk und Fernsehen 1994/95.* Baden-Baden: Nomos.

Kleinsteuber, Hans J., Denis McQuail, and Karen Siune. (Eds.). (1986). *Electronic Media and Politics in Western Europe.* Frankfurt am Main: Campus.

Kleinsteuber, Hans J., and Volkert Wiesner. (1988). Eigenproduktionen in Fernsehen und Hörfunk: Erfahrungen mit den Canadian Contents Rules, *Rundfunk und Fernsehen, 36*(3): 329–346.

Kleinwort Benson Securites. (1990). *New Media: Opportunity Knocks.* London: Author.

Köbberling, Ursel. (1984). Eskimo-Fernsehen im Norden Kanadas, *Rundfunk und Fernsehen, 32*(3): 341–347.

Kopetz, Dieter. (1993). Landesmedienanstalten und Lizenzvergabeverfahren in den neuen Bundesländern, pp. 87–96 in Otfried Jarren, Frank Marzinkowski, and Heribert Schatz (Eds.), *Landesmedienanstalten—Steuerung der Rundfunkentwicklung? Jahrbuch 1993 der Arbeitskreise "Politik und Kommunikation" der DVPW und der DGPuK.* Münster: LIT Verlag.

Krasnow, Erwin G., Lawrence D. Longley, and Herbert A. Terry. (1982). *The Politics of Broadcast Regulation,* 3rd ed. New York: St. Martin's Press.

Krüger, Udo Michael. (1991). Positionierung öffentlichrechtlicher und privater Fernsehveranstalter im dualen System—Programmanalyse 1990, *Media Perspektiven, 1991*(5): 303–332.

Kübler, Fiedrich. (1973). *Kommunikation und Verantwortung: Eine verfassungstheoretische und rechtspolitische Skizze zur Funkktion professioneller und kollegialer Autonomie in Presse, Funk und Hochschule.* Konstanz: Universitätsverlag.

Kübler, Friedrich. (1982). *Medienverflechtung.* Frankfurt am Main: Metzner.

Kübler, Friedrich. (1992). Regelungsprobleme der Medienverflechtung, pp. 43–55 in Heinz Hübner (Ed.), *Rechtsprobleme der privaten Rundfunkordnung. Vortragsveranstaltung vom 26. und 27. April 1991.* München: C. H. Beck.

Kuhn, Raymond. (1985a). The end of the government monopoly, pp. 47–82 in Raymond Kuhn (Ed.), *The Politics of Broadcasting.* London: Croom Helm.

Kuhn, Raymond. (Ed.). (1985b). *The Politics of Broadcasting.* London: Croom Helm.

Kuhn, Raymond. (1988). Satellite broadcasting in France, pp. 176–195 in Ralph

Negrine (Ed.), *Satellite Broadcasting: The Politics and Multiplications of the New Media.* London: Routledge.

Kull, Edgar. (1993). Dienende Freiheit—dienstbare Medien? pp. 663–674 in Peter Badura and Rupert Scholz (Eds.), *Wege und Verfahren im Verfassungsleben— Festschrift für Peter Lerche.* München: C.H. Beck.

Kunkel, Dale and Bruce Watkins. (1987). Evolution of children's television regulatory policy, *Journal of Broadcasting and Electronic Media, 31*(4): 367–389.

Ladeur, Karl-Heinz. (1989). Probleme der Koordination der Aufsicht über private Rundfunkveranstalter, *Archiv für Presserecht, 20*(4): 717–721.

Landesanstalt für Kommunikation (LfK). (1989). *Bericht an die Landesregierung gemäss 88 Landesmediengesetz Baden-Württemberg.* Stuttgart: Schriftenreihe LfK-Dialog.

Landesanstalt für Rundfunk Nordrhein-Westfalen (LfR). (1988). *Jahresbericht 1988.* Düsseldorf: Author.

Landesanstalt für Rundfunk Nordrhein-Westfalen (LfR). (Ed.). (1989). *Entscheidungsverfahren und Ergebnisse im Überblick—Landesanstalt für Rundfunk Nordrhein-Westfalen-Schriftenreihe,* Vol. 3. Düsseldorf: Author.

Landesanstalt für Rundfunk Nordrhein-Westfalen (LfR). (1990). *Jahresbericht 1990.* Düsseldorf: Author.

Landesmedienanstalten (DLM). (Ed.). (1990). *DLM Jahrbuch 89/90—Privater Rundfunk in Deutschland.* München: R. Fischer.

Lane, Patrik H. (1990). *An Introduction to the Australian Constitution,* 5th ed. Sydney: The Law Book Company. (Originally published 1974)

Lange, André. (1994). Das Rundfunksystem Frankreichs, pp. B64–75 in Hans-Bredow-Institut (Ed.), *Internationales Handbuch für Rundfunk und Fernsehen 1994/95.* Baden-Baden: Nomos.

Lange, André, and Jean-Luc Renaud. (1989). *The Future of the European Audiovisual Industry.* Manchester: The European Institute for the Media.

LaPointe, Alain, and Jean-Pierre Le Goff. (1988). Canadian television: An alternative to Caplan-Sauvageau, *Canadian Public Policy, 14*(3): 245–253.

Law Reform Commission of Canada. (1986). *Policy Implementation, Compliance and Administration Law.* Ottawa: Minister of Supply and Services: Canada.

Le Duc, Don R. (1973). *Cable Television and the FCC: A Crisis in Media Control.* Philadelphia: Temple University Press.

Le Duc, Don R. (1987). *Beyond Broadcasting: Pattern in Policy and Law.* New York: Longman.

Le Duc, Don R. (1988). Der Angriff auf die amerikanische Fairness-Doktrin: eine Schlacht in einem grösseren, Krieg, *Rundfunk unk Fernsehen, 36*(1): 56–66.

Lerche, Peter. (1974). *Fernsehabgabe und Bundeskompetenz: Kompetenzfragen im Hinblick auf eine von den Rundfunkanstalten zu entrichtende Filmhilfsabgabe mit einem materiell-rechtlichen Exkurs.* Frankfurt am Main: Metzner.

Lerche, Peter. (1979). Landesbericht Bundesrepublik Deutschland, pp. 22–107 in Martin Bullinger and Friedrich Kübler (Eds.), *Rundfunkorganisation und Kommunikationsfreiheit,* Baden-Baden: Nomos.

Lerche, Peter. (1984). *Presse und privater Rundfunk.* Berlin: Duncker & Humblot.

Lerg, Winfried B. (1980). *Rundfunk in Deutschland, Band 1: Rundfunkpolitik in Weimarer Republik.* München: DTV.

Levin, Harvey J. (1980). *Fact and Fancy in Television Regulation.* New York: Sage.

Lincoln, Anthony. (1979). Landesbericht Grossbritannien, pp. 125–142 in Martin Bullinger and Friedrich Kübler (Eds.), *Rundfunkorganisation und Kommunikationsfreiheit.* Baden-Baden: Nomos.

Lively, Donald E. (1992). *Essential Principles of Communication Law.* New York: Praeger.

Luppatsch, Michael. (1986). Vom "lokalen Gegenradio" zum "fond musicale": über die Entwicklung des privaten Lokalfunks in Frankreich seit 1981, *Media Perspektiven, 1986*(12): 789–797.

MacBride, Sean. (1980). *Many Voices, One World—Communication and Society Today and Tomorrow (MacBride Report), UNESCO 1980.* Colchester: Spottiswoode Bullentin Ltd.

Madge, Tim. (1989). *Beyond the BBC: Broadcasters and the Public in the 1980s.* London: Macmillan.

Magiera, Siegfried. (1981). Direct broadcasting by satellite and a new international information order, *German Yearbook of International Law, 24*: 288–307.

Mahle, Walter. (1984). *Kommerzielles Fernsehen in der Medienkonkurrenz, Band I: Grossbritannien, ein Modell für die Bundesrepublik?* Berlin: Volker Spiess.

Mallamud, Jonathan. (1973). The Broadcast licensee as fiduciary: Toward the enforcement of discretion, *Duke Law Journal, 23*(1): 89–133.

Malone, Donald M. (1972). Broadcasting: The reluctant dragon, *University of Michigan Journal of Law Reform, 5*: 194–234.

Marmont, Vince. (1990). TV industry structure: Market forces and the failure of regulation, *Media Information Australia, 1990*(56): 50–56.

Matte, Nicolas Matesco, and Ram S. Jakhu. (1987). *Law of International Telecommunications in Canada.* Baden-Baden: Nomos.

Mazzoleni, Gianpietro. (1992). Is there a question of vulnerable values in Italy? pp. 79–95 in Jay G. Blumler (Ed.), *Television and the Public Interest.* London: Sage.

McFayden, Stuart, Colin Hoskins, and David Gillen. (1980). *Canadian Broadcasting: Market Structure and Economic Performance.* Montreal: The Institute for Research on Public Policy.

McGregor, Michael A. (1989). Assessment of the renewal expectancy in FCC comparative renewal hearings, *Journalism Quarterly, 66*(2): 295–301.

McLaughlin, Margaret L. (Ed.). (1987). *Communication Yearbook 10.* Newbury Park, CA: Sage.

McPhail, Brenda. (1986). Canadian content regulations and the Canadian Charter of Rights and Freedom, *Canadian Journal of Communication, 12*(1): 41–53.

McPhail, Thomas L., and Brenda M. McPhail. (1990). *Communication: The Canadian Experience.* Toronto: Copp Clark Pitman.

Media Council of Australia. (1990). *Australian Advertising Co-Regulation: Procedures, Structures and Codes.* Sydney: Author.

Meier, Werner A. (1984). Pay-TV in Kanada: Grosse Kluft zwischen Anspruch und Wirklichkeit, *Media Perspektiven, 1984*(7): 528–535.

Meiklejohn, Alexander. (1948). *Free Speech and Its Relation to Self-Government.* New York: Harper & Row.

Meise, Martin. (1992). Zur Situation des französischen Fernsehens. Das Duale System im Spannungsfeld zwische Staat und Markt, *Media Perspektiven, 1992*(4): 236–255.

Meisel, John. (1989). Fanning the air: The Canadian state and broadcasting, pp. 1–21 in Royal Society of Canada (Eds.), *Symposium on "The State and the Arts," Quebec City, June 5, 1989*. Quebec: Author.

Melody, William H. (1983). Direktstrahlende Rundfunksatelliten: Die kanadischen Erfahrungen, *Rundfunk und Fernsehen, 31*(1): 5–11.

Menningen, Walter. (1981). Rundfunkarbeit als politisches Mandat? *Rundfunk und Fernsehen, 29*(2/3): 185–199.

Mestmäcker, Ernst-Joachim. (1988). In welcher Weise empfielt es sich, die Ordnung des Rundfunks und sein Verhältnis zu anderen Medien—auch unter dem Gesichtspunkt der Harmonisierung—zu regeln? pp. 9–37 in Deutscher Juristentag (Ed.), *Sitzungsbericht O zum 56. Deutschen Juristentag*. München: C. H. Beck.

Mestmäcker, Ernst-Joachm, Christoph Engel, Karin Gabriel-Bräutigam, and Martin Hoffmann. (1990). *Der Einfluss des europäischen Gemeinschaftsrechts auf die deutsche Rundfunkordnung*. Baden-Baden: Nomos.

Miller, Nod, and Cresta Norris. (Eds.). (1989). *Life after the Broadcasting Bill*. Manchester: Manchester Monograph.

Miller, Robert. (1973). The CRTC: Guardian of the Canadian identity, *Journal of Broadcasting, 17*(2): 189–199.

Minehan, Mike. (1993). Australiens Rundfunkgesetz und die Aufsicht über das private Fernsehen, *Rundfunk und Fernsehen, 41*(2): 212–222.

Missika, Jean-Louis. (1986). Die Deregulation der Audiovision in Frankreich, *Media Perspektiven, 1986*(8): 526–531.

Moderne, Frank. (1987). Les arrets et le contentieux de la concession de service public, *Revue Française de Droit Administratif, 3*(1): 11–19.

Montag, Helga. (1978). *Privater oder öffentlich-rechlicher Rundfunk?* Berlin: Verlag Volker Spiess.

Montgomery, Kathryn C. (1989). *Target: Prime Time—Advocacy Groups and the Struggle over Entertainment Television*. New York: Oxford University Press.

Morange, Jean. (1987). La Commission National de la Communication et des Libertés et le droit de la communication audiovisuelle, *Revue Française de Droit Administratif, 3*(3): 372–385.

Morange, J. (1989). Le Conseil Supérieur de l'Audiovisuel, *Revue Française de Droit Administratif, 5*(2): 235–251.

Mosco, Vincent. (1979). *Broadcasting in the United States: Innovative Challenge and Organizational Control*. Norwood, NJ: Ablex.

Müller-Römer, Frank. (1992). Rundfunkversorgung (Hörfunk und Fernsehen)—Verbreitung von Rundfunkprogrammen und neue Rundfunkdienste, pp. A 125–156 in Hans-Bredow-Institut (Ed.), *Internationales Handbuch für Hörfunk und Fernsehen 1992/93*. Baden-Baden: Nomos.

Müller-Römer, Frank. (1994). Rundfunkversorgung (Hörfunk und Fernsehen, pp. A 147–179 in Hans-Bredow-Institut (Ed.), *Internationales Handbuch für Rundfunk und Fernsehen 1994/95*. Baden-Baden: Nomos.

Murdock, Graham. (1992). Ausverkauf des Familiensilbers—Das kommerzielle Fernsehen in Grossbritannien nach der Lizenzauktion, *Media Perspektiven, 1992*(4): 222–235.

National Institute of Mental Health, Television and Behavior. (1982). *10 Years of Scientific Progress and Implications for the Eighties, Vol. 1: Summary Report.* Washington: Author.

Negrine, Ralph. (1989). *Politics and the Mass Media in Britain.* London: Routledge.

Nevoltry, Florence, and Bertrand, Delcros. (1989). *Le Conseil Supérieur de l'Audiovisuel: Fondement politique et analyse juridique—Loi du 17 janvier 1989.* Paris: Legipresse.

Nieuwenhuis, A. J. (1992). Media policy in the Netherlands—beyond the market? *European Journal of Communication, 8*(1): 195–218.

Nippon Hoso Kyokai/Broadcasting Culture Research Institute (NHK). (1991). *Studies of Broadcasting, Special Issue: Quality Assessment of Broadcast Programming.* Tokyo: Author.

Nippon Hoso Kyokai/Broadcasting Culture Research Institute (NHK). (1992). *Studies of Broadcasting, Special Issue: Quality Assessment of Broadcast Programming II.* Tokyo: Author.

Nippon Hoso Kyokai/Broadcasting Culture Research Institute (NHK). (1993). *Studies of Broadcasting, Special Issue: Quality Assessment of Broadcast Programming III.* Tokyo: Author.

Noam, Eli. (1991). *Television in Europe.* New York: Oxford University Press.

Noam, Eli M. (1993). *Television in Europe.* New York: Oxford University Press.

Noam, Eli M. (1994a). Beyond liberalisation II: The impending doom of common carriage, *Telecommunications Policy, 18*: 435–452.

Noble, Elisabeth. (1990). The mass, the market and the moral: The muddle of broadcasting policy in Australia, *Media Information Australia, 1990*(55): 21–27.

Noll, Roger G., Merton J. Peck, and John J. McGowan. (1973). *Economic Aspects of Television Regulation.* Washington, DC: The Brookings Institution.

Oehler, Thomas. (1988a). Lokale Radio-Networks in Frankreich, *Media Perspektiven, 1988*(6): 358–365.

Oehler, Thomas. (1988b). Vom Monopol des Staates zur Dominanz der Einschaltquoten, *Medien Bulletin, 7*(11–12): 34–41.

Office of Technology Assessment. (1993). The Internet, pp. 15–34 in Office of Technology Assessment (Ed.), *Advanced Network Technology—Background Paper,* Washington, DC: U.S. Government Printing Office.

Opitz, Gert. (1983). Das neue französische Mediengesetz; Bestandsaufnahme der französichen AV-Medienland-schaft, *Media Perspektiven, 1983*(2): 94–113.

Opitz, Gert. (1992). Das Rundfunksystem Frankreichs, pp. D 78-106 in Hans-Bredow-Institut (Ed.), *Internationales Handbuch für Rundfunk und Fernsehen 1992/93.* Baden-Baden: Nomos.

Ory, Stephan. (1988). Vom Teleshopping zum Elektronischen Anzeigenblatt? *Archiv für Presserecht, 19*(2): 120–124.

Oswin Report. (1984). *Localism in Australian Broadcasting: A Review of the Policy.* Canberra: Australian Government Publishing Service.

Ott, Michaela. (1990). *Die Liberalisierung des französichen Rundfunks unter Francois Mitterand. (1981–1988).* Frankfurt: Peter Lang.

Ott, Ursula. (1988). *Das freie Radio. Private Radios in Frankreich.* Marburg: Hitzeroth.

Owen, Bruce M., Jack H. Beebe, and Willard G. Manning, Jr. (1974). *Television Economics.* Lexington: D.C. Heath and Company.

Pace, Alessandro. (1983). *Stampa, geornalismo, radiotelevisione. Problemi costituzionali e indirizzi di giurisprudenza.* Padova: Cedam Padova.

Palmer, Michael, and Claude Sorbets. (1992). France, pp. 57–74 in Euromedia Research Group (Ed.), *The Media in Western Europe.* London: Sage.

Passler, R. G. (1990). Regulation of indecent radio broadcasts: George Carlin revisited—What does the future hold for the seven dirty words? *Tulane Law Review, 65*(1): 131–167.

Paulu, Burtun. (1981). *Television and Radio in the United Kingdom.* London: Macmillan.

Peacock Committee. (1986). *Report on the Committee on Financing the BBC.* London: Her Majesty's Stationery Office.

Peers, Frank W. (1969). *The Politics of Canadian Broadcasting 1920–1951.* Toronto: University of Toronto Press.

Peers, Frank W. (1979). *The Public Eye: Television and the Politics of Canadian Broadcasting 1952–1968.* Toronto: University of Toronto Press.

Peltzman, Sam. (1976). Toward a more general theory of regulation, *Journal of Law and Economics, 19*: 211–258.

Pengilley, Warren. (1990). Restrictive trade practices regulation of media, *Communications Law Bulletin, 10*(4): 5–10.

Petersen, Nikolaus. (1994). *Rundfunkfreiheit und EG-Vertrag.* Baden-Baden: Nomos.

Phillipe, Georges. (1991). Nouvelle reformé du contrôle de la publicité: Une revolution, *Angle Droit, 6*: 8.

Pilkington Report. (1962). *Report of the Committee on Broadcasting 1960.* London: Her Majesty's Stationery Office.

Plog, Jobst. (1981). Organisation und gesellschaftliche Kontrolle des Rundfunks, pp. 52–61 in Jörg Aufermann, Wilfried Scharf, and Otto Schlie (Eds.), *Fernsehen und Hörfunk für die Demokratie,* 2nd ed. Opladen: Westdeutscher Verlag.

Porter, Vincent, and Suzanne Hasselbach. (1991). *Pluralism, Politics and the Marketplace—The Reputation of German Broadcasting.* London: Routledge.

Postal and Telecommunications Department. (1976). *Australian Broadcasting (Green Report).* Canberra: Australian Government Publishing Service.

Postman, Neil. (1986). *Amusing Ourselves to Death.* New York: Penguin Books.

Potter, Jeremy. (1989). *Independent Television in Britain, Vol. 3: Politics and Control, 1968–80.* London: Macmillan.

Pragnell, Anthony. (1986). Authority view 2, pp. 295–328 in Henry Brian (Ed.), *British Television Advertising: The First Thirty Years.* London: David & Charles.

Prodoehl, Hans Gerd. (1987). Organisationsprobleme des lokalen Rundfunks—Das Zwei-Säulen-Modell im nordrhein-westfälischen Landesrundfunkgesetz, *Media Perspektiven, 1987*(4): 229–238.

Raboy, Marc. (1990). *Missed Opportunities: The Story of Canada's Broadcasting Policy.* Montreal: McGill-Queen's University Press.

Raboy, Marc. (1994). The Role of the public in broadcasting policy-making and regulation: Lesson for Europa from Canada, *European Journal of Communication, 9*(1): 5–23.

Ray, William B. (1990). *FCC: The Ups and Downs of Radio-TV Regulation.* Ames: Iowa University Press.

Reese, Jürgen. (1989). Nordrhein-Westfalen: Landesanstalt für Rundfunk (LfR), pp.

347–374 in Gerd-Michael Hellstern, Wolfgang Hoffmann-Riem, and Jürgen Reese (Eds.), *Rundfunkaufsicht—Vol. 16/I Begleitforschung des Landes Nordrhein-Westfalen zum Kabelpilotprojekt Dortmund.* Düsseldorf: Presse und Informationsamt der Landesregierung Nordrhein-Westfalen.

Reville, Nicholas. (1991). *Broadcasting—the New Law.* London: Butterworths.

Ricker, Reinhart. (1985). *Privatfunk-Besetze im Bundesstaat.* München: Beck.

Ridoux, George. (1984). Une nouvelle chaine de télévision, Canal Plus, *Revue de l'Union Européenne de Radiodiffusion, 35*(2): 46–49.

Ridoux, George. (1986). La 5 première chaine de télévision privée, *Revue de l'Union Européenne de Radiodiffusion, 37*(2): 42–45.

Riedel, Heide. (1977). *Hörfunk und Fernsehen in der DDR.* Köln: Literarischer Verlag Braun.

Ring, Wolf-Dieter. (1988). Entwicklung des privaten Rundfunk, pp. 1–3 in Direktorenkonferenz der Landesmedienbanstalten (DLM). (Ed.), *Jahrbuch 1988.* München: Neue Mediengesellschaft.

Ring, Wolf-Dieter. (1990). Pay TV ist Rundfunk! Entgegnung zum Aufsatz von C. Schwarz-Schilling "Pay-TV und doch kein Rundfunk," *Zeitschrift für Urheber und Medienrecht, 34*(6): 279–281.

Ring, Wolf-Dieter. (1992). Durchsetzung der gesetzlichen Programmanforderungen, pp. 35–42 in Heinz Hübner (Ed.), *Rechtsprobleme der privaten Rundfunkordnung. Vortragsveranstaltung vom 26. und 27. April 1991.* München: C. H. Beck.

Ring, Wolf-Dieter. (1993). Gefährdung der Rundfunkfreiheit. Neue Formen staatlicher Steuerung im dualen Rundfunksystem, pp. 707–719 in Peter Badura and Rupert Scholz (Hrsg.), *Wege und Verfahren des Verfassungslebens—Festschrift für Peter Lerche.* München: C. H. Beck.

Rinke Treuhand GmbH. (1989). Effekte von Kooperationsformen im lokalen Rundfunk in Nordrhein-Westfalen, pp. 219–270 in Landesanstalt für Rundfunk Nordrhein-Westfalen (Ed.), *Entscheidungsverfahren und Ergebnisse im Überblick, Landesanstalt für Rundfunk Nordrhein-Westfalen-Schriftenreihe Vol. 3.* Düsseldorf: Landesanstalt für Rundfunk Nordrhein-Westfalen.

Robertson, Geoffrey. (1990). *Media Law: The Rights of Journalists, Broadcasters and Publishers,* 2nd ed. London: Longman. (Originally published 1989)

Robinson, Glen O. (1978a). The Federal Communications Commission: An essay on regulatory watchdogs, *Virginia Law Review, 64*: 169–255.

Rödding, Gerhard. (1989). Die Aufgabe der Landesanstalt für Rundfunk Nordrhein-Westfalen, *Archiv für Presserecht, 20*(3): 648–651.

Romanow, Walter Ivan. (1976). A developing Canadian identity: A consequence of a defensive regulatory posture for broadcasting, *Gazette, 22*(1): 26–37.

Ronneberger, Franz. (1986). *Kommunikationspolitik III: Kommunikationspolitik als Medienpolitik.* Mainz: Von Hase & Koehler.

Röper, Horst. (1989). Der Stand der Verflechtung von privatem Rundfunk und Presse 1989, *Media Perspektiven, 1989*(9): 533–551.

Röper, Horst. (1990). Formationen deutscher Medienmultis, *Media Perspektiven, 1990*(12): 755–774

Röper, Horst. (1992). Formation deutscher Medienmultis 1992, *Media Perspektiven, 1992*(2): 2–22.

Röper, Horst. (1993). Formationen deutscher Medienmulties 1993, *Media Perspektiven, 1993*(2), 56–74.

Röper, Horst. (1994). Formationen deutscher Medienmultis 1993, *Media Perspektiven, 1994*(3): 125–144.

Rossen, Helge. (1994). Selbststeuerung im Rundfunk—Modell "FSK" für Kommeswiller Feusehen, *Zeitschuittuil Urhebes-und Medienecht* 38(4): 224–236.

Ross, Dieter. (1986). Der Rundfunk in Deutschland. Entwicklungen—Strukturen—Probleme, pp. B56–66 in Hans-Bredow-Institut (Ed.), *Internationales Handbuch für Rundfunk und Fernsehen 1986/87*. Baden-Baden: Nomos.

Rowan, Ford. (1984). *Broadcasting Fairness: Doctrine, Practices, Prospects*. White Plains, NY: Longman.

Royal Comission on the Press. (1977). *Final Report*. London: Her Majesty's Stationery Office.

Royal Society of Canada. (Eds.). (1989). *Symposium on "The State and the Arts." Quebec City, June 5, 1989*. Quebec City: Author.

Rozenblum, Serge Allain. (1984). Neue Fernsehangebote in Frankreich (Teil 1): Canal Plus—ein neuer Fernsehkanal, *Media Perspektiven, 1984*(2): 123 131.

Russel, Jim. (1993). Demystifying Canadian Content: Challenging the Television Broadcast Regulator to "7 Say What It Means and Mean What It Says," *Media and Communications Law Review, 3*(2): 171–208.

Rutherford, Paul. (1990). *When Television Was Young: Primetime Canada 1952–1967*. Toronto: University of Toronto Press.

Saunderson Report. (1988). *The Role and Functions of the Australian Broadcasting Tribunal—Parliamentary Paper No. 263/1988*. Canberra: The Parliament of the Commonwealth of Australia.

Saxer, Ulrich. (1994). Das Rundfunksystem der Schweiz, pp. B196–203 in Hans-Bredow-Institut (Ed.), *Internationales Handbuch für Rundfunk und Fernsehen 1994/95*. Baden-Baden: Nomos.

Schacht, Michael. (1981). Macht und Ohnmacht öffentlicher Kontrolle über privaten Rundfunk, *Media Perspektiven, 1981*(10): 689–699.

Scharpf, Fritz. (1970). *Die politischen Kosten des Rechtsstaats*. Tübingen: Mohr.

Scherer, Joachim. (1987). *Nachrichtenübertragung und Datenverarbeitung im Telekomminukationsrecht, Eine vergleichende Unterscheidung telekommunikationsrechtlicher Regelungsmodelle*. Baden-Baden: Nomos.

Schmidt, Benno C., Jr. (1978). Pluralistic programming and regulation of mass communications media, pp. 191–228 in Glen O. Robinson (Ed.), *Communications for Tomorrow*. New York: Praeger.

Schmidt, Hendrik. (1990). Mit dem Li-La-Launebär zum Erfolg, *Media Spectrum, 28*(11): 64–67.

Schmitt-Glaeser, W. (1979). *Kabelkommunikation und Verfassung*. Berlin: Duncker & Humblot.

Schneider, Hans. (1981). Expertenkommission neue Medien—Baden-Württemberg, *Die öffentliche Verwaltung, 39*(9): 334–337.

Schneider, Wolfgang. (1992). Über den Tellerrand—Zu den Werberichtlinien für den privaten Rundfunk, *Evangelischer Pressedienst/Kirche und Rundfunk, 1993*(13): 22–24.

Schröder, Hermann-Dieter. (1990). *Die Berichterstattung der Tagespresse in der*

Bundesrepublik Deutschland: Literaturdokumentation 1984–1988—Arbeitsberichte/Dokumentation 1990. Hamburg: Hans-Bredow-Institut.

Schröder, Hermann-Dieter, and Thorsten Sill. (1993). *Konstruktion und Realisierung des nordrhein-westfälischen Lokalfunkmodells.* Opladen: Leske & Budrich.

Schuler-Harms, Margarete. (1995). *Rundfunkaufsicht im Bundesstaat.* Baden-Baden: Nomos.

Schulz, Ferdinand F. (1990). Konzentrationstrend in Frankreichs Medienlandschaft, Daten zu den grossen Medienunternehmen, *Media Perspektiven, 1990*(3): 175–193.

Schurig, Christian. (1988). Programmzulieferung zwischen Wirtschaftlichkeit und Vielfalt unter Berücksichtigung baden-württembergischer Verhältnisse, pp. 50–55 in Direktorenkonferenz der Landesmedienanstalten (DLM) (Ed.), *Jahrbuch 88.* München: Neue Mediengesellschaft.

Schuster, Detlev. (1990). *Meinungsvielfalt in der dualen Rundfunkordnung.* Berlin: Duncker & Humblot.

Scott, Sheridan. (1990). The new broadcasting act: An analysis, *Media and Communications Law Review, 1*(1): 25–58.

Sendall, Bernard. (1982). *Independent Television in Britain, Vol. 1: Origin and Foundation, 1946–1962.* London: Macmillan.

Seufert, Wolfgang. (1988). *Struktur und Entwicklung des Rundfunk-Werbemarktes.* Düsseldorf: Presse- und Informationsamt der Landesregierung.

Seymour-Ure, Collin. (1987). Media Policy in Britain: Now You See It, Now You Don't, *European Journal of Communication, 2*(4): 269–288.

Shaw, Paul W. (1975). Purging Madison Avenue from Canadian cable television, *Law and Policy in International Business, 10*(4): 655–671.

Shedd, M. S., Elizabeth A. Wilman and R. Douglas Burch. (1990). An economic analysis of Canadian content regulations and a new proposal, *Canadian Public Policy, 16*(1): 60–72.

Simmons, Steven J. (1978). *The Fairness Doctrine and the Media.* Berkeley: University of California Press.

Simon, Jean-Paul. (1988). New media policies and the politics of central–local relations in France, pp. 161–184 in Kenneth Dyson and Peter Humphreys (Eds.), *Broadcasting and New Media Policies in Western Europe.* London: Routledge.

Smith, Anthony. (1986). Licenses and liberty: Public service broadcasting in Britain, pp. 1–29 in Colin MacCabe and Olivia Stewart (Eds.), *The BBC and Public Service Broadcasting.* Manchester: Manchester University Press.

Smith, Quentin. (1990). BBC rethinks hard news case after legal threat, *Broadcast, 4.5.1990*: 1.

Social and Liberal Democrats. (1988). *The Future of British Television.* Hebden Bridge: Hebden Royd Publications.

Sola Pool, Ithiel de. (1983). *Technologies of Freedom: On Free Speech in an Electronic Age.* Cambridge, MA: Belknap Press.

Sparks, Colin. (1988). Very slow progress: Deregulation and community radio in the United Kingdom, *RTV Theory and Practice, Special Issue No. 3./1988*: 145–160.

Special Broadcasting Service (SBS). (1989). *Annual Report 1988–89.* Canberra: Author.

Spicers Consulting Group. (1989). *The ITV—Network in Jeopardy? (Media Bulletin No. 1).* London: Author.

Spitz, Bernard. (1991). *Pour en finir avec le Pagialle Audiovisuelle Française.* Paris: Notes de la Fondation Saint-Simon.

Starowicz, Mark. (1990). The Americanisation of TV in Canada, *Combroad, 89*: 6–10.

Steinmaurer, Thomas. (1994). Das Rundfunksystem Österreichs, pp. B160–171 in Hans-Bredow-Institut (Ed.), *Internationales Handbuch für Rundfunk und Fernsehen 1994/95,* Baden-Baden: Nomos.

Stevenson, Wilf, and Nick Smedley. (Eds.). (1989). *Responses to the White Paper.* London: British Film Institute.

Stevenson, Wilf, and Nick Smedley. (1990). *Responses to the White Paper.* London: British Film Institute.

Stock, Martin. (1987). Ein fragwürdiges Konzept dualer Rundfunksysteme, *Rundfunk und Fernsehen 35*(1): 5–24.

Stoler, Andrew L. (1979). The border broadcasting dispute: A unique case under Section 301, *Osgoode Hall Law Journal, 17*(1): 39–54.

Street, H. (1982). *Freedom, the Individual and the Law,* 5th ed. Harmondsworth: Penguin Books.

Streeter, Thomas. (1983). Policy discourse and broadcast practice: The FCC, the US broadcast networks and the discourse of the marketplace, *Media, Culture and Society, 1983*(5): 247–262.

Suich, Max. (1991). What did the Labor Party do for you? *The Independent Monthly, March 1991*: 21–23.

Swinton, Katherine. (1977). Advertising and Canadian cable television—a problem in international communications law, *Osgoode Hall Law Journal, 15*(3): 543–590.

Task Force. (1986). *Report of the Task Force on Broadcasting Policy.* Ottawa: Minister of Supply and Services Canada.

Task Force. (1991). *The Economic Status of Canadian Television: Report of the Task Force.* Ottawa: Minister of Supply and Services Canada.

Tedford, Tomas L. (1993). *Feedom of Speech in the United States.* New York: McGraw-Hill.

Teeter, Dwight L., and Don R. Le Duc. (1992). *Law of Mass Communication—Freedom and Control of Print and Broadcast Media,* 7th ed. Westbury, NY: Foundation Press. (Originally published 1969)

Teichert, Will, and Peter Steinborn. (1990). *Werbemarkt Hamburg: Gutachten zu den wirtschaftlichen und marktpsychologischen Voraussetzungen für werbefinanzierte Hörfunkprogramme im Agglomerationsraum Hamburg.* Hamburg: Schriftenreihe der HAM.

Teidelt, Irene. (1986). Mediengesetzgebung in Frankreich: Trendwende nach dem Regierungswechsel, *Media Perspektiven, 1986*(8): 531–540.

Téléfilm Canada. (1990). *Rapport annuel 1989–1990.* Montréal: Service des Communications de Téléfilm Canada.

Thaenert, Wolfgang. (1990). Programm- und Konzentrationskontrolle privater Rundfunkveranstalter, pp. 31–51 in Direktorenkonferenz der Landesmedienanstalten (DLM) (Ed.), *DLM Jahrbuch 89/90—Privater Rundfunk in Deutschland.* München: DLM.

Theunert, Helga. (1992). Zwischen Vergnügen und Angst—Fernsehen im Alltag von

Kindern. Schriftenreihe der Hamburgischen Anstalt für neue Medien. Berlin: Vistas.

Truchet, Didier. (1987). *Regime juridique de la communication audiovisuelle, juris-classeur administratif, Vol. 3, Fasc. 273.* Paris.

Truchet, Didier. (1989). La loi du 17 janvier 1989 sur la communication audiovisuelle au la fin d'une illusion lyrique, *Revue Française de Droit Administratif, 5*(2): 208–214.

Trudel, Pierre. (1984). *Droit de l'information et de la communication. Notes et Documents, Université de Montréal, Faculté de droit.* Montréal: Les Éditions Thémis.

Trudel, Pierre. (1989). Le standard de programmation de haute qualité dans la législation sur la radio et la télévision, *McGill Law Journal, 34*(2): 203–232.

Trudel, Pierre. (1990). *L'audiovisuel Canadien: du discours de la législation à la réglementation du discours. (Rapport présenté aux entretiens Poitiers-Montréal "La communicatique et le droit," Université de Montréal, Faculté de droit.)* Montréal: Les Éditions Thémis.

Trudel, Pierre, and France Abran. (1991). *Droit de la Radio et de la Télévision.* Montréal: Les Éditions Thémis.

Tucker, David E., and Jeffrey Saffelle. (1982). The Federal Communications Commission and the regulation of children's television, *Journal of Broadcasting, 26*(3): 657–669.

Tunstall, Jeremy. (1983). *The Media in Britain.* London: Constable.

Tunstall, Jeremy. (1986). Great Britain, pp. 110–134 in Hans J. Kleinsteuber, Denis McQuail, and Karen Siune (Eds.), *Electronic Media and Politics in Western Europe.* Frankfurt am Main: Campus.

Tunstall, Jeremy. (1992). The United Kingdom, pp. 238–255 in Euromedia Research Group (Ed.), *The Media in Western Europe.* London: Sage.

Turpin, Dominique. (1988). Neue Entwicklungen im Recht der audiovisuellen Kommunikation in Frankreich, *Zeitschrift für Urheber- und Medienrecht/Film und Recht, 32*(3): 101–120.

Unabhängige Landesanstalt für das Rundfunkwesen Schleswig-Holstein. (ULR). (1989). *Jahresbericht der ULR 1989.* Kiel: Author.

Unabhängige Landesanstalt für das Rundfunkewesen Schleswig-Holstein (ULR). (1990). *Rechenschaftsbericht 1989.* Kiel: Author.

Vardy, Jill, and Richard Siklos. (1990). Rivals applaud as CBC slashes budget, *Financial Post, December 6, 1990*: 3.

Veljanovski, Cento. (1990). *The Media in Britain Today.* London: New International.

Vipond, Mary. (1989). *The Mass Media in Canada.* Toronto: James Wiesand.

Vogel, Paul O. (1983). Tödliches Ende auf Parteisohlen, *Evangelischer Pressedienst/Kirche und Rundfunk, 1983*(65): 1–6.

Wagner, Christoph. (1990a). *Die Landesmedienanstalten.* Baden-Baden: Nomos.

Wagner, Christoph. (1990b). Konzentrationskontrolle im privaten Rundfunk, *Rundfunk und Fernsehen, 38*(2): 165–182.

Walker, Sally. (1989). *The Law of Journalism in Australia.* Sydney: Law Book Company.

Wassermann, Rudolf. (Ed.). (1989). *Kommentar zum Grundgesetz für die Bundesrepublik Deutschland,* 2nd ed. Neuwied: Luchterhand. (Originally published 1983)

Weber, Werner. (1973). *Innere Pressefreiheit als Verfassungsproblem.* Berlin: Duncker & Humblot.

Weelwright, T., and K. Buckley. (Eds.). (1987). *Communications and the Media in Australia.* Sydney: Allen & Unwin.

Weibull, Lennart, and Britt Börjesson. (1992). The Swedish media accountability system: A research perspective, *European Journal of Communication, 7*(1): 121–139.

Weiss, Frederic A., David Ostroff, and Charles E. Clift III. (1980). Station license revocations and denials of renewal, 1970–78, *Journal of Broadcasting, 24*(1): 69–77.

Widlok, Peter. (1984). Kanadische Identität oder "American way of life?" Zum US-Einfluss auf den öffentlichen Rudnfunk und die Neuen Medien in Kanada, *Media Perspektiven, 1984*(5): 399–407.

Wiedemann, Verena A. M. (1989). *Law of International Telecommunications in the United Kingdom: Regulation of Electronic Media.* Baden-Baden: Nomos.

Wiesner, Volkert. (1988a). Community radio in Kanada. Das Beispiel Vancouver Cooperative Radio, *Rundfunk und Fernsehen, 36*(2): 229–246.

Wiesner, Volkert. (1988b). Das Rundfunksystem Kanadas zwischen öffentlichem Programmauftrag und Kommerzialisierung, *Media Perspektiven, 1988*(11): 705–714.

Wiesner, Volkert. (1991). *Rundfunkpolitik und kulturelle Identität in Canada. Eine Analyse staatlicher Steuerungsmöglichkeiten durch Normensetzung und Regulierungs-Institutionen im dualen Rundfunksystem.* Münster: LIT Verlag.

Wiley, Richard E. (1977). Family viewing: A balancing of interests, *Journal of Communications, 27*(1): 188–192.

Williams, Frederick, and John V. Pavlik. (Eds.). (1994). *The People's Right to Know: Media, Democracy and the Information Highway.* Hillsdale, NJ: Lawrence Erlbaum.

Wilson, H. (1989). *Australian Communications and the Public Sphere.* Melbourne: Macmillan Australia.

Wilson, J. Q. (1980). *The Politics of Regulation.* New York: Basic Books.

Woldt, Runar. (1988). Teleshopping. Aktuelle Entwicklungen in 4 Ländern, *Media Perspektiven, 1988*(7): 421–436.

Wolton, D. (1990). *Eloge du grand public. Une thèorie critique de la television.* Paris: Flammarion.

Wöste, Marlene. (1989). Networkbildung durch die Hintertür? Programmzulieferer für privaten Hörfunk in der Bundesrepublik, *Media Perspektiven, 1989*(1): 9–22.

Wöste, Marlene. (1990). Nur knapp die Hälfte für Lizensierung und Kontrolle: Die Einnahmen und Ausgaben der Landesmedienanstalten 1985–1990. *Media Perspektiven, 1990*(5): 281–304.

Wöste, Marlene. (1991). Programmquellen privater Radios in Deutschland—Rahmenprogramm, Beitragsanbieter und PR-Audioagenturen, *Media Perspektiven, 1991*(9): 561–567.

Wyman, Kenneth. (1983). Rationales for cable regulation: A Canadian perspective, pp. 171–179 in Oscar H. Gandy, Jr., Paul Espinosa, and Janusz A. Ordoner (Eds.), *Proceedings from the Tenth Annual Telecommunications Policy Research Conference.* Norwood, NJ: Ablex.

Ziethen, Michael P. (1989). Rechtliche Spielräume der Lizensierung und Kontrolle: Ausgewählte Regelungsfelder, pp. 59–164 in Gerd-Michael Hellstern, Wolfgang Hoffmann-Riem, and Jürgen Reese (Eds.), *Rundfunkaufsicht—Vol. 16/III Begleitforschung des Landes Nordrhein-Westfalen zum Kabelpilotprojekt Dortmund.* Düsseldorf: Presse und Informationsamt der Landesregierung Nordrhein-Westfalen.

Zuckman, Harvey L., Martin J. Gaynes, Barton T. Carter, and Juliet Lushbough Dee. (1988). *Mass Communications Law,* 3rd ed. St. Paul, MN: West.

Index